Convergence of Probability Measures

WILEY SERIES IN PROBABILITY AND STATISTICS
PROBABILITY AND STATISTICS SECTION

Established by WALTER A. SHEWHART and SAMUEL S. WILKS

A complete list of the titles in this series appears at the end of this volume.

Convergence of Probability Measures

Second Edition

PATRICK BILLINGSLEY
The University of Chicago
Chicago, Illinois

A Wiley-Interscience Publication
JOHN WILEY & SONS, INC.
New York • Chichester • Weinheim • Brisbane • Singapore • Toronto

This text is printed on acid-free paper. ♾

For ordering and customer service, call 1-800-CALL-WILEY.

Library of Congress Cataloging in Publication Data:

Billingsley, Patrick.
Convergence of probability measures / Patrick Billingsley. — 2nd ed.
p. cm. — (Wiley series in probability and statistics. Probability and statistics)
"Wiley-Interscience publication."
Includes bibliographical references and indexes.
ISBN 0-471-19745-9 (alk. paper)
1. Probability measures. 2. Metric spaces. 3. Convergence.
I. Title. II. Series: Wiley series in probability and statistics. Probability and statistics.
QA273.6.B55 1999
519.2—dc21 99-30372
CIP

Printed in the United States of America

14 13 12 11

PREFACE

From the preface to the first edition. Asymptotic distribution theorems in probability and statistics have from the beginning depended on the classical theory of weak convergence of distribution functions in Euclidean space—convergence, that is, at continuity points of the limit function. The past several decades have seen the creation and extensive application of a more inclusive theory of weak convergence of probability measures on metric spaces. There are many asymptotic results that can be formulated within the classical theory but require for their proofs this more general theory, which thus does not merely study itself. This book is about weak-convergence methods in metric spaces, with applications sufficient to show their power and utility.

The second edition. A person who read the first edition of this book when it appeared thirty years ago could move directly on to the periodical literature and to research in the subject. Although the book no longer takes the reader to the current boundary of what is known in this area of probability theory, I think it is still useful as a textbook one can study before tackling the industrial-strength treatises now available. For the second edition I have reworked most of the sections, clarifying some and shortening others (most notably the ones on dependent random variables) by using discoveries of the last thirty years, and I have added some new topics. I have written with students in mind; for example, instead of going directly to the space $D[0,\infty)$, I have moved, in what I hope are easy stages, from $C[0,1]$ to $D[0,1]$ to $D[0,\infty)$. In an earlier book of mine, I said that I had tried to follow the excellent example of Hardy and Wright, who wrote their *Introduction to the Theory of Numbers* with the avowed aim, as they say in the preface, of producing an interesting book, and I have again taken them as my model.

Chicago,
January 1999

Patrick Billingsley

For mathematical information and advice, I thank Richard Arratia, Peter Donnelly, Walter Philipp, Simon Tavaré, and Michael Wichura. For essential help on Teχ and the figures, I thank Marty Billingsley and Mitzi Nakatsuka.

PB

CONTENTS

★ Starred topics can be omitted on a first reading

INTRODUCTION

The De Moivre-Laplace limit theorem says that, if

$$(1) \qquad F_n(x) = \mathsf{P}\Big[\frac{S_n - np}{\sqrt{npq}} \le x\Big]$$

is the distribution function of the normalized number of successes in n Bernoulli trials, and if

$$(2) \qquad F(x) = \frac{1}{\sqrt{2\pi}} \int_{-\infty}^{x} e^{-u^2/2} du$$

is the standard normal distibution function, then

$$(3) \qquad F_n(x) \to F(x)$$

for all x ($n \to \infty$, the probability p of success fixed).

We say of arbitrary distribution functions F_n and F on the line that F_n *converges weakly to* F, which we indicate by writing $F_n \Rightarrow F$, if (3) holds at every continuity point x of F. Thus the De Moivre-Laplace theorem says that (1) converges weakly to (2); since (2) is everywhere continuous, the proviso about continuity points is vacuous in this case. If F_n and F are defined by

$$(4) \qquad F_n(x) = I_{[n^{-1},\infty)}(x)$$

(I for the indicator function) and

$$(5) \qquad F(x) = I_{[0,\infty)}(x),$$

then again $F_n \Rightarrow F$, and this time the proviso does come into play: (3) fails at $x = 0$.

For a better understanding of this notion of weak convergence, which underlies a large class of limit theorems in probability, consider the probability measures P_n and P generated by arbitrary distribution functions F_n and F. These probability measures, defined on the class of Borel subsets of the line, are uniquely determined by the requirements

$$P_n(-\infty, x] = F_n(x), \quad P(-\infty, x] = F(x).$$

Since F is continuous at x if and only if the set $\{x\}$ consisting of x alone has P-measure 0, $F_n \Rightarrow F$ means that the implication

$$P_n(-\infty, x] \to P(-\infty, x] \quad \text{if } P\{x\} = 0 \tag{6}$$

holds for each x.

Let ∂A denote the boundary of a subset A of the line; ∂A consists of those points that are limits of sequences of points in A and are also limits of sequences of points outside A. Since the boundary of $(\infty, x]$ consists of the single point x, (6) is the same thing as

$$P_n(A) \to P(A) \quad \text{if } P(\partial A) = 0, \tag{7}$$

where we have written A for $(-\infty, x]$. The fact of the matter is that $F_n \Rightarrow F$ holds if and only if the implication (7) is true for *every* Borel set A—a result proved in Chapter 1.

Let us distinguish by the term *P-continuity set* those Borel sets A for which $P(\partial A) = 0$, and let us say that P_n *converges weakly to* P, and write $P_n \Rightarrow P$, if $P_n(A) \to P(A)$ for each P-continuity set A—that is, if (7) holds. As just asserted, $P_n \Rightarrow P$ if and only if the corresponding distribution functions satisy $F_n \Rightarrow F$.

This reformulation clarifies the reason why we allow (3) to fail if F has a jump at x. Without this exemption, (4) would not converge weakly to (5), but this example may appear artificial. If we turn our attention to probability measures P_n and P, however, we see that $P_n(A) \to P(A)$ may fail if $P(\partial A) > 0$ even in the De Moivre-Laplace theorem. The measures P_n and P generated by (1) and (2) satisfy

$$P_n(A) = \mathsf{P}\left[\frac{S_n - np}{\sqrt{npq}} \in A\right] \tag{8}$$

and

$$P(A) = \frac{1}{\sqrt{2\pi}} \int_A e^{-u^2/2} du \tag{9}$$

for Borel sets A. Now if A consists of the countably many points

$$\frac{k-np}{\sqrt{npq}}, \quad n = 1, 2, \ldots, \quad k = 0, 1, \ldots, n,$$

then $P_n(A) = 1$ for all n and $P(A) = 0$, so that $P(A_n) \to P(A)$ is impossible. Since ∂A is the entire real line, this does not violate (7).

Although the concept of weak convergence of distribution functions is tied to the real line (or to Euclidean space, at any rate), the concept of weak convergence of probability measures can be formulated for the general metric space, which is the real reason for preferring the latter concept. Let S be an arbitrary metric space, let $\mathcal{S}$ be the class of Borel sets ($\mathcal{S}$ is the σ-field generated by the open sets), and consider probability measures P_n and P defined on $\mathcal{S}$. Exactly as before, we define weak convergence $P_n \Rightarrow P$ by requiring the implication (7) to hold for all Borel sets A. In Chapter 1 we investigate the general theory of this concept and see what it reduces to in various special cases. We prove there, for example, that P_n converges weakly to P if and only if

$$\int_S f\,dP_n \to \int_S f\,dP \tag{10}$$

holds for all bounded, continuous real-valued functions on S. (In order to conform with general mathematical usage, we take (10) as the defintion of weak convergence, so that (7) becomes a necessary and sufficient condition instead of a definition.)

Chapter 2 concerns weak convergence on the space $C = C[0,1]$ with the uniform topology; C is the space of all continuous real functions on the closed unit interval $[0,1]$, metrized by taking the distance between two functions $x = x(t)$ and $y = y(t)$ to be

$$\rho(x, y) = \sup_{0 \le t \le 1} |x(t) - y(t)|. \tag{11}$$

An example of the sort of application made in Chapter 2 will show why it is both interesting and useful to develop a general theory of weak convergence—one that goes beyond the Euclidean case. Let $\xi_1, \xi_2, \ldots$ be a sequence of independent, identically distributed random variables defined on some probability space $(\Omega, \mathcal{F}, \mathsf{P})$. If the ξ_n have mean 0 and variance σ^2, then, by the Lindeberg-Lévy central limit theorem, the distribution of the normalized sum

$$\frac{1}{\sigma\sqrt{n}} S_n = \frac{1}{\sigma\sqrt{n}}(\xi_1 + \cdots + \xi_n) \tag{12}$$

converges weakly, as n tends to infinity, to the normal distribution defined by (9).

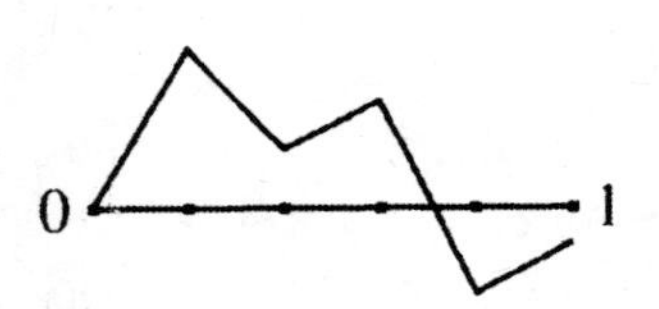

We can formulate a refinement of the central limit theorem by proving weak convergence of the distributions of certain random functions constructed from the partial sums S_n. For each integer n and each sample point ω, construct on the unit interval the polygonal function that is linear on each of the subintervals $[(i-1)/n, i/n]$, $i = 1, 2, \ldots, n$, and has the value $S_i(\omega)/\sigma\sqrt{n}$ at the point i/n ($S_0(\omega) = 0$). In other words, construct the function $X^n(\omega)$ whose value at a point t of $[0, 1]$ is

(13)
$$X_t^n(\omega) = \frac{1}{\sigma\sqrt{n}} S_{i-1}(\omega) + \frac{t - (i-1)/n}{1/n} \frac{1}{\sigma\sqrt{n}} \xi_i(\omega), \quad \text{if } t \in \left[\frac{i-1}{n}, \frac{i}{n}\right].$$

For each ω, $X^n(\omega)$ is an element of the space C, a random function. Let P_n be the distribution of $X^n(\omega)$ on C, defined for Borel subsets A of C—Borel sets relative to the metric (11)—by

$$P_n(A) = \mathsf{P}[\omega\colon X^n(\omega) \in A]$$

(the definition is possible because the mapping $\omega \to X^n(\omega)$ turns out to be measurable in the right way). In Chapter 2 we prove Donsker's theorem, which says that

$$P_n \Rightarrow W, \tag{14}$$

where W is Wiener measure. We also prove the existence in C of Wiener measure, which describes the probability distribution of the path traced out by a particle in Brownian motion.

If $A = [x\colon x(1) \le \alpha]$, then, since the value of the function $X^n(\omega)$ at $t = 1$ is $X_1^n(\omega) = S_n(\omega)/\sigma\sqrt{n}$,

$$P_n(A) = \mathsf{P}\left[\omega\colon \frac{1}{\sigma\sqrt{n}} S_n(\omega) \le \alpha\right].$$

It turns out that $W(\partial A) = 0$, so that (14) implies

$$\mathsf{P}\left[\omega\colon \frac{1}{\sigma\sqrt{n}} S_n(\omega) \le \alpha\right] \to W[x\colon x(1) \le \alpha].$$

It also turns out that

$$W[x\colon x(1) \le 1] = \frac{1}{\sqrt{2\pi}} \int_{-\infty}^{x} e^{-u^2/2} du,$$

so that (14) does contain the Lindeberg-Lévy theorem.

If ξ_i takes the values $+1$ and -1 with probability $\frac{1}{2}$ each, S_n can be interpreted as the position at time n in a symmetric random walk. The central limit theorem says that this position, normalized by $\sqrt{n}$ ($\sigma = 1$), is, for n large, approximately distributed as the position at time $t = 1$ of a particle in Brownian motion. The relation (14) says that the entire path of the random walk during the first n steps is, for n large, distributed approximately as the path up to time $t = 1$ of a particle in Brownian motion.

To see in a concrete way that (14) contains information going beyond the central limit theorem, consider the set

$$A = \left[x\colon \sup_{0 \le t \le 1} x(t) \le \alpha\right].$$

Again it turns out that $W(\partial A) = 0$, so that (14) implies

$$\text{(15)} \quad \lim_{n\to\infty} \mathsf{P}\left[\omega\colon \frac{1}{\sigma\sqrt{n}} \max_{1\le k\le n} S_k(\omega) \le \alpha\right] = \lim_{n\to\infty} P_n(A) = W\left[x\colon \sup_{0\le t\le 1} x(t) \le \alpha\right].$$

We can evaluate the rightmost member of (15) by finding the limit on the left for the special case of symmetric random walk, which is easy to analyze. And then we have a limit theorem for the distribution of $\max_{k\le n} S_k$ under the hypothesis of the Lindeberg-Lévy theorem.

For another example involving $X^n(\omega)$, take A to be the set of x in C for which the set $[t\colon x(t) > 0]$ has Lebesgue measure at most α (where $0 \le \alpha \le 1$). As before, $P_n(A) \to P(A)$. Since the Lebesgue measure of $[t\colon X_t^n(\omega) > 0]$ is essentially the fraction of the partial sums $S_1, S_2, \ldots, S_n$ that exceed 0, this argument leads to an arc sine law under the hypotheses of the Lindeberg-Lévy theorem. Chapter 2 contains the details of these derivations.

We can in this way use the theory of weak convergence in C to obtain a whole class of limit theorems for functions of the partial sums $S_1, S_2, \ldots, S_n$. The fact that Wiener measure W is the weak limit of the distribution over C of the random function $X^n(\omega)$ can also be used to prove theorems about W, and W is interesting in its own right.

Chapter 3 specializes the theory of weak convergence to another space of functions on $[0,1]$—the space $D[0,1]$ of functions having only discontinuities of the first kind (jump discontinuities). This is the natural space in which to analyze the Poisson process and other processes with paths that are necessarily discontinuous. We also study spaces of discontinuous functions on $[0,\infty)$ and on certain other sets that play the role of a generalized "time"—for example, the set of convex subsets of the unit square. Chapter 4 concerns weak convergence of the distributions of random functions derived from various sequences of dependent random variables. Chapter 5 has to do with other asymptotic properties of random functions; there we prove Strassens's theorem, a far-reaching generalization of the law of the iterated logarithm.

Many of the conclusions in Chapters 2 through 5, although not requiring function-space concepts for their statement, could hardly have been derived without function-space methods. Standard measure-theoretic probability and metric-space topology are used from the beginning of the book. Although the point of view throughout is that of functional analysis (a function is a point in a space), nothing of functional analysis is assumed (beyond an initial willingness to view a function as a point in a space). All function-analytic results needed are proved in the text or else in Appendix M at the end of the book. This appendix also gathers together for easy reference some results in metric-space topology, analysis, and probability.

CHAPTER 1

WEAK CONVERGENCE IN METRIC SPACES

SECTION 1. MEASURES ON METRIC SPACES

We begin by studying probability measures on the general metric space. Denote the space by S, and let $\mathcal{S}$ be the *Borel σ-field*, the one generated by the open sets;[†] its elements are the *Borel sets*. A probability measure on $\mathcal{S}$ is a nonnegative, countably additive set function P satisfying $PS = 1$.

If probability measures P_n and P satisfy[‡] $P_n f \to Pf$ for every bounded, continuous real function f on S, we say that P_n *converges weakly* to P and write $P_n \Rightarrow P$. In Chapter 1 we study the basic, general theory of weak convergence, together with the associated concept of convergence in distribution. We first derive some properties of individual measures on $(S, \mathcal{S})$. Although S is sometimes assumed separable or complete, most of the theorems in this chapter hold for all metric spaces.

Measures and Integrals

Theorem 1.1. *Every probability measure P on $(S, \mathcal{S})$ is regular; that is, for every $\mathcal{S}$-set A and every ϵ there exist a closed set F and an open set G such that $F \subset A \subset G$ and $P(G - F) < \epsilon$.*

PROOF. Denote the metric on S by $\rho(x, y)$ and the distance from x to A by $\rho(x, A)$ [M1].[§] If A is closed, we can take $F = A$ and $G = A^\delta = [x : \rho(x, A) < \delta]$ for some δ, since the latter sets decrease to A as $\delta \downarrow 0$. Hence we need only show that the class $\mathcal{G}$ of $\mathcal{S}$-sets with

---

[†] Or, for example, by the closed sets. All unindicted sets and functions are assumed measurable with respect to $\mathcal{S}$. Another convention: all ϵ's and δ's are positive.

[‡] Write Pf for $\int_S f dP$.

[§] A reference [Mn] is to paragraph n of the Appendix starting on p. 236.

the asserted property is a σ-field. Given sets A_n in $\mathcal{G}$, choose closed sets F_n and open sets G_n such that $F_n \subset A_n \subset G_n$ and $P(G_n - F_n) < \epsilon/2^{n+1}$. If $G = \bigcup_n G_n$, and if $F = \bigcup_{n \le n_0} F_n$, with n_0 so chosen that $P(\bigcup_n F_n - F) < \epsilon/2$, then $F \subset \bigcup_n A_n \subset G$ and $P(G - F) < \epsilon$. Thus $\mathcal{G}$ is closed under the formation of countable unions; since it is obviously closed under complementation, $\mathcal{G}$ is a σ-field. □

Theorem 1.1 implies that P is completely determined by the values of PF for closed sets F. The next theorem shows that P is also determined by the values of Pf for bounded, continuous f. The proof depends on approximating the indicator I_F by such an f, and the function $f(x) = (1 - \rho(x, F)/\epsilon)^+$ works. It is bounded, and it is continuous, even uniformly continuous, because $|f(x) - f(y)| \le \rho(x,y)/\epsilon$. And $x \in F$ implies $f(x) = 1$, while $x \notin F^\epsilon$ implies $\rho(x, F) \ge \epsilon$ and hence $f(x) = 0$. Therefore,

$$I_F(x) \le f(x) = (1 - \rho(x, F)/\epsilon)^+ \le I_{F^\epsilon}(x). \tag{1.1}$$

Theorem 1.2. *Probability measures P and Q on $\mathcal{S}$ coincide if $Pf = Qf$ for all bounded, uniformly continuous real functions f.*

PROOF. For the bounded, uniformly continuous f of (1.1), $PF \le Pf = Qf \le QF^\epsilon$. Letting $\epsilon \downarrow 0$ gives $PF \le QF$, provided F is closed. By symmetry and Theorem 1.1, $P = Q$. □

Because of theorems like this, it is possible to work with measures PA or with integrals Pf, whichever is simpler or more natural. We defined weak convergence in terms of the convergence of integrals of functions, and in the next section we characterize it in terms of the convergence of measures of sets.

Tightness

The following notion of tightness plays a fundamental role both in the theory of weak convergence and in its applications. A probability measure P on $(S, \mathcal{S})$ is *tight* if for each ϵ there exists a compact set K such that $PK > 1 - \epsilon$. By Theorem 1.1, P is tight if and only if PA is, for each A in $\mathcal{S}$, the supremum of PK over the compact subsets K of A.

Theorem 1.3. *If S is separable and complete, then each probability measure on $(S, \mathcal{S})$ is tight.*

PROOF. Since S is separable, there is, for each k, a sequence $A_{k1}, A_{k2}, \ldots$ of open $1/k$-balls covering S. Choose n_k large enough that

$P(\bigcup_{i \le n_k} A_{ki}) > 1 - \epsilon/2^k$. By the completeness hypothesis, the totally bounded set $\bigcap_{k \ge 1} \bigcup_{i \le n_k} A_{ki}$ has compact closure K. But clearly $PK > 1 - \epsilon$. □

Some Examples

Here are four metric spaces, together with some facts about them we need further on. Define a subclass $\mathcal{A}$ of $\mathcal{S}$ to be a *separating class* if two probability measures that agree on $\mathcal{A}$ necessarily agree also on the whole of $\mathcal{S}$: The values of PA for A in $\mathcal{A}$ are enough to separate P from all the other probability measures on $\mathcal{S}$. For example, by Theorem 1.1 the closed sets form a separating class. Recall that a class is a π-system if it is closed under the formation of finite intersections and that $\mathcal{A}$ is a separating class if it is a π-system generating the σ-field $\mathcal{S}$ [PM.42].†

Example 1.1. Let R^k denote k-dimensional Euclidean space with the ordinary metric $|x - y| = \sqrt{[\sum_{i=1}^k (x_i - y_i)^2]}$, and denote by $\mathcal{R}^k$ the corresponding class of Borel sets—the k-dimensional Borel sets. The distribution function corresponding to a probability measure P on $\mathcal{R}^k$ is

$$F(x_1, \dots, x_k) = P[y \colon y_i \le x_i,\ i \le k]. \tag{1.2}$$

Since the sets on the right here form a π-system that generates $\mathcal{R}^k$, they constitute a separating class. Therefore, F completely determines P.

By Theorem 1.3, each probability measure on $(R^k, \mathcal{R}^k)$ is tight. But tightness is in this case obvious because the space is σ-compact—is a countable union of compact sets ($B(0,n)^- \uparrow R^k$, for example). □

Example 1.2. Let R^∞ be the space of sequences $x = (x_1, x_2, \dots)$ of real numbers—the product of countably many copies of R^1. If $b(\alpha, \beta) = 1 \wedge |\alpha - \beta|$, then b is a metric on R^1 equivalent to the usual one, and under it, R^1 is complete as well as separable [M4]. Metrize R^∞ by $\rho(x, y) = \sum_i b(x_i, y_i)/2^i$. Obviously,‡ $\rho(x^n, x) \to_n 0$ implies $b(x_i^n, x_i) \to_n 0$ for each i; but by the M-test [PM.543], the reverse implication also holds. Under ρ, therefore, R^∞ has the topology of pointwise convergence: $x^n \to_n x$ if and only if $x_i^n \to_n x_i$ for each i.

† These are page references to *Probability and Measure*, third edition, for any who may find them useful.

‡ As often in the book, the superscript is not a power: $x^n = (x_1^n, x_2^n, \dots)$.

Let $\pi_k: R^\infty \to R^k$ be the natural projection: $\pi_k(x) = (x_1, \ldots, x_k)$. Since convergence in R^∞ implies coordinatewise convergence, π_k is continuous, and therefore the sets

$$N_{k,\epsilon}(x) = [y: |y_i - x_i| < \epsilon,\ i = 1, \ldots, k] \tag{1.3}$$

are open. Moreover, $y \in N_{k,\epsilon}(x)$ implies $\rho(x,y) < \epsilon + 2^{-k}$. Given a positive r, choose ϵ and k so that $\epsilon + 2^{-k} < r$; then $N_{k,\epsilon}(x) \subset B(x,r)$. This means that sets (1.3) form a base for the topology of R^∞. It follows that the space is separable: one countable, dense subset consists of those points having only finitely many nonzero coordinates, each of them rational. If $\{x^n\}$ is fundamental, then each $\{x_i^n\}$ is fundamental and hence converges to some x_i, and of course x^n converges to the point with coordinates x_i. Therefore, R^∞ is also complete. (These facts are proved in a more general setting in [M6]).

Since R^∞ is separable and complete, it follows by Theorem 1.3 that each probability measure on $\mathcal{R}^\infty$ is tight.

In the case of R^k, tightness follows directly from the fact that the space is σ-compact, an argument which cannot work here, since R^∞ is not σ-compact. Indeed, if $y_i^n = x_i$ for $i \le k$ and $y_i^n = n$ for $i > k$, then the sequence $\{y^n\}$ is contained in (1.3) but has no convergent subsequence. This means that the closure of (1.3) is not compact, and therefore no closed ball $B(x,\epsilon)^-$ is compact. This of course implies that R^∞ is not locally compact. Furthermore, if K is compact, then an arbitrary open ball B must share some point x with K^c ($B \subset K$ being impossible), and so (since $B \cap K^c$ is open) there is an ϵ such that $B(x,\epsilon) \subset B \cap K^c$. Therefore, every compact set is nowhere dense. Finally, by Baire's category theorem [M7], this implies that R^∞ is not σ-compact.

Let $\mathcal{R}_f^\infty$ be the class of *finite-dimensional* sets, that is to say, the sets of the form $\pi_k^{-1}H$ for $k \ge 1$ and $H \in \mathcal{R}^k$. Since π_k is continuous, it is measurable $\mathcal{R}^\infty/\mathcal{R}^k$ [M10], and so $\mathcal{R}_f^\infty \subset \mathcal{R}^\infty$. Moreover, since $\pi_k^{-1}H = \pi_{k+1}^{-1}(H \times R^1)$, the set of indices in the specification of an $\mathcal{R}_f^\infty$-set can always be enlarged, and it follows that two sets A and A' in $\mathcal{R}_f^\infty$ can be represented as $A = \pi_k^{-1}H$ and $A' = \pi_k^{-1}H'$ for the same value of k. And now $A \cap A' = \pi_k^{-1}(H \cap H')$ makes it clear that $\mathcal{R}_f^\infty$ is a π-system (even a field). Further, since the sets (1.3) form a basis and each lies in $\mathcal{R}_f^\infty$, it follows by separability that each open set is a countable union of sets in $\mathcal{R}_f^\infty$, which therefore generates the Borel σ-field $\mathcal{R}^\infty$: *$\mathcal{R}_f^\infty$ is a separating class.*

If P is a probability measure on $(R^\infty, \mathcal{R}^\infty)$, its *finite-dimensional distributions* are the measures $P\pi_k^{-1}$ on $(R^k, \mathcal{R}^k)$, $k \geq 1$, and since $\mathcal{R}_f^\infty$ is a separating class, these measures completely determine P. □

Example 1.3. Let $C = C[0,1]$ be the space of continuous functions $x = x(\cdot)$ on $[0,1]$. Define the *norm* of x as $\|x\| = \sup_t |x(t)|$, and give C the *uniform metric*

$$\rho(x, y) = \|x - y\| = \sup_t |x(t) - y(t)|. \tag{1.4}$$

The random-walk polygons of the Introduction lie in C, which will be studied systematically in Chapter 2. Since $\rho(x_n, x) \to 0$ means that x_n converges to x uniformly, it implies pointwise convergence. But of course the converse is false: Consider the function z_n that increases linearly from 0 to 1 over $[0, n^{-1}]$, decreases linearly from 1 to 0 over $[n^{-1}, 2n^{-1}]$, and stays at 0 to the right of $2n^{-1}$; that is,

$$z_n(t) = ntI_{[0,n^{-1}]}(t) + (2 - nt)I_{(n^{-1},2n^{-1}]}(t). \tag{1.5}$$

This z_n convereges pointwise to the 0-function, while $\rho(z_n, 0) = 1$.

The space C is separable. For let D_k be the set of polygonal functions that are linear over each subinterval $I_{ki} = [(i-1)/k, i/k]$ and have rational values at the endpoints. Then $\bigcup_k D_k$ is countable. To show that it is dense, for given x and ϵ choose k so that $|x(t) - x(i/k)| < \epsilon$ for $t \in I_{ki}$, $1 \leq i \leq k$, which is possible by uniform continuity, and then choose a y in D_k so that $|y(i/k) - x(i/k)| < \epsilon$ for each i. Now $y(i/k)$ is within 2ϵ of $x(t)$ for $t \in I_{ki}$, and similarly for $y((i-1)/k)$. Since $y(t)$ is a convex combination of $y((i-1)/k)$ and $y(i/k)$, it too is within 2ϵ of $x(t)$: $\rho(x, y) \leq 2\epsilon$.

And C is also complete: If x_n is fundamental, which means that $\epsilon_n = \sup_{m>n} \rho(x_n, x_m) \to_n 0$, then, for each t, $\{x_n(t)\}$ is fundamental on the line and hence has a limit $x(t)$. Letting $m \to \infty$ in the inequality $|x_n(t) - x_m(t)| \leq \epsilon_n$ gives $|x_n(t) - x(t)| \leq \epsilon_n$; therefore, $x_n(t)$ converges uniformly to $x(t)$, x is continuous, and $\rho(x_n, x) \to 0$.

Since C is separable and complete, it follows, again by Theorem 1.3, that each probability measure on the Borel σ-field $\mathcal{C}$ is tight.

Like R^∞, C is not σ-compact. To see this, consider the function (1.4) again. No subsequence of ϵz_n can converge, because if $\rho(\epsilon z_{n_i}, z) \to_i 0$, then z must be the 0-function, while $\rho(\epsilon z_{n_i}, 0) = \epsilon$. Therefore, the closed ball $B(0,\epsilon)^-$ is not compact, and in fact no closed ball $B(x,\epsilon)^-$ is compact (consider the points $x + \epsilon z_n$). It follows as before that every compact set is nowhere dense and that C is not σ-compact.

For $0 \le t_1 < \cdots < t_k \le 1$, define the natural projection from C to R^k by $\pi_{t_1 \cdots t_k}(x) = (x(t_1), \ldots, x(t_k))$. In C the *finite-dimensional sets* are those of the form $\pi^{-1}_{t_1 \cdots t_k} H$, $H \in \mathcal{R}^k$, and they lie in $\mathcal{C}$ because the projections are continuous. As in the preceding example, the index set defining a finite-dimensional set can always be enlarged. For suppose we want to enlarge t_1, t_2 to t_1, s, t_2 (where $t_1 < s < t_2$). For the projection ψ from R^3 to R^2 defined by $\psi(u, v, w) = (u, w)$, we have $\pi_{t_1 t_2} = \psi \pi_{t_1 s t_2}$ and hence $\pi^{-1}_{t_1 t_2} H = \pi^{-1}_{t_1 s t_2} \psi^{-1} H$, and of course $\psi^{-1} H \in \mathcal{R}^3$ if $H \in \mathcal{R}^2$. The proof for the general case involves more notation, but the idea is the same. It follows as before that the class $\mathcal{C}_f$ of finite-dimensional sets is a π-system. Furthermore, we have $B(x, \epsilon)^- = \bigcap_r [y: |y(r) - x(r)| \le \epsilon]$, where r ranges over the rationals in $[0, 1]$. Therefore, the σ-field $\sigma(\mathcal{C}_f)$ generated by $\mathcal{C}_f$ contains the closed balls, hence the open balls, and hence (separability) the open sets. Since $\mathcal{C}_f$ is a π-system and $\sigma(\mathcal{C}_f) = \mathcal{C}$, *$\mathcal{C}_f$ is a separating class.* □

The final example clarifies several technical points. Let $\mathcal{S}_0$ be the σ-field generated by the open balls; we can call it the *ball σ-field.* Of course, $\mathcal{S}_0 \subset \mathcal{S}$. If S is separable, then each open set is a countable union of open balls, and therefore $\mathcal{S}_0 = \mathcal{S}$, a fact we used in each of the preceding two examples. In the nonseparable case, $\mathcal{S}_0$ is usually[†] smaller than $\mathcal{S}$.

Example 1.4. Let S be an uncountable discrete space ($\rho(x, y) = 1$ for $x \ne y$); S is complete but not separable. Since the open balls are the singletons and S itself, $\mathcal{S}_0$ consists of the countable and the cocountable sets. But since every set is open, $\mathcal{S} = 2^S$, and so $\mathcal{S}_0$ is strictly smaller than $\mathcal{S}$.

Suppose there exists on $\mathcal{S}$ a probability measure P that is not tight. If S_0 consists of all the x for which $P\{x\} > 0$, then S_0 is countable[‡] and hence $PS_0 < 1$ (since otherwise P would be tight). But then, if $\mu A = P(A \cap S_0^c)$ for A in $\mathcal{S}$, μ is a finite, nontrivial measure (countably additive) on the class 2^S, and $\mu\{x\} = 0$ for each x. If S has the power of the continuum, then, assuming the axiom of choice and the continuum hypothesis, one can show [PM.46] that this is impossible. Thus Theorem 1.3 sometimes holds in the nonseparable case. □

The ball σ-field will play a role only in Sections 6 and 15.

[†] For some nonseparable spaces, the two σ–fields coincide—by accident, so to speak. See Talagrand [64].

[‡] Since P is finite, $\mathcal{S}$ cannot contain an uncountable, disjoint collection of sets of positive P-measure [PM.162].

Problems

A simple assertion is understood to be prefaced by "show that." See Some Notes on the Problems, p. 264.

1.1. The open finite-dimensional sets in R^∞ form a basis for the topology. Is this true in C?

1.2. Are $\mathcal{R}_f^\infty$ and $\mathcal{C}_f$ σ-fields?

1.3. Show that, if A and B are at positive distance, then they can be separated by a uniformly continuous f, in the sense that $I_A \le f \le I_{B^c}$. If A and B have disjoint closures but are at distance zero, this holds for a continuous f but not for a uniformly continuous one.

1.4. If S is a Banach space, then either (i) no closed ball of positive radius is compact or else (ii) they all are. Alternative (i) holds for R^∞ and C. Alternative (ii) holds if and only if S has a finite basis; see Liusternik and Sobolev [44], p. 69.

1.5. Suppose only that P is finitely additive. Show that, if each PA is the supremum of PK over the compact subsets K of A, then P is countably additive after all.

1.6. A real function on a metric space S is by definition a Borel function if it is measurable with respect to $\mathcal{S}$. It is by definition a Baire function if it is measurable with respect to the σ-field generated by the continuous functions. Show (what is not true for the general topological space) that the two concepts coincide.

1.7. Inequivalent metrics can give rise to the same class of Borel sets.

1.8. Try to reverse the roles of F and G in Theorem 1.1. When is it true that, for each ϵ, there exist an open G and a closed F such that $G \subset A \subset F$ and $P(F - G) < \epsilon$?

1.9. If S is separable and locally compact, then it is σ-compact, which implies of course that each probability measure on $\mathcal{S}$ is tight. Euclidean space is an example.

1.10. Call a class $\mathcal{F}$ of bounded, continuous functions a separating class if $Pf = Qf$ for all f in $\mathcal{F}$ implies that $P = Q$. The functions (1.1) form a separating class. Show that $\mathcal{F}$ is a separating class if each bounded, continuous function is the uniform limit of elements of $\mathcal{F}$.

1.11. Suppose that S is separable and locally compact. Since S is then σ-compact, each compact set in S is contained in the interior of another compact set. Suppose that $Pf = Qf$ for all continuous f with compact support; show that P and Q agree for compact sets, for closed sets, for Borel sets: The continuous functions with compact support form a separating class in the sense of the prededing problem.

1.12. Completeness can be replaced by topological completeness [M4] in Theorem 1.3, and separability can be replaced by the hypothesis that P has separable support.

1.13. The hypothesis of completeness (or topological completeness) cannot be suppressed in Theorem 1.3. Let S be a subset of $[0,1]$ with inner and outer Lebesgue measures 0 and 1: $\lambda_*(S) = 0$, $\lambda^*(S) = 1$. Give S the usual metric, and let P be the restriction of λ^* to $\mathcal{S}$ ([M10] & [PM.39]). If K is compact, then $PK = 0$, and so P is not tight.

1.14. If S consists of the rationals with the relative topology of the line, then each P on S is tight, even though S is not topologically complete (use the Baire category theorem).

1.15. If $r \le \epsilon/(1+\epsilon)2^k$, then (see (1.3)) $B(x,r) \subset N_{k,\epsilon}(x)$.

1.16. If A is nowhere dense, then $A^\circ = \emptyset$, but the converse is false. Find an A that is everywhere dense even though $A^\circ = \emptyset$.

1.17. Every locally compact subset of C is nowhere dense.

1.18. In connection with Example 1.3, consider the space $C_b(T)$ of bounded, continuous functions on the general metric space T; metrize it by (1.4) (we specify boundedness because otherwise $\|x\|$ may not be finite). Show that $C_b(T)$ is complete. Show that it need not be separable, even if T is totally bounded.

SECTION 2. PROPERTIES OF WEAK CONVERGENCE

We have defined $P_n \Rightarrow P$ to mean that $P_n f \to Pf$ for each bounded, continuous real f on S. Note that, since the integrals Pf completely determine P (Theorem 1.2), a single sequence $\{P_n\}$ cannot converge weakly to each of two different limits. Although it is not important to the subject at this point, it is easy to topologize the space of probability measures on $(S, \mathcal{S})$ in such a way that weak convergence is convergence in this topology: Take as the basic neighborhoods the sets of the form $[Q\colon |Qf_i - Pf_i| < \epsilon, i \le k]$, where the f_i are bounded and continuous. If S is separable and complete, this topology can be defined by a metric, the Prohorov metric; see Section 6.

Weak convergence is the subject of the entire book. We start with a pair of simple examples to illustrate the ideas lying behind the definition.

Example 2.1. On an arbitrary S, write δ_x for the *unit mass at* x, the probability measure on $\mathcal{S}$ defined by $\delta_x(A) = I_A(x)$. If $x_n \to x_0$ and f is continuous, then

$$\delta_{x_n} f = f(x_n) \to f(x_0) = \delta_{x_0} f, \tag{2.1}$$

and therefore $\delta_{x_n} \Rightarrow \delta_{x_0}$. On the other hand, if $x_n \not\to x_0$, then there is an ϵ such that $\rho(x_0, x_n) > \epsilon$ for infinitely many n. If f is the function in (1.1) for $F = \{x_0\}$, then $f(x_0) = 1$ and $f(x_n) = 0$ for infinitely many n, and so (2.1) fails: $\delta_{x_n} \not\Rightarrow \delta_{x_0}$. Therefore, $\delta_{x_n} \Rightarrow \delta_{x_0}$ if and only if $x_n \to x_0$. This simplest of examples is useful when doubtful conjectures present themselves. □

Example 2.2. Let S be $[0,1]$ with the usual metric, and for each n, suppose that x_{nk}, $0 \le k < r_n$, are r_n points of $[0,1]$. Suppose that these points are asymptotically uniformly distributed, in the sense that, for each subinterval J,

$$\frac{1}{r_n}\#[k: x_{nk} \in J] \to |J|, \tag{2.2}$$

where $|J|$ denotes length: As $n \to \infty$, the proportion of the points x_{nk} that lie in an interval is asymptotically equal to its length. Take P_n to have a point-mass of $1/r_n$ at each x_{nk} (if several of the x_{nk} coincide, let the masses add), and let P be Lebesgue measure restricted to $[0,1]$. If (2.2) holds, then $P_n \Rightarrow P$. For suppose that f is continuous on $[0,1]$. Then it is Riemann integrable, and for any given ϵ, there is a finite decomposition of $[0,1]$ into subintervals J_i such that, if v_i and u_i are the supremum and infimum of f over J_i, then the upper and lower Darboux sums $\sum v_i|J_i|$ and $\sum u_i|J_i|$ are within ϵ of the Riemann intergal $Pf = \int_0^1 f(x)\,dx$. By(2.2),

$$P_n f = \sum_k \frac{1}{r_n} f(x_{nk}) \le \sum_i v_i \frac{1}{r_n}\#[k: x_{nk} \in J_i] \to \sum_i v_i|J_i| \le Pf + \epsilon.$$

This, together with the symmetric bound from below, shows that $P_n f \to Pf$. Therefore, $P_n \Rightarrow P$; this fact, as Theorem 2.1 will show, contains information going beyond (2.2).

As a special case, take $r_n = 10^n$ and let $x_{nk} = k10^{-n}$ for $0 \le k < r_n$. If $J = (a,b]$, then (2.2) holds because the set there consists of those k satisfying $\lfloor a10^n \rfloor < k \le \lfloor b10^n \rfloor$. That $P_n \Rightarrow P$ holds in this case is an expression of the fact that one can produce approximately uniformly distributed observations by generating a stream of random digits (somehow) and for (a suitably chosen) large n breaking them into groups of n with decimal points to the left.

As a second special case, take x_{nk} to be the fractional part of $k\theta$, $0 \le k < r_n = n$. If θ is irrational, then (2.2) holds [PM.328], which is expressed by saying that the integer multiples of θ are uniformly distributed modulo 1. □

The Portmanteau Theorem

The following theorem provides useful conditions equivalent to weak convergence; any of them could serve as the definition. A set A in S whose boundary ∂A satisfies $P(\partial A) = 0$ is called a *P-continuity set* (note that ∂A is closed and hence lies in $\mathcal{S}$). Let P_n, P be probability measures on $(S, \mathcal{S})$.

Theorem 2.1. *These five conditions are equivalent:*

(i) $P_n \Rightarrow P$.
(ii) $P_n f \to Pf$ *for all bounded, uniformly continuous* f.
(iii) $\limsup_n P_n F \le PF$ *for all closed* F.
(iv) $\liminf_n P_n G \ge PG$ *for all open* G.
(v) $P_n A \to PA$ *for all* P*-continuity sets* A.

To see the significance of these conditions, return to Example 2.1. Suppose that $x_n \to x_0$, so that $\delta_{x_n} \Rightarrow \delta_{x_0}$, and suppose further that the x_n are all distinct from x_0 (take $x_0 = 0$ and $x_n = 1/n$ on the line, for example). Then the inequality in part (iii) is strict if $F = \{x_0\}$, and the inequality in (iv) is strict if $G = \{x_0\}^c$. If $A = \{x_0\}$, then convergence does not hold in (v); but this does not contradict the theorem, because the limit measure of $\partial\{x_0\} = \{x_0\}$ is 1, not 0.

And suppose in Example 2.2 that (2.2) holds, so that $P_n \Rightarrow P$. If A is the set of all the x_{nk}, for all n and k, then A is countable and supports each P_n, so that $P_n A = 1 \not\to PA = 0$; but of course, $\partial A = S$ in this case. By regularity (Theorem 1.1), there is an open G such that $A \subset G$ and $PG < \frac{1}{2}$ (say); for this G the inequality in part (iv) is strict. To end on a positive note: If (2.2) holds for intervals J, then by part (v) of the theorem, it also holds for a much wider class of sets—those having boundary of Lebesgue measure 0.

PROOF OF THEOREM 2.1. Of course, the implication (i) $\to$ (ii) is trivial.

Proof that (ii) $\to$ (iii). The f of (1.1) is bounded and uniformly continuous. By the two inequalities in (1.1), condition (ii) here implies $\limsup_n P_n F \le \limsup_n P_n f = Pf \le PF^\epsilon$. If F is closed, letting $\epsilon \downarrow 0$ gives the inequality in (iii).

The equivalence of (iii) and (iv) follows easily by complementation.

Proof that (iii) & (iv) $\to$ (v). If A° and A^- are the interior and closure of A, then conditions (iii) and (iv) together imply

$$PA^- \ge \limsup_n P_n A^- \ge \limsup_n P_n A \ge \liminf_n P_n A \ge \liminf_n P_n A^\circ \ge PA^\circ. \tag{2.3}$$

If A is a P-continuity set, then the extreme terms here coincide with PA, and (v) follows.

Proof that (v) $\to$ (i). By linearity we may assume that the bounded f satisfies $0 < f < 1$. Then $Pf = \int_0^\infty P[f > t]\,dt = \int_0^1 P[f > t]\,dt$, and similarly for $P_n f$. If f is continuous, then $\partial[f > t] \subset [f = t]$, and hence $[f > t]$ is a P-continuity set except for countably many t. By condition (v) and the bounded convergence theorem,

$$P_n f = \int_0^1 P_n[f > t]\,dt \to \int_0^1 P[f > t]\,dt = Pf. \qquad \square$$

Other Criteria

Weak convergence is often proved by showing that $P_n A \to PA$ holds for the sets A of some advantageous subclass of $\mathcal{S}$.

Theorem 2.2. *Suppose* (i) *that $\mathcal{A}_P$ is a π-system and* (ii) *that each open set is a countable union of $\mathcal{A}_P$-sets. If $P_n A \to PA$ for every A in $\mathcal{A}_P$, then $P_n \Rightarrow P$.*

PROOF. If $A_1, \ldots A_r$ lie in $\mathcal{A}_P$, then so do their intersections; hence, by the inclusion-exclusion formula,

$$P_n\Big(\bigcup_{i=1}^r A_i\Big) = \sum_i P_n A_i - \sum_{ij} P_n A_i A_j + \sum_{ijk} P_n A_i A_j A_k - \cdots$$

$$\to \sum_i PA_i - \sum_{ij} PA_i A_j + \sum_{ijk} PA_i A_j A_k - \cdots = P\Big(\bigcup_{i=1}^r A_i\Big).$$

If G is open, then $G = \bigcup_i A_i$ for some sequence $\{A_i\}$ of sets in $\mathcal{A}_P$. Given ϵ, choose r so that $P(\bigcup_{i\le r} A_i) > PG - \epsilon$. By the relation just proved, $PG - \epsilon \le P(\bigcup_{i\le r} A_i) = \lim_n P_n(\bigcup_{i\le r} A_i) \le \liminf_n P_n G$. Since ϵ was arbitrary, condition (iv) of the preceding theorem holds. $\square$

The next result transforms condition (ii) above in a useful way.

Theorem 2.3. *Suppose* (i) *that $\mathcal{A}_P$ is a π-system and* (ii) *that S is separable and, for every x in S and positive ϵ, there is in $\mathcal{A}_P$ an A for which $x \in A^\circ \subset A \subset B(x, \epsilon)$. If $P_n A \to PA$ for every A in $\mathcal{A}_P$, then $P_n \Rightarrow P$.*

PROOF. The hypothesis implies that, for each point x of a given open set G, $x \in A_x^\circ \subset A_x \subset G$ holds for some A_x in $\mathcal{A}_P$. Since S is separable, there is a countable subcollection $\{A_{x_i}^\circ\}$ of $\{A_x^\circ : x \in G\}$ that covers G (the Lindelöf property [M3]), and then $G = \bigcup_i A_{x_i}^\circ$. Therefore, the hypotheses of Theorem 2.2 are satisfied. $\square$

Call a subclass $\mathcal{A}$ of $\mathcal{S}$ a *convergence-determining class* if, for every P and every sequence $\{P_n\}$, convergence $P_n A \to PA$ for all P-continuity sets in $\mathcal{A}$ implies $P_n \Rightarrow P$. A convergence-determining class is obviously a separating class in the sense of the preceding section. To ensure that a given $\mathcal{A}$ is a convergence-determining class, we need conditions implying that, whatever P may be, the class $\mathcal{A}_P$ of P-continuity sets in $\mathcal{A}$ satisfies the hypothesis of Theorem 2.3. For given $\mathcal{A}$, let $\mathcal{A}_{x,\epsilon}$ be the class of $\mathcal{A}$-sets satisfying $x \in A^\circ \subset A \subset B(x, \epsilon)$, and let $\partial\mathcal{A}_{x,\epsilon}$ be the class of their boundaries. If $\partial\mathcal{A}_{x,\epsilon}$ contains uncountably many disjoint sets, then at least one of them must have P-measure 0, which provides a usable condition.

Theorem 2.4. *Suppose* (i) *that $\mathcal{A}$ is a π-system and* (ii) *that S is separable and, for each x and ϵ, $\partial\mathcal{A}_{x,\epsilon}$ either contains $\emptyset$ or contains uncountably many disjoint sets. Then $\mathcal{A}$ is a convergence-determining class.*

Since $\partial B(x,r) \subset [y\colon \rho(x,y) = r]$, the finite intersections of open balls satisfy the hypotheses.

PROOF. Fix an arbitrary P and let $\mathcal{A}_P$ be the class of P-continuity sets in $\mathcal{A}$. Since

$$\partial(A \cap B) \subset (\partial A) \cup (\partial B), \tag{2.4}$$

$\mathcal{A}_P$ is a π-system. Suppose that $P_n A \to PA$ for every A in $\mathcal{A}$ satisfying $P(\partial A) = 0$, that is, for every A in $\mathcal{A}_P$. If $\partial\mathcal{A}_{x,\epsilon}$ does not contain $\emptyset$, then it must contain uncountably many distinct, pairwise disjoint sets; in either case, it contains a set of P-measure 0. This means that each $\mathcal{A}_{x,\epsilon}$ contains an element of $\mathcal{A}_P$, which therefore satisfies the hypothesis of Theorem 2.3. Since $P_n A \to PA$ for each A in $\mathcal{A}_P$, it follows that $P_n \Rightarrow P$. □

Example 2.3. Consider R^k, as in Example 1.1, and let $\mathcal{A}$ be the class of rectangles, the sets $[y\colon a_i < y_i \le b_i,\ i \le k]$. Since it obviously satisfies the hypotheses of Theorem 2.4, $\mathcal{A}$ is a convergence-determining class.

The class of sets $Q_x = [y\colon y_i \le x_i,\ i \le k]$ is also a convergence-determining class. For suppose that $P_n Q_x \to PQ_x$ for each x such that $P(\partial Q_x) = 0$. The set E_i of t satisfying $P[y\colon y_i = t] > 0$ is at most countable, and so the set $D = (\bigcup_i E_i)^c$ is dense. Let $\mathcal{A}_P$ be the class of rectangles such that each coordinate of each vertex lies in D. If $A \in \mathcal{A}_P$, then, since $\partial Q_x \subset \bigcup_i [y\colon y_i = x_i]$, Q_x is a P-continuity set for each vertex x of A, and it follows by inclusion-exclusion that $P_nA \to PA$. From this and the fact that D is dense, it follows that $\mathcal{A}_P$ satisfies the hypothesis of Theorem 2.3.

That these sets Q_x form a convergence-determining class can be restated as a familiar fact. Let $F(x) = P(Q_x)$ and $F_n(x) = P_n(Q_x)$ be the distribution functions for P and P_n. Since F is continuous at x if and only if Q_x is a P-continuity set, $P_n \Rightarrow P$ if and only if $F_n(x) \to F(x)$ for all continuity points x of F. □

Example 2.4. In Example 1.2 we showed that the class $\mathcal{R}_f^\infty$ of finite-dimensional sets is a separating class. It is also a convergence-determining class: Given x and ϵ, choose k so that $2^{-k} < \epsilon/2$ and consider the finite-dimensional sets $A_\eta = [y: |y_i - x_i| < \eta,\ i \le k]$ for $0 < \eta < \epsilon/2$. Then $x \in A_\eta^\circ = A_\eta \subset B(x, \epsilon)$. Since ∂A_η consists of the points y such that $|y_i - x_i| \le \eta$ for all $i \le k$, with equality for some i, these boundaries are disjoint. And since R^∞ is separable, Theorem 2.4 applies: *$\mathcal{R}_f^\infty$ is a convergence-determining class*, and $P_n \Rightarrow P$ if and only if $P_nA \to PA$ for all finite-dimensional P-continuity sets A. □

Example 2.5. In Example 1.3, the sequence of functions z_n defined by (1.5) shows that the space C is not σ-compact, but it also shows something much more important: Although the class $\mathcal{C}_f$ of finite-dimensional sets in C is a separating class, it is *not* a convergence-determining class. For let $P_n = \delta_{z_n}$ and let $P = \delta_0$ be the unit mass at the 0-function. Then $P_n \not\Rightarrow P$, because $z_n \not\to 0$.

On the other hand, if $2n^{-1}$ is less than the smallest nonzero t_i, then $\pi_{t_1\cdots t_k}(z_n) = \pi_{t_1\cdots t_k}(0) = (0, \cdots, 0)$, and so $P_n\pi_{t_1\cdots t_k}^{-1}H = P\pi_{t_1\cdots t_k}^{-1}H$ for all H. In this example, $P_nA \to PA$ for all sets A in $\mathcal{C}_f$ (including those that are not P-continuity sets, as it happens), even though $P_n \not\Rightarrow P$. In the space C, the arguments and results involving weak convergence go far beyond the finite-dimensional theory. □

Theorem 2.2 has a corollary used in Section 4. Recall that $\mathcal{A}$ is a semiring if it is a π-system containing $\emptyset$ and if $A, B \in \mathcal{A}$ and $A \subset B$ together imply that there exist finitely many disjoint $\mathcal{A}$-sets C_i such that $B - A = \bigcup_{i=1}^m C_i$.

Theorem 2.5. *Suppose that* (i) *$\mathcal{A}$ is a semiring and* (ii) *each open set is a countable union of $\mathcal{A}$-sets. If $PA \le \liminf_n P_nA$ for each A in $\mathcal{A}$, then $P_n \Rightarrow P$.*

PROOF. If $A_1, \cdots, A_r$ lie in $\mathcal{A}$, then, since $\mathcal{A}$ is a semiring, $\bigcup_{i=1}^r A_i$ can be represented as a disjoint union $\bigcup_{j=1}^s B_j$ of other sets in $\mathcal{A}$ [PM.168], and it follows that

$$P\left(\bigcup_{i=1}^r A_i\right) = P\left(\bigcup_{j=1}^s B_j\right) = \sum_{j=1}^s PB_j \le \liminf_n \sum_{j=1}^s P_n B_j$$
$$= \liminf_n P_n\left(\bigcup_{i=1}^r A_i\right).$$

The proof is completed as before. □

A further simple condition for weak convergence:

Theorem 2.6. *A necessary and sufficient condition for $P_n \Rightarrow P$ is that each subsequence $\{P_{n_i}\}$ contain a further subsequence $\{P_{n_{i(m)}}\}$ converging weakly ($m \to \infty$) to P.*

PROOF. The necessity is easy (but not useful). As for sufficiency, if $P_n \not\Rightarrow P$, then $P_n f \not\to Pf$ for some bounded, continuous f. But then, for some positive ϵ and some subsequence P_{n_i}, $|P_{n_i} f - Pf| > \epsilon$ for all i, and no further subsequence can converge weakly to P. □

The Mapping Theorem

Suppose that h maps S into another metric space S', with metric ρ' and Borel σ-field $\mathcal{S}'$. If h is measurable $\mathcal{S}/\mathcal{S}'$ [M10], then each probability P on $(S, \mathcal{S})$ induces on $(S', \mathcal{S}')$ a probability Ph^{-1} defined as usual by $Ph^{-1}(A) = P(h^{-1}A)$. We need conditions under which $P_n \Rightarrow P$ implies $P_n h^{-1} \Rightarrow Ph^{-1}$. One such condition is that h is continuous: If f is bounded and continuous on S', then fh is bounded and continuous on S, and by change of variable [PM.216], $P_n \Rightarrow P$ implies

$$(2,5) \quad \int_{S'} f(y) P_n h^{-1}(dy) = \int_S f(h(x)) P_n(dx)$$
$$\to \int_S f(h(x)) P(dx) = \int_{S'} f(y) Ph^{-1}(dy).$$

Example 2.6. Since the natural projections π_k from R^∞ to R^k are continuous, if $P_n \Rightarrow P$ holds on R^∞, then $P_n \pi_k^{-1} \Rightarrow P\pi_k^{-1}$ holds on R^k for each k. As the following argument shows, the converse implication is a consequence of the fact that the class $\mathcal{R}_f^\infty$ of finite-dimensional sets in R^∞ is a convergence-determining class (Example 2.4).

From the continuity of π_k it follows easily that $\partial \pi_k^{-1} H \subset \pi_k^{-1} \partial H$ for $H \subset R^k$. Using special properties of the projections we can prove inclusion in the other direction. If $x \in \pi_k^{-1} \partial H$, so that $\pi_k x \in \partial H$, then there are points $\alpha^{(u)}$ in H and points $\beta^{(u)}$ in H^c such that $\alpha^{(u)} \to \pi_k x$ and

$\beta^{(u)} \to \pi_k x$ $(u \to \infty)$. Since the points $(\alpha_1^{(u)}, \cdots \alpha_k^{(u)}, x_{k+1}, \cdots)$ lie in $\pi_k^{-1}H$ and converge to x, and since the points $(\beta_1^{(u)}, \cdots \beta_k^{(u)}, x_{k+1}, \cdots)$ lie in $(\pi_k^{-1}H)^c$ and also converge to x, it follows that $x \in \partial(\pi_k^{-1}H)$. Therefore, $\partial\pi_k^{-1}H = \pi_k^{-1}\partial H$.

If $A = \pi_k^{-1}H$ is a finite-dimensional P-continuity set, then we have $P\pi_k^{-1}(\partial H) = P(\pi_k^{-1}\partial H) = P(\partial\pi_k^{-1}H) = P(\partial A) = 0$, and so H is a $P\pi_k^{-1}$-continuity set. This means that, if $P_n\pi_k^{-1} \Rightarrow P\pi_k^{-1}$ for all k, then $P_nA \to PA$ for each P-continuity set A in $\mathcal{R}_f^\infty$, and hence (since $\mathcal{R}_f^\infty$ is a convergence-determining class) it means that $P_n \Rightarrow P$. Therefore: *$P_n \Rightarrow P$ if and only if $P_n\pi_k^{-1} \Rightarrow P\pi_k^{-1}$ for all k.* This is essentially just a restatement of the fact that the finite-dimensional sets form a convergence-determining class. The theory of weak convergence in R^∞ is applied in Section 4 to the study of some problems in number theory and combinatorial analysis. □

Example 2.7. Because of the continuity of the natural projections $\pi_{t_1\cdots t_k}$ from C to R^k, if $P_n \Rightarrow P$ for probabilities on C, then $P_n\pi_{t_1\cdots t_k}^{-1} \Rightarrow P\pi_{t_1\cdots t_k}^{-1}$ for all k and all k–tuples $t_1, \ldots, t_k$. But the converse is false, because, as Example 2.5 shows, $\mathcal{C}_f$ is not a convergence-determining class. In fact, for P_n and P as in that example, $P_n \not\Rightarrow P$, even though for $2n^{-1}$ less than the smallest nonzero t_i we have $P_n\pi_{t_1\cdots t_k}^{-1} = P\pi_{t_1\cdots t_k}^{-1}$ (a unit mass at the origin of R^k). Again: Weak-convergence theory in C goes beyond the finite-dimensional case in an essential way. The space C is studied in detail in Chapter 2. □

By (2.5), $P_n \Rightarrow P$ implies $P_nh^{-1} \Rightarrow Ph^{-1}$ if h is a continuous mapping from S to S', but the continuity assumption can be weakened. Assume only that h is measurable $\mathcal{S}/\mathcal{S}'$, and let D_h be the set of its discontinuities; D_h lies in $\mathcal{S}$ [M10]. The *mapping theorem*:

Theorem 2.7. *If $P_n \Rightarrow P$ and $PD_h = 0$, then $P_nh^{-1} \Rightarrow Ph^{-1}$.*

PROOF. If $x \in (h^{-1}F)^-$, then $x_n \to x$ for some sequence $\{x_n\}$ such that $hx_n \in F$; but then, if $x \in D_h^c$, hx lies in F^-. Therefore, $D_h^c \cap (h^{-1}F)^- \subset h^{-1}(F^-)$. If F is a closed set in S', it therefore follows from $PD_h^c = 1$ that

$$(2.6)\quad \limsup_n P_n(h^{-1}F) \le \limsup_n P_n(h^{-1}F)^- \le P(h^{-1}F)^- \\ = P(D_h^c \cap (h^{-1}F)^-) \le P(h^{-1}(F^-)) = P(h^{-1}F).$$

Condition (iii) of Theorem 2.1 holds.[†] □

Example 2.8. Let F be a distribution function on the line, and let φ be the corresponding quantile function: $\varphi(u) = \inf[x: u \le F(x)]$ for $0 < u < 1$ (put $\varphi(0) = \varphi(1) = 0$, say). If P is Lebesgue measure restricted to $[0,1]$, as in Example 2.2, then $P\varphi^{-1}$ is the probability measure having distribution function F, since $P\varphi^{-1}(-\infty, x] = P[u: \varphi(u) \le x] = P[u: u \le F(x)] = F(x)$. Since φ has at most countably many disconinuities, $PD_\varphi = 0$. If P_n is also defined as in Example 2.2, then $P_n \Rightarrow P$, and it follows that $P_n\varphi^{-1} \Rightarrow P\varphi^{-1}$. Consider the case where $x_{nk} = k10^{-n}$. If φ is calculated for each approximately uniformly distributed observation as it is generated, this gives a sequence of observations approximately distributed according to F. □

Example 2.9. If $S_0 \in \mathcal{S}$, then [M10] the Borel σ-field for S_0 in the relative topology consists of the $\mathcal{S}$-sets contained in S_0. Suppose that P_n and P are probability measures on $\mathcal{S}$ and that $P_nS_0 \equiv PS_0 = 1$. Let Q_n and Q be the restrictions of P_n and P to $\mathcal{S}_0$. The identity map h from S_0 to S is continuous, and $P_n = Q_nh^{-1}$, $P = Qh^{-1}$. Therefore, by the mapping theorem, $Q_n \Rightarrow Q$ implies $P_n \Rightarrow P$.

The converse holds as well. The general open set in S_0 is $G \cap S_0$, where G is open in S. But $Q_n(G \cap S_0) = P_n(G \cap S_0) = P_nG$, and similarly for Q. If $P_n \Rightarrow P$, then $\liminf_n Q_n(G \cap S_0) = \liminf_n P_nG \ge PG = Q(G \cap S_0)$. Therefore:

If $P_nS_0 \equiv PS_0 = 1$, then $P_n \Rightarrow P$ (on S) if and only if $Q_n \Rightarrow Q$ (on S_0). □

Example 2.10. Suppose again that $S_0 \in \mathcal{S}$ and $P_nS_0 \equiv PS_0 = 1$. And suppose that the measurable map $h: S \to S'$ is continuous when restricted to S_0, in the sense that, if points x_n of S_0 converge to a point x of S_0, then $hx_n \to hx$. If $P_n \Rightarrow P$, then the Q_n and Q of the preceding example satisfy $Q_n \Rightarrow Q$. The restriction h_0 of h from S to S_0 is a continuous map from S_0 to S', and the mapping theorem gives $(S_0 \cap h^{-1}A' = h_0^{-1}A')$ $P_nh^{-1} = Q_nh_0^{-1} \Rightarrow Qh_0^{-1} = Ph^{-1}$. Therefore:

If $P_nS_0 \equiv PS_0 = 1$ and h is continuous when restricted to S_0, then $P_n \Rightarrow P$ implies $P_nh^{-1} \Rightarrow Ph^{-1}$.[‡] □

[†] For a different approach to the mapping theorem, see Problem 2.10.

[‡] See Problem 2.2.

Product Spaces

Assume that the product $T = S' \times S''$ is separable, which implies that S' and S'' are separable and that the three Borel σ-fields are related by $\mathcal{T} = \mathcal{S}' \times \mathcal{S}''$ [M10]. Denote the marginal distributions of a probability measure P on $\mathcal{T}$ by P' and P'': $P'(A') = P(A' \times S'')$ and $P''(A'') = P(S' \times A'')$. Since the projections $\pi'(x', x'') = x'$ and $\pi''(x', x'') = x''$ are continuous, and since $P' = P(\pi')^{-1}$ and $P'' = P(\pi'')^{-1}$, it follows by the mapping theorem that $P_n \Rightarrow P$ implies $P_n' \Rightarrow P'$ and $P_n'' \Rightarrow P''$.

The reverse implication is false.[†] But consider the π-system $\mathcal{A}$ of measurable rectangles $A' \times A''$ ($A' \in \mathcal{S}'$ and $A'' \in \mathcal{S}''$). If we take the distance between (x', x'') and (y', y'') to be $\rho(x', y') \vee \rho(x'', y'')$, then the open balls in T have the form [M10]

$$B_t((x', x''), r)) = B_{\rho'}(x', r) \times B_{\rho''}(x'', r). \tag{2.7}$$

These balls lie in $\mathcal{A}$, and since the $\partial B_t((x', x''), r)$ are disjoint for different values of r, $\mathcal{A}$ satisfies the hypothesis of Theorem 2.4 and is therefore a convergence-determining class.

There is a related result that is more useful. Let $\mathcal{A}_P$ be the class of $A' \times A''$ in $\mathcal{A}$ such that $P'(\partial A') = P''(\partial A'') = 0$. Applying (2.4) in S' and in S'' shows that $\mathcal{A}_P$ is a π-system. And since

$$\partial(A' \times A'') \subset ((\partial A') \times S'') \cup (S' \times (\partial A'')), \tag{2.8}$$

each set in $\mathcal{A}_P$ is a P-continuity set. Since the $B_{\rho'}(x', r)$ in (2.7) have disjoint boundaries for different values of r, and since the same is true of the $B_{\rho''}(x'', r)$, there are arbitrarily small r for which (2.7) lies in $\mathcal{A}_P$. It follows that Theorem 2.3 applies to $\mathcal{A}_P$: $P_n \Rightarrow P$ if and only if $P_n A \to PA$ for all A in $\mathcal{A}_P$. Therefore, we have the following theorem, in which (ii) is an obvious consequence of (i).

Theorem 2.8. (i) *If $T = S' \times S''$ is separable, then $P_n \Rightarrow P$ if and only if $P_n(A' \times A'') \to P(A' \times A'')$ for each P'-continuity set A' and each P''-continuity set A''.*

(ii) *If T is separable, then $P_n' \times P_n'' \Rightarrow P' \times P''$ if and only if $P_n' \Rightarrow P'$ and $P_n'' \Rightarrow P''$.*

Problems

2.1. According to Examples 2.4 and 2.5, $\mathcal{R}_f^\infty$ is a convergence-determining class, while $\mathcal{C}_f$ is not. What essential difference between R^∞ and C makes this possible? (See Problem 1.1.)

[†] See Problem 2.7.

2.2. Show that the assertion in Example 2.10 is false without the assumption that $P_n S_0 \equiv 1$, which points up the distinction between continuity of h at each point of S_0 and continuity of h when it is restricted to S_0.

2.3. If S is countable and discrete, then $P_n \Rightarrow P$ if and only if $P_n\{x\} \to P\{x\}$ for each singleton. Show that in this case $\sup_{A \in \mathcal{S}} |P_n A - PA| \to 0$.

2.4. In connection with Example 2.3, show for $k = 1$ that F has at most countably many discontinuities. Show for $k = 2$ that, if F has at least one discontinuity, then it has uncountably many. Show that, if $F_n(x) \to F(x)$ fails for one x, then it fails for uncountably many.

2.5. The class of P-continuity sets (P fixed) is a field but may not be a σ-field.

2.6. If f is bounded and upper semicontinuous [M8], then $P_n \Rightarrow P$ implies that $\limsup_n P_n f \le Pf$. Show that this contains part (iii) of Theorem 2.1 as a special case. Generalize part (iv) in the same way.

2.7. The uniform distribution on the unit square and the uniform distribution on its diagonal have identical marginal distributions. Relate this to Theorem 2.8.

2.8. Show that, if $\delta_{x_n} \Rightarrow P$, then $P = \delta_x$ for some x.

2.9. If f is $\mathcal{S}$-measurable and $P|f| < \infty$, then for each ϵ there is a bounded, uniformly continuous g such that $P|f - g| < \epsilon$.

2.10. **(a)** Without using the mapping theorem, show that $P_n \Rightarrow Pf$ if $P_n f \to P$ for all bounded, continuous f such that $PD_f = 0$. (See the proof of (v)$\to$(i) in Theorem 2.1.)

(b) Now give a second proof of the mapping theorem. (Use (2.5) and the fact that $D_{fh} \subset D_h$.)

SECTION 3. CONVERGENCE IN DISTRIBUTION

The theory of weak convergence can be paraphrased as the theory of convergence in distribution. When stated in terms of this second theory, which involves no new ideas, many results assume a compact and perspicuous form.

Random Elements

Let X be a mapping from a probability space $(\Omega, \mathcal{F}, \mathsf{P})$ to a metric space S. We call X a *random element* if it is measurable $\mathcal{F}/\mathcal{S}$; we say that it is defined *on* its domain Ω and *in* its range S, and we call it a random element *of* S. We call X a random *variable* if S is R^1, a random *vector* if S is R^k, a random *sequence* if S is R^∞, and a random *function* if S is C or some other function space.[†]

The *distribution* of X is the probability measure $P = \mathsf{P}X^{-1}$ on $(S, \mathcal{S})$ defined by

$$PA = \mathsf{P}(X^{-1}A) = \mathsf{P}[\omega: X(\omega) \in A] = \mathsf{P}[X \in A]. \tag{3.1}$$

[†] Often in the literature an arbitrary random element is called a random variable.

This is also called the *law* of X and denoted $\mathcal{L}(X)$. In the case $S = R^k$, there is also the associated *distribution function* of $X = (X_1, \ldots, X_k)$, defined by

$$F(x_1, \ldots, x_k) = P[y: y_i \le x_i,\ i \le k] = \mathsf{P}[X_i \le x_i,\ i \le k]. \tag{3.2}$$

Note that P is a probability measure on an arbitrary measurable space, whereas P is always defined on the Borel σ-field of a metric space. The distribution P contains the essential information about the random element X. For example, if f is a real measurable function on S (measurable $\mathcal{S}/\mathcal{R}^1$), then by change of variable,

$$\mathsf{E}[f(X)] = \int_\Omega f(X(\omega))\, \mathsf{P}(d\omega) = \int_S f(x) P(dx) = Pf, \tag{3.3}$$

in the sense that both integrals exist or neither does, and they have the same value if they do exist.

Each probability measure on each metric space is the distribution of some random element on some probability space. In fact, given P on $(S, \mathcal{S})$, we can simply take $(\Omega, \mathcal{F}, \mathsf{P}) = (S, \mathcal{S}, P)$ and take X to be the identity, so that $X(\omega) = \omega$ for $\omega \in \Omega = S$:

$$(\Omega, \mathcal{F}, \mathsf{P}) = (S, \mathcal{S}, P), \quad X(\omega) = \omega \text{ for } \omega \in \Omega = S. \tag{3.4}$$

Then X is a random element on Ω with values in S (measurable $\mathcal{F}/\mathcal{S}$), and it has P as its distribution.

Convergence in Distribution

We say a sequence $\{X_n\}$ of random elements *converges in distribution* to the random element X if $P_n \Rightarrow P$, where P_n and P are the distributions of X_n and X. In this case we write $X_n \Rightarrow X$. (This double use of the double arrow causes no confusion.) Thus $X_n \Rightarrow X$ means that $\mathcal{L}(X_n) \Rightarrow \mathcal{L}(X)$. Although this definition of course makes no sense unless the image space S (the range) and the topology on it are the same for all the $X, X_1, X_2, \cdots$, the underlying probability spaces (the domains)—$(\Omega, \mathcal{F}, \mathsf{P})$ and $(\Omega_n, \mathcal{F}_n, \mathsf{P}_n)$, say—may all be distinct. These spaces ordinarily remain offstage; we make no mention of them because their structures enter into the argument only by way of the distributions on S they induce. For example, if we write E_n for integrals with respect to P_n, then, by (3.3), $P_n f \to Pf$ if and only if $\mathsf{E}_n[f(X_n)] \to \mathsf{E}[f(X)]$. But we simply write P in place of P_n and E in place of E_n: P and E will refer to whatever probability space the random element in question is defined on. Thus $X_n \Rightarrow X$ if and only if $\mathsf{E}[f(X_n)] \to \mathsf{E}[f(X)]$ for all bounded, continuous f on S.

Theorem 2.1 asserts the equivalence of the following five statements. Call a set A in $\mathcal{S}$ an X-continuity set if $\mathsf{P}[X \in \partial A] = 0$.

(i) $X_n \Rightarrow X$.
(ii) $\mathsf{E}[f(X_n)] \to \mathsf{E}[f(X)]$ *for all bounded, uniformly continuous* f.
(iii) $\limsup_n \mathsf{P}[X_n \in F] \le \mathsf{P}[X \in F]$ *for all closed* F.
(iv) $\liminf_n \mathsf{P}[X_n \in G] \ge \mathsf{P}[X \in G]$ *for all open* G.
(v) $\mathsf{P}[X_n \in A] \to \mathsf{P}[X \in A]$ *for all* X*-continuity sets* A.

This requires no proof; it is just a matter of translating the terms. Each theorem about weak convergence can be recast in the same way. Suppose $h: S \to S'$ is measurable $\mathcal{S}/\mathcal{S}'$ and let D_h be the set of its discontinuity points. If X has distribution P, then $h(X)$ has distribution Ph^{-1}. Therefore, the mapping theorem becomes: $X_n \Rightarrow X$ (on S) implies $h(X_n) \Rightarrow h(X)$ (on S') if $\mathsf{P}[X \in D_h] = 0$.

The following hybrid terminology is convenient. If X_n and X are random elements of S, and if P_n and P are their distributions, then $X_n \Rightarrow X$ means $P_n \Rightarrow P$. But we can just as well write $X_n \Rightarrow P$ or $P_n \Rightarrow X$. Thus there are four contexts for the double arrow:

$$\left\{\begin{array}{l} P_n \Rightarrow P, \\ X_n \Rightarrow X, \\ X_n \Rightarrow P, \\ P_n \Rightarrow X. \end{array}\right. \tag{3.5}$$

The last three relations are defined by the first. If X_n are random variables having asymptotically the standard normal distribution, this fact is expressed as $X_n \Rightarrow N$, and one can interpret N as the standard normal distribution on the line or (better) as any random variable having this distribution. In all that follows, N will be such a random variable—normally distributed with mean 0 and variance 1.

Example 3.1. If $S_0 \in \mathcal{S}$, then the Borel σ-field of S_0 for the relative topology is $\mathcal{S}_0 = [A \cap S_0 : A \in \mathcal{S}]$ [M10] and $\mathcal{S}_0 \subset \mathcal{S}$. If $X: \Omega \to S_0$, then X is a random element of S_0 (measurable $\mathcal{F}/\mathcal{S}_0$) if and only if it is a random element of S (measurable $\mathcal{F}/\mathcal{S}$), the common requirement being that the set $[\omega: X(\omega) \in A \cap S_0] = [\omega: X(\omega) \in A]$ lie in $\mathcal{F}$ for every A in $\mathcal{S}$.

Suppose now that X_n and X are random elements of S_0. The general open set in S_0 is $G \cap S_0$ with G open in S. Therefore, there is convergence in distribution in the sense of S_0 if and only if there is convergence in distribution in the sense of S, the common condition being that $\liminf_n \mathsf{P}[X_n \in G \cap S_0] \geq \mathsf{P}[X \in G \cap S_0]$ for every open set G in S. This is essentially the convergence-in-distribution version of Example 2.9. □

Convergence in Probability

If, for an element a of S,

$$\mathsf{P}[\rho(X_n, a) < \epsilon] \to 1 \tag{3.6}$$

for each ϵ, we say X_n *converges in probability* to a. Conceive of a as a constant-valued random element on an arbitrary space, and suppose that (3.6) holds. If G is open and $a \in G$, then, for small enough ϵ, $\liminf_n \mathsf{P}[X_n \in G] \geq \lim_n \mathsf{P}[\rho(X_n, a) < \epsilon] = 1 = \mathsf{P}[a \in G]$, whereas if $a \notin G$, then $\liminf_n \mathsf{P}[X_n \in G] \geq 0 = \mathsf{P}[a \in G]$. Thus (3.6) implies $X_n \Rightarrow a$, in the sense of the second line of (3.5). On the other hand, if $X_n \Rightarrow a$ in this sense, then (3.6) follows because $[x\colon \rho(x, a) < \epsilon]$ is open. For this reason, we express convergence in probability as

$$X_n \Rightarrow a. \tag{3.7}$$

This can also be interpreted by the third line of (3.5): Identify a with δ_a. Again, the random elements X_n in (3.7) can be defined on different probability spaces. By the mapping theorem, $X_n \Rightarrow a$ implies $h(X_n) \Rightarrow h(a)$ if h is continuous at a.[†]

Suppose that (X_n, Y_n) is a random element of $S \times S$; this implies (since the projections $(x, y) \to x$ and $(x, y) \to y$ are continuous) that X_n and Y_n are random elements of S.[‡] If ρ is the metric on S, then $\rho(x, y)$ maps $S \times S$ continuously to the line, and so it makes sense to speak of the distance $\rho(X_n, Y_n)$—the random variable with value $\rho(X_n(\omega), Y_n(\omega))$ at ω. The next two theorems generalize much-used results about random variables.

Theorem 3.1. *Suppose that (X_n, Y_n) are random elements of $S \times S$. If $X_n \Rightarrow X$ and $\rho(X_n, Y_n) \Rightarrow 0$, then $Y_n \Rightarrow X$.*

[†] For X and X_n *all defined on the same* Ω, $\mathsf{P}[\rho(X_n, X) < \epsilon] \to_n 1$ defines convergence of X_n to X in probability, but this concept is not used in the book until Chapter 5 (except for Problem 3.8). Before that, X will always be a constant and the X_n may be defined on different spaces.

[‡] The reverse implication holds if S is separable, but not in complete generality [M10].

PROOF. Apply the convergence-in-distribution version of Theorem 2.1 twice. If F_ϵ is defined as $[x: \rho(x, F) \le \epsilon]$, then

$$\mathsf{P}[Y_n \in F] \le \mathsf{P}[\rho(X_n, Y_n) \ge \epsilon] + \mathsf{P}[X_n \in F_\epsilon].$$

Since F_ϵ is closed, the hypotheses imply

$$\limsup_n \mathsf{P}[Y_n \in F] \le \limsup_n \mathsf{P}[X_n \in F_\epsilon] \le \mathsf{P}[X \in F_\epsilon].$$

If F is closed, then $F_\epsilon \downarrow F$ as $\epsilon \downarrow 0$. □

Take $X_n \equiv X$:

Corollary. *Suppose that (X, Y_n) are random elements of $S \times S$. If $\rho(X, Y_n) \Rightarrow 0$, then $Y_n \Rightarrow X$.*

Theorem 3.2. *Suppose that (X_{un}, X_n) are random elements of $S \times S$. If*† *$X_{un} \Rightarrow_n Z_u \Rightarrow_u X$ and*

$$\lim_u \limsup_n \mathsf{P}[\rho(X_{un}, X_n) \ge \epsilon] = 0 \tag{3.8}$$

for each ϵ, then $X_n \Rightarrow_n X$.

PROOF. For F_ϵ defined as before, we have

$$\mathsf{P}[X_n \in F] \le \mathsf{P}[X_{un} \in F_\epsilon] + \mathsf{P}[\rho(X_{un}, X_n) \ge \epsilon].$$

Since $X_{un} \Rightarrow_n Z_u$ and F_ϵ is closed,

$$\limsup_n \mathsf{P}[X_n \in F] \le \mathsf{P}[Z_u \in F_\epsilon] + \limsup_n \mathsf{P}[\rho(X_{un}, X_n) \ge \epsilon].$$

And since $Z_u \Rightarrow_u X$, (3.8) gives

$$\limsup_n \mathsf{P}[X_n \in F] \le \mathsf{P}[X \in F_\epsilon].$$

Let $\epsilon \downarrow 0$, as before. □

Other standard results can be extended beyond the Euclidean case in a routine way:

---

† $X_{un} \Rightarrow_n Z_u$ indicates weak convergence as $n \to \infty$ with u fixed.

Example 3.2. Assume of the random elements (X_n, Y_n) of $S \times S$ that X_n and Y_n are independent in the sense that the events $[X_n \in A]$ and $[Y_n \in B]$ are independent for $A, B \in \mathcal{S}$; assume that X and Y are independent as well. If S is separable, then (Theorem 2.8) $X_n \Rightarrow X$ and $Y_n \Rightarrow Y$ imply $(X_n, Y_n) \Rightarrow (X, Y)$.

If $S = R^1$, we can go further and conclude from the mapping theorem that, for example, $X_n Y_n \Rightarrow XY$. If $X_n \equiv \alpha_n$ and $X \equiv \alpha$, then $X_n \Rightarrow X$ is the same thing as $\alpha_n \to \alpha$ (Example 2.1), and the independence condition is necessarily satisfied: $\alpha_n \to \alpha$ and $Y_n \Rightarrow Y$ imply $\alpha_n Y_n \Rightarrow \alpha Y$. □

Local vs. Integral Laws

Suppose P_n and P have densities f_n and f with respect to a measure μ on $(S, \mathcal{S})$. If

$$f_n(x) \to f(x) \tag{3.9}$$

outside a set of μ-measure 0, then, by Scheffé's theorem [PM.215],

$$\sup_{A \in \mathcal{S}} |PA - P_n A| \le \int_S |f(x) - f_n(x)|\, \mu(dx) \to 0. \tag{3.10}$$

Thus (3.9), a *local* limit theorem, implies $P_n \Rightarrow P$, an *integral* limit theorem. But the reverse implication is false:

Example 3.3. Let $P = \mu$ be Lebesgue measure on $S = [0, 1]$. Take f_n to be n^2 times the indicator of the set $\bigcup_{k=0}^{n-1}(kn^{-1}, kn^{-1}+n^{-3})$. Then $\mu[f_n > 0] = n^{-2}$, and by the Borel–Cantelli lemma, $f_n(x) \to 0$ outside a set B of μ-measure 0; redefine f_n as 0 on B, so that $f_n(x) \to 0$ everywhere. If P_n has density f_n with respect to μ, then $|P_n[0, x] - x| \le 1/n$, and it follows that $P_n \Rightarrow P$, a limit theorem for integral laws. With respect to μ, P has density $f(x) \equiv 1$, and so (3.9) does not hold for any x at all: There is no local limit theorem. □

Local laws imply integral laws also in the case where $S = R^k$, P has a density with respect to Lebesgue measure, and P_n is supported by a lattice. Let $\delta(n) = (\delta_1(n), \ldots, \delta_k(n))$ be a point of R^k with positive coordinates, let $\alpha(n) = (\alpha_1(n), \ldots, \alpha_k(n))$ be an arbitrary point of R^k, and denote by L_n the lattice consisting of the points of the form $(u_1\delta_1(n) - \alpha_1(n), \ldots, u_k\delta_k(n) - \alpha_k(n))$, where $u_1, \ldots, u_k$ range independently over the integers, positive and negative. If x is a point of L_n, then

$$[y\colon x_i - \delta_i(n) < y_i \le x_i,\ i \le k], \quad x \in L_n \tag{3.11}$$

is a cell of volume $v_n = \delta_1(n) \cdots \delta_k(n)$, and R^k is the union of this countable collection of cells.

Suppose now that P_n and P are probability measures on R^k, where P_n is supported by L_n and P has density p with respect to Lebesgue measure. For $x \in L_n$, let $p_n(x)$ be the P_n-mass (possibly 0) at that point.

Theorem 3.3. *Suppose that*

$$\delta_1(n) \vee \cdots \vee \delta_k(n) \to_n 0 \tag{3.12}$$

and that, if x_n is a point of L_n varying with n in such a way that $x_n \to x$, then

$$p_n(x_n)/v_n \to p(x). \tag{3.13}$$

Then $P_n \Rightarrow P$.

PROOF. Define a probability density q_n on R^k by setting $q_n(y) = p_n(x)/v_n$ if y lies in the cell (3.11). Since $x_n \to x$ implies (3.13), it follows by (3.12) that $q_n(x) \to p(x)$. Let X_n have the density q_n, and define Y_n on the same probability space by setting $Y_n = x$ if X_n lies in the cell (3.11). We are to prove that $Y_n \Rightarrow P$. Since $|X_n - Y_n| \leq |\delta(n)|$, this will follow by (3.12) and Theorem 3.1 if we prove that $X_n \Rightarrow P$. But since q_n converges pointwise to p, this is a consequence of Scheffé's theorem. □

Example 3.4. If S_n is the number of successes in n Bernoulli trials and $v_n = 1/\sqrt{npq}$, then [†]

$$\mathsf{P}[S_n = k]/v_n = \binom{n}{k} p^k q^{n-k} \sqrt{npq} \to_n \frac{1}{\sqrt{2\pi}} e^{-x^2/2}, \tag{3.14}$$

provided k varies with n in such a way that $(k - np)/\sqrt{npq} \to_n x$. Therefore, Theorem 3.3 applies to the lattice of points of the form $(k - np)/\sqrt{npq}$: P_n is the distribution of $(S_n - np)/\sqrt{npq}$ and P is the standard normal distribution. This gives the central limit theorem for Bernoulli trials: $(S_n - np)/\sqrt{npq} \Rightarrow N$. □

Integration to the Limit

If $X_n \Rightarrow X$ for random variables, when does $\mathsf{E}X_n \to \mathsf{E}X$ hold?

[†] Feller [28], p. 184.

Theorem 3.4. *If* $X_n \Rightarrow X$, *then* $\mathsf{E}|X| \le \liminf_n \mathsf{E}|X_n|$.

PROOF. By the mapping theorem, $|X_n| \Rightarrow |X|$, and therefore $\mathsf{P}[|X_n| > t] \to \mathsf{P}[|X| > t]$ for all but countably many t. By Fatou's lemma on the line,

$$\mathsf{E}|X| = \int_0^\infty \mathsf{P}[|X| > t]\,dt \le \liminf_n \int_0^\infty \mathsf{P}[|X_n| > t]\,dt = \liminf_n \mathsf{E}|X_n|.$$

□

The X_n are by definition *uniformly integrable* if

$$\lim_\alpha \sup_n \int_{|X_n| \ge \alpha} |X_n|\,d\mathsf{P} = 0. \tag{3.15}$$

This obviously holds if the X_n are uniformly bounded. If α is large enough that the supremum in (3.15) is at most 1, then $\sup_n \mathsf{E}|X_n| \le 1 + \alpha < \infty$.

Theorem 3.5. *If* X_n *are uniformly integrable and* $X_n \Rightarrow X$, *then* X *is integrable and* $\mathsf{E}X_n \to \mathsf{E}X$.

PROOF. Since the $\mathsf{E}|X_n|$ are bounded, Theorem 3.4 implies that X is integrable. And since $X_n^+ \Rightarrow X^+$ and $X_n^- \Rightarrow X^-$ by the mapping theorem, the variables can be assumed nonnegative, in which case

$$\mathsf{E}X_n = \int_0^\alpha \mathsf{P}[t < X_n < \alpha]\,dt + \int_{X_n \ge \alpha} X_n\,d\mathsf{P} \tag{3.16}$$

and

$$\mathsf{E}X = \int_0^\alpha \mathsf{P}[t < X < \alpha]\,dt + \int_{X \ge \alpha} X\,d\mathsf{P}. \tag{3.17}$$

By uniform integrability, for given ϵ there is an α such that the second term on the right in each equation is less than ϵ, and it is therefore enough to show that the first term on the right in (3.16) converges to the corresponding term in (3.17). But α can be chosen in such a way that $\mathsf{P}[X = \alpha] = 0$, and then this follows by the bounded convergence theorem applied in the interval $[0, \alpha]$. □

A simple condition for uniform integrability is that

$$\sup_n \mathsf{E}[|X_n|^{1+\epsilon}] < \infty \tag{3.18}$$

for some ϵ; in this case, (3.15) follows from

$$\int_{|X_n|\geq\alpha} |X_n| d\mathsf{P} \leq \frac{1}{\alpha^\epsilon}\mathsf{E}[|X_n|^{1+\epsilon}].$$

Theorem 3.6. *If X and the X_n are nonnegative and integrable, and if $X_n \Rightarrow X$ and $\mathsf{E}X_n \to \mathsf{E}X$, then the X_n are uniformly integrable.*

PROOF. From the hypothesis and (3.16) and (3.17) it follows that

$$\int_{X_n\geq\alpha} X_n\, d\mathsf{P} \to \int_{X\geq\alpha} X\, d\mathsf{P}$$

if $\mathsf{P}[X = \alpha] = 0$. Choose α so that the limit here is less than a given ϵ. Then, for n beyond some n_0, $\int_{X_n\geq\alpha} X_n\, d\mathsf{P} < \epsilon$. Increase α enough to take care of each of the integrable random variables $X_1, X_2, \ldots, X_{n_0}$.□

Relative Measure*

Let P_T be the probability measure on $(R^1, \mathcal{R}^1)$ corresponding to a uniform distribution over $[-T, T]$:

$$\mathsf{P}_T A = \frac{1}{2T}|A \cap [-T, T]|, \quad A \in \mathcal{R}^1, \tag{3.19}$$

where the bars refer to Lebesgue measure. Now define $\mathsf{P}_\infty A$ as

$$\mathsf{P}_\infty A = \lim_{T\to\infty} \mathsf{P}_T A, \tag{3.20}$$

provided the limit exists; this is called the *relative measure* of A. Note that, since $\mathsf{P}_\infty A = 0$ if A is bounded, P_∞ is not countably additive on its domain of definition. For Borel functions f, define

$$\mathsf{E}_\infty f = \lim_{T\to\infty} \mathsf{E}_T f = \lim_{T\to\infty} \frac{1}{2T}\int_{-T}^{T} f(\omega)\, d\omega, \tag{3.21}$$

provided the integrals and the limit exist; $\mathsf{E}_\infty f$ is the *mean value* of f. To write $\mathsf{P}_\infty A$ or $\mathsf{E}_\infty f$ is to assert or assume that the corresponding limit exists. If f is bounded and has period T_0, then it has a mean value,† and in fact, $\mathsf{E}_\infty f = \mathsf{E}_{T_0} f$. And if a set A has period T_0, in the sense that its indicator function does, then $\mathsf{P}_\infty A = \mathsf{P}_{T_0} A$.

† Almost periodic functions also have means; see Bohr [10].

Suppose that $X: R^1 \to S$ is measurable $\mathcal{R}^1/\mathcal{S}$. Then X, regarded as a random element defined on $(R^1, \mathcal{R}^1, \mathsf{P}_T)$, has $\mathsf{P}_T X^{-1}$ as its distribution over S. If[†]

$$\mathsf{P}_T X^{-1} \Rightarrow_T P \tag{3.22}$$

for some probability measure on $\mathcal{S}$, then P is called the *distribution* of X. By definition, this means that $\mathsf{E}_\infty[f(X)] = Pf$ for every bounded, continuous f on S, and by Theorem 2.1, it holds if and only if, for every P-continuity set A, $\mathsf{P}_\infty[X \in A] = PA$. By the mapping theorem, if X has distribution P, and if h is a measurable map from S to S' for which $PD_h = 0$, then $h(X)$ has distribution Ph^{-1}. We can derive the distributions of some interesting random elements by using standard methods of probability theory. For example, by the continuity theorem for characteristic functions, if $f(\omega)$ is real, then it has distribution P if and only if $\mathsf{E}_\infty[\exp(itf(\omega))]$ coincides with the characteristic function of P.

If λ is a nonzero real number, then $\cos \lambda\omega$ has period $T_\lambda = 2\pi/|\lambda|$, and its distribution is described by

$$\begin{aligned} \mathsf{P}_\infty[\cos \lambda\omega \le x] &= \mathsf{P}_{T_\lambda}[\cos \lambda\omega \le x] \\ &= 1 - \frac{1}{\pi} \arccos x, \quad -1 \le x \le +1, \end{aligned} \tag{3.23}$$

where for the arc cosine we use the continuous version on $[-1, +1]$ that takes the values π and 0 at -1 and $+1$. By using the simple relation

$$\mathsf{E}_\infty[e^{is\omega}] = \begin{cases} 1 & \text{if } s = 0, \\ 0 & \text{if } s \ne 0, \end{cases} \tag{3.24}$$

we can calculate some moments and characteristic functions. Since E_∞ is obviously linear, it follows by (3.24) and the assumption $\lambda \ne 0$ that

$$\begin{aligned} \mathsf{E}_\infty[\cos^r \lambda\omega] &= \mathsf{E}_\infty\left[\left(\frac{e^{i\lambda\omega} + e^{-i\lambda\omega}}{2}\right)^r\right] \\ &= \frac{1}{2^r} \sum_{j=0}^{r} \binom{r}{j} \mathsf{E}_\infty[e^{i(2j-r)\lambda\omega}] = \begin{cases} \frac{1}{2^r}\binom{r}{r/2}, & r \text{ even}, \\ 0, & r \text{ odd}. \end{cases} \end{aligned} \tag{3.25}$$

We have

$$\mathsf{E}_T[e^{it\cos\lambda\omega}] = \sum_{r=0}^{\infty} \frac{(it)^r}{r!} \mathsf{E}_T[\cos^r \lambda\omega],$$

[†] In the theory of weak convergence, T can obviously go to infinity continuously just as well as through the integers.

where the expected value can be taken inside the sum because the series on the right converges absolutely. And by the M-test, we can let T go to infinity on each side:

$$\varphi(t) := \mathsf{E}_\infty[e^{it\cos\lambda\omega}] = \sum_{r=0}^{\infty} \frac{(it)^r}{r!} \mathsf{E}_\infty[\cos^r \lambda\omega]. \tag{3.26}$$

Because of (3.23), this characteristic function is the same for all λ.

A two-dimensional version of the same argument shows that

$$\begin{aligned}\mathsf{E}_\infty[&\exp(it_1\cos\lambda_1\omega + it_2\cos\lambda_2\omega)] \\ &= \sum_{r_1=0}^{\infty}\sum_{r_2=0}^{\infty} \frac{(it_1)^{r_1}}{r_1!}\frac{(it_2)^{r_2}}{r_2!} \mathsf{E}_\infty[\cos^{r_1}\lambda_1\omega \cdot \cos^{r_2}\lambda_2\omega].\end{aligned} \tag{3.27}$$

Suppose now that λ_1 and λ_2 are incommensurable (neither is 0 and λ_1/λ_2 is irrational). We can show in this case that

$$\mathsf{E}_\infty[\cos^{r_1}\lambda_1\omega \cdot \cos^{r_2}\lambda_2\omega] = \mathsf{E}_\infty[\cos^{r_1}\lambda_1\omega] \cdot \mathsf{E}_\infty[\cos^{r_2}\lambda_2\omega]. \tag{3.28}$$

The argument is like that for (3.25):

$$\begin{aligned}\mathsf{E}_\infty[&\cos^{r_1}\lambda_1\omega \cdot \cos^{r_2}\lambda\omega] \\ &= \mathsf{E}_\infty\left[\left(\frac{e^{i\lambda_1\omega}+e^{-i\lambda_1\omega}}{2}\right)^{r_1}\left(\frac{e^{i\lambda_2\omega}+e^{-i\lambda_2\omega}}{2}\right)^{r_2}\right] \\ &= \frac{1}{2^{r_1}}\frac{1}{2^{r_2}}\sum_{j_1=0}^{r_1}\sum_{j_2=0}^{r_2}\binom{r_1}{j_1}\binom{r_2}{j_2} \\ &\qquad\times \mathsf{E}_\infty[\exp\{i((2j_1-r_1)\lambda_1+(2j_2-r_2)\lambda_2)\omega\}].\end{aligned} \tag{3.29}$$

By (3.24) and the assumption that λ_1 and λ_2 are incommensurable, this last expected value is 1 or 0 according as $2j_1 - r_1 = 0 = 2j_2 - r_2$ or not. But the product

$$\mathsf{E}_\infty[\exp\{i(2j_1-r_1)\lambda_1\}] \cdot \mathsf{E}_\infty[\exp\{i(2j_2-r_2)\lambda_2\}]$$

is also 1 or 0 according as $2j_1-r_1=0=2j_2-r_2$ or not, and substituting this product into (3.29) gives (3.28)—see (3.25). Finally, (3.26), (3.27), and (3.28) together imply

$$\mathsf{E}_\infty[\exp(it_1\cos\lambda_1\omega + it_2\cos\lambda_2\omega)] = \varphi(t_1)\varphi(t_2). \tag{3.30}$$

Let $\eta_1, \eta_2, \ldots$ be independent random variables, each having the characteristic function φ and the distribution function in (3.23). By the continuity theorem for characteristic functions in two dimensions, the random element $(\cos\lambda_1\omega, \cos\lambda_2\omega)$ of R^2 has the same distribution (in the sense of (3.22)) as the random vector (η_1, η_2), provided λ_1 and λ_2 are incommensurable. Now suppose that $\lambda_1, \lambda_2 \ldots$ are linearly independent in the sense that $m_1\lambda_1 + \cdots + m_k\lambda_k = 0$ for integers m_i can hold only if $m_i \equiv 0$. Then the preceding argument extends from R^2 to R^n. Since the η_k have mean 0 and variance $\frac{1}{2}$ by (3.25), it is probabilistically natural to replace them by $\xi_k = \sqrt{2}\eta_k$. By (3.23),

$$\text{(3.31)} \qquad \mathsf{P}[\xi_k \le x] = 1 - \frac{1}{\pi}\arccos\frac{x}{\sqrt{2}}, \quad -\sqrt{2} \le x \le +\sqrt{2}.$$

Since the distribution of $(\xi_1, \ldots, \xi_n)$ is absolutely continuous with respect to Lebesgue measure in R^n, we arrive at the following theorem.

Theorem 3.7. *Suppose that* $\lambda_1, \lambda_2, \ldots$ *are linearly independent and that* $\xi_1, \xi_2, \ldots$ *are independent, each distributed according to* (3.31). *Then* $(\sqrt{2}\cos\lambda_1\omega, \ldots, \sqrt{2}\cos\lambda_n\omega)$ *has the distribution of* $(\xi_1, \ldots, \xi_n)$ *for each* n:

$$\text{(3.32)} \quad \mathsf{P}_\infty[(\sqrt{2}\cos\lambda_1\omega, \ldots, \sqrt{2}\cos\lambda_n\omega) \in A] = \mathsf{P}[(\xi_1, \ldots, \xi_n) \in A]$$

if ∂A *has Lebesgue measure* 0.

Since the ξ_k have mean 0 and variance 1, the Lindeberg-Lévy theorem implies a result of Kac and Steinhaus:

$$\text{(3.33)} \quad \lim_{n\to\infty} \mathsf{P}_\infty\left[x \le \sqrt{\frac{2}{n}}\sum_{k=1}^{n}\cos\lambda_k\omega \le y\right] = \frac{1}{\sqrt{2\pi}}\int_x^y e^{-u^2/2}\,du.$$

This approximates the relative time a superposition of vibrations with incommensurable frequencies spends between x and y. For a refinement, see Theorem 11.2.

Return to the random variables η_k, and let $\psi(x) = (2\pi)^{-1}\arccos x$, with the arc cosine defined as in (3.23). Then the distribution function on the right in (3.23) is $F(x) = 1 - 2\psi(x)$ for $-1 \le x \le +1$, and since $F(\eta_k)$ is uniformly distributed over $[0, 1]$ (the probability transformation), $\beta_k = \psi(\eta_k)$ is uniformly distributed over $[0, \frac{1}{2}]$. And for every x, $\psi(\cos 2\pi x)$ is $\langle x\rangle$, the distance from x to the nearest integer. From the linear independence of $\{\lambda_k\}$ follows that of $\{2\pi\lambda_k\}$, and so the $\cos 2\pi\lambda_k\omega$ are distributed like the η_k. By the mapping theorem, we can apply ψ to each $\cos 2\pi\lambda_k$ if we relace each η_k by β_k. Therefore: Suppose that $\lambda_1, \lambda_2, \ldots$ are linearly independent and that $\beta_1, \beta_2, \ldots$ are independent random variables, each uniformly distributed over $[0, \frac{1}{2}]$. Then, for each n,

$$\mathsf{P}_\infty[(\langle\lambda_1\omega\rangle, \ldots, \langle\lambda_n\omega\rangle) \in A] = \mathsf{P}[(\beta_1, \ldots, \beta_n) \in A] \tag{3.34}$$

if ∂A has Lebesgue measure 0, a connection with Diophantine approximation.

Suppose that $0 < \alpha < \frac{1}{2}$ and take A to be the set of $(x_1, \ldots, x_n)$ such that $0 \le x_i \le \alpha$ for exactly k values of i. If $U_n(\omega, \alpha)$ is the number of i, $1 \le i \le n$, for which $\langle\lambda_i\omega\rangle \le \alpha$, then the event on the left in (3.34) is $[U_n(\omega, \alpha) = k]$ (for the corresponding A, ∂A has Lebesgue measure 0), and the right side is a binomial probability:

$$\mathsf{P}_\infty[U_n(\omega, \alpha) = k] = \binom{n}{k}\alpha^k(1-\alpha)^{n-k}. \tag{3.35}$$

We can apply the binomial central limit theorem here—either the local or the integral one. Or take a Poisson limit:

$$\lim_{n\to\infty} \mathsf{P}_\infty\left[U_n\left(\omega, \frac{\alpha}{n}\right) = k\right] = e^{-\alpha}\frac{\alpha^k}{k!}. \tag{3.36}$$

Three Lemmas*

The rest of the section is needed only for the proofs of Theorems 14.4 and 14.5 and in Section 17.

Theorem 3.8. *If (X_n, X) is a random element of $S \times S$ and $\rho(X_n, X) \Rightarrow 0$, and if A is an X-continuity set, then it follows that $\mathsf{P}([X_n \in A]\Delta[X \in A]) \to 0$.*

PROOF. For each positive ϵ,

$$\mathsf{P}[X_n \in A,\ X \notin A]) \le \mathsf{P}[\rho(X_n, X) \ge \epsilon] + \mathsf{P}[\rho(X, A) < \epsilon,\ X \notin A].$$

This, the same inequality with A^c in place of A, and the assumption $\rho(X_n, X) \Rightarrow 0$, together imply

$$\limsup_n \mathsf{P}([X_n \in A]\Delta[X \in A]) \le \mathsf{P}[\rho(X, A) < \epsilon,\ X \notin A]] + \mathsf{P}[\rho(X, A^c) < \epsilon,\ X \in A].$$

If A is an X-continuity set, then the right side goes to 0 as $\epsilon \to 0$. □

Let (X', X'') and (X'_n, X''_n) be random elements of $T = S' \times S''$. If S is separable, then, by Theorem 2.8(i),

$$(X'_n, X''_n) \Rightarrow (X', X'') \tag{3.37}$$

holds if and only if

$$\mathsf{P}[X'_n \in A',\ X''_n \in A''] \to \mathsf{P}[X' \in A,\ X'' \in A''] \tag{3.38}$$

holds for all X'-continuity sets A' and all X''-continuity sets A''.

Theorem 3.9. *Suppose that T is separable. If $X'_n \Rightarrow X'$ and $X''_n \Rightarrow a''$, then $(X'_n, X''_n) \Rightarrow (X', a'')$.*

PROOF. We must verify (3.38) for $X'' \equiv a''$. Suppose that A' is an X'-continuity set and that $a'' \notin \partial A''$. If $a'' \in A''$, then $\mathsf{P}[X''_n \notin A''] \to 0$, and (3.38) follows from $X'_n \Rightarrow X'$ and

$$\mathsf{P}[X'_n \in A'] - \mathsf{P}[X''_n \notin A''] \le \mathsf{P}[X'_n \in A',\ X''_n \in A''] \le \mathsf{P}[X'_n \in A'].$$

If $a'' \notin A''$, then (3.38) follows from

$$\mathsf{P}[X'_n \in A',\ X''_n \in A''] \le \mathsf{P}[X''_n \in A''] \to 0. \qquad \square$$

Suppose that Y'' and all the (X'_n, X''_n) are all defined on the same $(\Omega, \mathcal{F}, \mathsf{P})$. Suppose that $\mathcal{F}_0$ is a field in $\mathcal{F}$ and let $\mathcal{F}_1 = \sigma(\mathcal{F}_0)$.

Theorem 3.10. *Suppose that T is separable, X' and X'' are independent, and X'' has the same distribution as Y''. If $\rho(X''_n, Y'') \Rightarrow 0$, if*

$$\mathsf{P}([X'_n \in A'] \cap E) \to \mathsf{P}[X' \in A']\mathsf{P}(E) \tag{3.39}$$

for each X'-continuity set A' and each E in $\mathcal{F}_0$, and if each X'_n is measurable $\mathcal{F}_1/S'$, then $(X'_n, X''_n) \Rightarrow (X', X'')$.

PROOF. Fix an X'-continuity set A' and an X''-continuity set A''. It follows from the hypothesis and Rényi's theorem on mixing sequences [M21] that (3.39) holds for every E in $\mathcal{F}$, and therefore,

$$\mathsf{P}[X'_n \in A',\ Y'' \in A''] \to \mathsf{P}[X' \in A']\mathsf{P}[Y'' \in A'']. \tag{3.40}$$

We are to prove (3.38), which, since X' and X'' are independent and the latter has the same distribution as Y'', reduces to

$$\mathsf{P}[X'_n \in A',\ X''_n \in A''] \to \mathsf{P}[X' \in A']\mathsf{P}[Y'' \in A''].$$

Since $\rho(X''_n, Y'') \Rightarrow 0$, it follows by Theorem 3.8 that this is the same thing as (3.40). $\square$

Problems

3.1. Let $X_1, X_2, \ldots$ be independent random elements having a common distribution P on S. Let $P_{n,\omega}$ be the empirical measure corresponding to the observation $(X_1(\omega), \ldots, X_n(\omega))$:

$$P_{n,\omega}(A) = \frac{1}{n}\sum_{k=1}^{n} I_A(X_k(\omega)).$$

Show that, if S is separable, then $P_{n,\omega} \Rightarrow_n P$ with probability 1. Use the strong law of large numbers and Theorem 2.3.

3.2. If X' and X'' are random elements of S' and S'', then $(X', X''): \Omega \to T = S' \times S''$ is measurable $\mathcal{F}/\mathcal{S}' \times \mathcal{S}''$. And (X', X'') is a random element of T if T is separable but may not be a random element in the nonseparable case [M10].

3.3. Let X_n and X be random elements of a separable S, defined on $(\Omega, \mathcal{F}, \mathsf{P})$. Then $[\omega: X_n(\omega) \to X(\omega)]$ lies in $\mathcal{F}$; if it has probability 1, then $\rho(X_n, X) \Rightarrow 0$, and hence $X_n \Rightarrow X$.

3.4. Prove directly that $P_n \Rightarrow \delta_a$ if and only if $P_n(B(a, \epsilon)) \to 1$ for positive ϵ. Prove directly from the definition (3.6) that $X_n \Rightarrow a$ implies $h(X_n) \Rightarrow h(X)$ if h is continuous at a.

3.5. Apply the method of Example 3.4 to the hypergeometric distribution (see Problem 10 on p. 194 of Feller [28]).

3.6. Suppose there is a nonnegative random variable Y such that $\mathsf{P}[|X_n| \geq t] \leq \mathsf{P}[Y \geq t]$ for all n and all positve t, and $\mathsf{E}Y < \infty$ (Y and the X_n can be defined on different spaces). Show that $X_n \Rightarrow X$ implies $\mathsf{E}X_n \to \mathsf{E}X$. This corresponds to the dominated convergence theorem.

3.7. Prove (in the notation of (3.34)) that

$$\lim_{n\to\infty} \mathsf{P}\left[\frac{1}{\sqrt{48n}}\sum_{k=1}^{n}(\langle\lambda_k\omega\rangle - \tfrac{1}{4}) \leq x\right] = \frac{1}{\sqrt{2\pi}}\int_{-\infty}^{x} e^{-u^2/2}du. \tag{3.41}$$

3.8. For random variables η_n and η on the same probability space, write $\eta_n \to_P \eta$ if $\eta_n - \eta \Rightarrow 0$ in the sense of (3.6). Let S_n be the partial sums of independent and identically distributed random variables with mean 0 and variance 1. By the Lindeberg-Lévy theorem, $S_n/\sqrt{n} \Rightarrow N$. Show that $S_n/\sqrt{n} \to_P \eta$ cannot hold for any random variable η. But see Chapter 5.

SECTION 4. LONG CYCLES AND LARGE DIVISORS*

Here we study the cycle structure of random permutations on n letters, obtaining in particular the joint limiting distribution ($n \to \infty$) of the greatest cycle lengths. And we derive analogous results on the limiting distribution of the largest prime divisors of an integer drawn at random from among the first n integers. These distributions are described by a certain measure—the Poisson-Dirichlet measure—on a certain space of sequences. We study weak convergence in this space, obtaining results that have applications to population biology as well as to long cycles and large divisors. The essential tool is the mapping theorem.

Long Cycles

Every permutation can be written as a product of cycles. For example, the permutation

$$(1\,4\,2\,7)(3)(5\,6) \tag{4.1}$$

on the seven "letters" $1, \ldots, 7$ sends $1 \to 4 \to 2 \to 7 \to 1$, $3 \to 3$, and $5 \to 6 \to 5$. To standarize the representation, start the first cycle with 1 and start each successive cycle with the smallest integer not yet encountered.

Of the $n!$ permutations on $1, \ldots, n$, for how many does the leftmost cycle have length i $(1 \le i \le n)$? The answer is

$$(n-1)(n-2)\cdots(n-i+1) \times (n-i)! = (n-1)! \tag{4.2}$$

To see this, note that what we want is the number of products $\alpha\beta$, where α is an i-long cyclic permutation of 1 together with $i-1$ other letters and β is a permutation of the remaining $n-i$ letters. The number of ways of choosing α is the product to the left of the $\times$ in (4.2); and, α having been chosen, the number of ways of then choosing β is $(n-i)!$

Suppose now that a permutation is chosen at random, all $n!$ possibilities having the same probability, and suppose that it is written as a product of cycles in *standard order*, on the pattern of (4.1). That is to say, the first cycle starts with 1, the second cycle starts with the smallest element not contained in the first cycle, and so on. Let $C_1^n, C_2^n, \ldots$ be the lengths of the successive cycles; if the number of cycles is u, set $C_v^n = 0$ for $v > u$ $(\sum_v C_v^n = n)$. It follows by (4.2) that $\mathsf{P}[C_1^n = i] = 1/n$ for $1 \le i \le n$. Suppose that $C_1^n = i$ and that in fact the initial cycle consists of i specified letters (including 1, of course). Then there remain $n-i$ letters that make up the rest of the permutation, and conditionally, the chance that $C_2^n = j$ is just the probability that the first cycle in a random permutation on the remaining $n-i$ letters has length j, and this is $1/(n-i)$ for $1 \le j \le n-i$. Therefore,

$$\mathsf{P}[C_1^n = i] = \frac{1}{n}, \quad 1 \le i \le n, \tag{4.3}$$

$$\mathsf{P}[C_1^n = i, C_2^n = j] = \frac{1}{n} \cdot \frac{1}{n-i}, \quad 2 \le i+j \le n,$$

$$\mathsf{P}[C_1^n = i, C_2^n = j, C_3^n = k] = \frac{1}{n} \cdot \frac{1}{n-i} \cdot \frac{1}{n-i-j}, \quad 3 \le i+j+k \le n,$$

and so on.

Let $L_u^n = C_u^n/n$ be the *relative* cycle lengths, and let $B_1, B_2, \cdots$ be independent random variables, each uniformly distributed over $[0,1]$. (The uniform distribution has the beta–(1,1) density [M11]; hence the notation B_i.) Since L_1^n is uniformly distributed over its possible values i/n, it should for large n be distributed approximately as B_1. That this is true can be shown by the argument in Example 2.2: If f is continuous over $[0,1]$, then $\mathsf{E}[f(L_1^n)] = n^{-1}\sum_{i\le n} f(i/n)$ is a Riemann sum converging to $\int_0^1 f(x)\,dx = \mathsf{E}[f(B_1)]$. Therefore, $L_1^n \Rightarrow B_1$.

If $C_1^n = i$, then C_2^n is, conditionally, uniformly distributed over the values j for $j \le n - i = n - C_1^n$. Therefore, conditionally on $L_1^n = i/n$, L_2^n is uniformly distributed over the values j/n for $j/n \le 1 - i/n = 1 - L_1^n$. This makes it plausible that $(L_1^n, L_2^n/(1-L_1^n))$ will be, in the limit, uniformly distributed over the unit square—like (B_1, B_2). This follows by a two-dimensional Riemann-sum argument: If f is continuous over the square $[0,1]^2$, then

$$\begin{aligned}\mathsf{E}\Big[f\Big(L_1^n, \frac{L_2^n}{1-L_1^n}\Big)\Big] &= \frac{1}{n^2}\sum_{i+j\le n}\frac{1}{1-i/n} f\Big(i/n, \frac{j/n}{1-i/n}\Big)\\ &\to \int\!\!\int_{x_1+x_2\le 1} f\Big(x_1, \frac{x_2}{1-x_1}\Big)\frac{dx_1\,dx_2}{1-x_1}\\ &= \int\!\!\int_{[0,1]^2} f(y_1,y_2)\,dy_1\,dy_2.\end{aligned}$$

Therefore, $(L_1^n, L_2^n/(1-L_1^n)) \Rightarrow (B_1, B_2)$, and it follows by the mapping theorem that $(L_1^n, L_2^n) \Rightarrow (B_1, (1-B_1)B_2)$.

Define

$$G_1 = B_1, \quad G_i = (1-B_1)\cdots(1-B_{i-1})B_i, \quad i > 1. \tag{4.4}$$

A three-dimensional version of the argument given above shows that

$$\Big(L_1^n, \frac{L_2^n}{1-L_1^n}, \frac{L_3^n}{1-L_1^n-L_2^n}\Big) \Rightarrow (B_1, B_2, B_3),$$

which, by the mapping theorem, implies $(L_1^n, L_2^n, L_3^n) \Rightarrow (G_1, G_2, G_3)$. This extends to the general r: $(L_1^n, \ldots, L_r^n) \Rightarrow_n (G_1, \ldots, G_r)$. We can regard $(L_1^n, L_2^n, \ldots)$ and $(G_1, G_2, \ldots)$ as random elements of R^∞ (Example 2.6), and since the finite-dimensional sets in R^∞ form a convergence-determining class, we have

$$L^n = (L_1^n, L_2^n, \ldots) \Rightarrow_n G = (G_1, G_2, \ldots). \tag{4.5}$$

This describes the limiting distribution of the relative cycle lengths of a random permutation when it is written as a product of cycles in standard order.

Induction shows that $\sum_{i=1}^k G_i = 1-\prod_{i=1}^k(1-B_i)$, and since, by the Borel-Cantelli lemma, there is probability 1 that $B_i > \frac{1}{2}$ holds infinitely often, it follows that $\sum_{i=1}^\infty G_i = 1$ with probability 1. The random sequence G can be viewed as a description of a random dissection of the unit interval into an infinite sequence of subintervals. A piece of length B_1 is broken off at the left, which leaves a piece of length $1-B_1$. From this, a piece of length $(1-B_1)B_2$ is broken off, which leaves a piece of length $(1-B_1)(1-B_2)$, and so on. What (4.5) says is that the splitting of a random permutation into cycles is, in the limit, like this random dissection of an interval.

Let $L^n_{(\cdot)} = (L^n_{(1)}, L^n_{(2)}, \ldots)$ be L^n with the components *ranked by size*: $L^n_{(1)} \geq L^n_{(2)} \geq \cdots$. The details of how this is to be done are taken up below. To study the asymptotic properties of the longest cycles of a random permutation is to study the limiting distribution of $L^n_{(\cdot)}$. Let $G_{(\cdot)}$ be G with the components ranked by size. After the appropriate groundwork, we will be in a position to deduce $L^n_{(\cdot)} \Rightarrow G_{(\cdot)}$ from (4.5) by means of the mapping theorem, as well as to describe the finite-dimensional distributions of $G_{(\cdot)}$. It will be very much worth our while to do this in a more general context.

The Space Δ

Consider the subset

$$\Delta = \left[x \in R^\infty : x_1, x_2, \ldots \geq 0, \sum\nolimits_{i=1}^\infty x_i = 1\right] \tag{4.6}$$

of R^∞ with the relative topology, that of coordinatewise convergence. Let $A_{\epsilon,k}$ be the set of x in R^∞ for which $x_1, \ldots, x_k \geq 0$ and $1 - \epsilon \leq \sum_{i=1}^k x_i \leq 1$; it is closed in R^∞, and since $\Delta = \bigcap_\epsilon \bigcup_{k>1/\epsilon} A_{\epsilon,k}$ (intersection over the positive rationals), (4.6) is a Borel set, a member of $\mathcal{R}^\infty$. From $\sum_i C_i^n = n$ follows $L^n \in \Delta$, and since $\sum_i G_i = 1$ with probability 1, $\mathsf{P}[G \in \Delta] = 1$. This leads us to study weak convergence in the space Δ.

If Y^n and Y are random elements of Δ, then they are also random elements of R^∞; see Example 3.1. And $Y^n \Rightarrow Y$ holds in the sense of Δ if and only if it holds in the sense of R^∞, which in turn holds if and only if $(Y_1^n, \ldots, Y_r^n) \Rightarrow (Y_1, \ldots, Y_r)$ holds for each r. Therefore:

$Y^n \Rightarrow Y$ holds for random elements of Δ if and only if $(Y_1^n, \ldots, Y_r^n) \Rightarrow (Y_1, \ldots, Y_r)$ holds for each r.

Change the G in (4.5) on a set of probability 0 in such a way that $\sum_i G_i = 1$ holds identically, so that G becomes a random element of Δ. Now (4.5) is to be interpreted as convergence in distribution on Δ rather than on R^∞.

We turn next to *ranking*. The *ranking function* $\rho\colon \Delta \to \Delta$ is defined this way: If $x \in \Delta$, then $x_i \to_i 0$, and so there is a maximum x_i, and this is the first component of $y = \rho x$. There may of course be ties for the maximum (finitely many), and if there are j components x_i of maximum size, take $y_1, \ldots, y_j$ to have this common value. Now take y_{j+1} to be the next-largest component of x, and so on: ρ is then defined in an unambiguous way.

Lemma 1. *The ranking function $\rho\colon \Delta \to \Delta$ is continuous.*

PROOF. To prove that $x^n \to x$ implies $\rho x^n \to \rho x$, we can start by applying an arbitrary permutation to the coordinates of x and applying the same perutation to the coordinates of each x^n: This does not change the hypothesis, which requires convergence $x_i^n \to_n x_i$ in each coordinate individually, and it does not change the conclusion because it leaves ρx and the ρx^n invariant. Use the permutation that ranks the x_i. Assume then that $x_1 \geq x_2 \geq \cdots$ (and hence $\rho x = x$) and that $x^n \to x$; we are to prove that $\rho x^n \to x$. Since $\sum_i x_i^n = \sum_i x_i = 1$ and the coordinates are all nonnegative, it follows by Scheffé's theorem that $\sum_i |x_i^n - x_i| \to_n 0$. Suppose that

$$a_1 = x_1 = \cdots = x_{i_1} > a_2 = x_{i_1+1} = \cdots = x_{i_2} > a_3 = x_{i_2+1} = \cdots.$$

For n beyond some n_ϵ, the coordinates $x_1^n, \ldots, x_{i_1}^n$ all lie in the range $a_1 \pm \epsilon$, $x_{i_1+1}^n, \ldots, x_{i_2}^n$ all lie in the range $a_2 \pm \epsilon$, and $x_{i_2+1}^n, x_{i_2+2}^n, \ldots$ are all less than $a_3 + \epsilon$. If $a_1 - \epsilon > a_2 + \epsilon > a_2 - \epsilon > a_3 + \epsilon$, then (for $n > n_\epsilon$) the first i_1 coordinates of ρx^n are some permutation of $x_1^n, \ldots, x_{i_1}^n$, all of which are within ϵ of a_1; and the next i_2 coordinates are some permutation of $x_{i_1+1}^n, \ldots, x_{i_2}^n$, all of which are within ϵ of a_2. A simple extension of this argument proves continuity. □

An immediate consequence of Lemma 1 and the mapping theorem:

Theorem 4.1. *If Y^n and Y are random elements of Δ, and if $Y^n \Rightarrow_n Y$, then $\rho Y^n \Rightarrow_n \rho Y$.*

Since (4.5) now means weak convergence in Δ, we can conclude that the vector $L_{(\cdot)}^n$ of ranked relative cycle lengths satisfies

$$L_{(\cdot)}^n = (L_{(1)}^n, L_{(2)}^n, \ldots) = \rho L^n \Rightarrow_n \rho G = G_{(\cdot)} = (G_{(1)}, G_{(2)}, \ldots). \tag{4.7}$$

For this result to have real significance, we need detailed information about the distribution of $G_{(\cdot)}$ over Δ.

The Poisson-Dirichlet Distribution

Let $B_1, B_2, \ldots$ be independent random variables on $(\Omega, \mathcal{F}, \mathsf{P})$, each having the beta–$(1,\theta)$ density $\theta(1-x)^{\theta-1}$ over $[0,1]$, where $\theta > 0$. If $\theta = 1$, the B_i are uniformly distributed. Define $G = (G_1, G_2, \ldots)$ by (4.4), as in the special case above. For the general θ, we still have $\sum_i G_i = 1$ with probability 1; alter G on a set of probability 0 so as to make it a random element of Δ. The distribution $\mathsf{P}G^{-1}$ of G, a probability measure on Δ, is called the *GEM distribution* with parameter θ. And as before, let $G_{(\cdot)} = \rho G = (G_{(1)}, G_{(2)}, \ldots)$ be the ranked version of G. The distribution $\mathsf{P}G_{(\cdot)}^{-1}$ of $G_{(\cdot)}$ on Δ is the *Poisson-Dirichlet distribution* with parameter θ.

For $\theta = 1$, the GEM distribution describes the random splitting of the unit interval into a sequence of subintervals, and the Poisson-Dirichlet distribution describes the random splitting after the pieces have been ordered by size (the longest piece to the left). And, also for $\theta = 1$, by (4.5) the GEM distribution describes the asymptotic distribution of the relative cycle lengths of a random permutation when the cycles are written in standard order, and by (4.7) the Poisson-Dirichlet distribution describes the asymptotic distribution when the cycles are written in order of size (the longest one first).

We can get a complete specification of the Poisson-Dirichlet distribution for the general θ by an indirect method that comes from population genetics. Imagine a population of individuals (genes) divided into n types (alleles),† the proportion or relative frequency of individuals of the ith type being $Z_i, i = 1, \ldots, n$. In the diffusion approximation to the Wright-Fisher model for the evolution of the population, the stationary distribution for the relative frequency vector $(Z_1, \ldots, Z_{n-1})$ has the symmetric Dirichlet density [M13]

$$(4.8) \quad \phi_n(z_1, \ldots, z_{n-1}) = \frac{\Gamma(n\alpha)}{\Gamma^n(\alpha)} z_1^{\alpha-1} \cdots z_{n-1}^{\alpha-1}(1 - z_1 - \cdots z_{n-1})^{\alpha-1}$$

$$\text{for} \begin{cases} z_1, \ldots, z_{n-1} > 0, \\ z_1 + \cdots + z_{n-1} < 1. \end{cases}$$

† In the genetics literature, the number of types is denoted by K. For a general treatment of mathematical population genetics, see Ewens [27]. The account here is self-contained but ignores the biology.

Here α represents the mutation rate per type. To study the limiting case of a population in which all mutations produce new types, there being infinitely many such types available, let n tend to infinity while the total mutation rate $\theta = n\alpha$ is held fixed. This forces the individual components to 0, and so we focus on the largest components—the components corresponding to types having appreciable proportions in the limiting population—which means studying the first r components $Z_{(1)}, \ldots, Z_{(r)}$ of the *ranked* relative frequencies.

The program is first to derive the limiting distribution of the ranked frequencies, and then later to show that this is in fact the Poisson-Dirichlet distribution with parameter θ. Some of the arguments involve considerable uphill calculation, and these are put at the end of the section, so as not to obscure the larger structure of the development.

Changing the notation, let Z^n be a random element of Δ for which $(Z_1^n, \ldots, Z_{n-1}^n)$ is distributed according to the density (4.8), $Z_n^n = 1 - Z_1^n - \cdots - Z_{n-1}^n$, and $Z_i^n = 0$ for $i > n$. And let $Z_{(\cdot)}^n = \rho Z^n$. Here n will go to infinity, α will be a positive function of n such that $n\alpha \to \theta > 0$, and r will be fixed. Define

$$M_r = \left[z \in R^r : 1 > z_1 > \cdots > z_r > 0, \ \sum_{i=1}^r z_i < 1 \right]. \tag{4.9}$$

Let $g_m(\cdot\,;\alpha)$ be the m-fold convolution of the density $\alpha x^{\alpha-1}$ on $(0,1)$, and write (assume $n > r+2$) $(n)_r = n(n-1)\cdots(n-r+1)$.

Lemma 2. *The density of $Z_{(1)}^n, \ldots, Z_{(r)}^n$ at points of M_r is*

(4.10)

$$(n)_r \frac{\Gamma(n\alpha)}{\Gamma^n(\alpha)} \alpha^{-(n-r)} z_1^{\alpha-1} \cdots z_{r-1}^{\alpha-1} z_r^{(n-r+1)\alpha-2} g_{n-r}\left(\frac{1 - z_1 - \cdots - z_r}{z_r}; \alpha \right).$$

For the proof, see the end of the section. To carry the calculation further, we need the limit of $g_m(\cdot\,;\alpha)$. Suppose that $\theta > 0$. For $k \ge 1$, let $C_k(x)$ be the $(s_1, \ldots, s_k)$-set where $s_1, \ldots, s_k > 1$ and $\sum_{i=1}^k s_i < x$, and define

$$J_k(x;\theta) = \int_{C_k(x)} \left(x - \sum_{i=1}^k s_i \right)^{\theta-1} \frac{ds_1 \cdots ds_k}{s_1 \cdots s_k}; \tag{4.11}$$

note that, for $x \le k$, $C_k(x) = \emptyset$ and $J_k(x;\theta) = 0$. Define $J_0(x;\theta) = x^{\theta-1}$.

Lemma 3. *For $0 < x < m$, we have*

$$(4.12)\quad g_m(x;\alpha) = \sum_{0\le k<x} (-1)^k \binom{m}{k} \frac{\Gamma^{m-k}(\alpha)\alpha^m}{\Gamma((m-k)\alpha)} \times \int_{C_k(x)} s_1^{\alpha-1}\cdots s_k^{\alpha-1}\Big(x-\sum_{i=1}^k s_i\Big)^{(m-k)\alpha-1} ds_1\cdots ds_k,$$

where for $k=0$ the integral is to be replaced by $x^{m\alpha-1}$. If $m\to\infty$ and $m\alpha\to\theta>0$, then $g_m(x;\alpha)$ converges to

$$(4.13)\qquad g_\theta(x) = \sum_{0\le k<x} \frac{1}{\Gamma(\theta)e^{\gamma\theta}} \frac{(-1)^k}{k!}\theta^k J_k(x;\theta)$$

for each positive x; g_θ is a probability density on $(0,\infty)$ having Laplace transform

$$(4.14)\qquad \int_0^\infty e^{-tx} g_\theta(x)\,dx = e^{-\gamma\theta}t^{-\theta}\exp[-\theta E_1(t)] = \exp\Big[-\theta\int_0^t (1-e^{-u})\frac{du}{u}\Big].$$

Although g_θ is integrable, term-by-term integration in (4.13) gives $\sum \pm\infty$. Here γ is Euler's constant and $E_1(t)$ is $\int_t^\infty e^{-u}u^{-1}du$, the exponential integral function. The sum in (4.13) can be extended to all $k\ge 0$, since the summand vanishes for $k\ge x$. Again the proof is deferred.

Theorem 4.2. *As $n\to\infty$ and $n\alpha\to\theta$ (r fixed), the density (4.10) of $Z_{(1)}^n,\ldots,Z_{(r)}^n$ converges to the probability density*

$$(4.15)\qquad \theta^r\Gamma(\theta)e^{\gamma\theta}z_1^{-1}\cdots z_{r-1}^{-1}z_r^{\theta-2}g_\theta\Big(\frac{1-z_1-\cdots-z_r}{z_r}\Big)$$

on the set M_r defined by (4.9).

Let $d_r(z_r;\theta)$ be (4.15) with $z_1,\ldots,z_{r-1}$ integrated out (for fixed z_r, $(z_1,\ldots,z_{r-1},z_r)$ ranges over M_r), so that $d_r(\,\cdot\,;\theta)$ is the rth marginal density.

Theorem 4.3. *The rth marginal density of (4.15) is*

$$(4.16)\qquad d_r(x;\theta) = x^{\theta-2}\sum_{0\le k<x^{-1}-r} \frac{(-1)^k\theta^{r+k}}{k!(r-1)!}J_{k+r-1}\Big(\frac{1-x}{x};\theta\Big)$$

for $0 < x < r^{-1}$. *The corresponding moments are*

$$(4.17)\quad \int_0^{r^{-1}} x^m d_r(x;\theta)\,dx = \frac{\Gamma(\theta+1)}{\Gamma(\theta+m)} \int_0^\infty y^{m-1} e^{-y} \frac{(\theta E_1(y))^{r-1}}{(r-1)!} e^{-\theta E_1(y)}\,dy$$

for $m = 0, 1, 2, \ldots$.

The two preceding theorems are proved at the end of the section.

Size-Biased Sampling

Let $X = (X_1, X_2, \ldots)$ be a random element of Δ. The *size-biased* version of X is another random element $\hat{X} = (\hat{X}_1, \hat{X}_2, \ldots)$ of Δ defined informally in the following way. Take $\hat{X}_1 = X_i$ with probability X_i. That done, take $\hat{X}_2 = X_j$ $(j \neq i)$ with the conditional probability $X_j/(1 - X_i)$; to put it another way, take $\hat{X}_1 = X_i$, $\hat{X}_2 = X_j$ $(i \neq j)$ with probability $X_i \cdot X_j/(1 - X_i)$. Continue in this way. We must set up a probability mechanism for making the choices, and we must avoid divisions by 0.

Replace the probability space on which X is defined by its product with another space, in such a way that the enlarged space supports random variables $\xi_1, \xi_2, \ldots$ that are independent of X and of each other and each is uniformly distributed over $[0, 1]$; arrange that $0 < \xi_u(\omega) < 1$ for every u and ω. Let $S_i = \sum_{h=1}^i X_h$ $(S_0 = 0)$; since X is a random element of Δ, $S_i \uparrow 1$. Let $\varsigma_u = i$ if $S_{i-1} < \xi_u \le S_i$. Then, conditionally on X, the ς_u are independent and each assumes the value i with probability X_i (if $X_i = 0$, then $[\varsigma_u = i] = \emptyset$). Let $\tau_1 = \varsigma_1$; let $\tau_2 = \varsigma_u$, where u is the smallest index for which ς_u is distinct from τ_1; let $\tau_3 = \varsigma_v$, where v is the smallest index for which ς_v is distinct from τ_1 and τ_2 (v necessarily exceeds u); and so on. If we define $\hat{X}_u = X_{\tau_u}$, then $\hat{X} = (\hat{X}_1, \hat{X}_2, \ldots)$ will have the structure we want, provided we accommodate the case where we run out new values for the τ_u.

Suppose that the procedure just described defines $\tau_1, \ldots, \tau_u$ unambiguously but that $X_{\tau_1} + \cdots + X_{\tau_u} = 1$. Then τ_{u+1} is not defined, because there is no ς_v distinct from $\tau_1, \ldots, \tau_u$. In this case, take $\tau_v = \infty$ and $\hat{X}_v = 0$ for $v > u$. Under this definition, if X has infinitely many positive components, then all the components of $\hat{X}$ are positive, and they are a (random) permutation of the positive components of X; and if X has exactly u positive components, then $\hat{X}_1, \ldots, \hat{X}_u$ are positive, a permutation of the positive components of X, and $\hat{X}_v = 0$ for $v > u$. And $\sum_{u=1}^{\infty} \hat{X}_u = \sum_{u=1}^{\infty} X_u = 1$ in any case, so that $\hat{X}$ is a random element of Δ (verify in succession the measurability of ς_u, τ_u, $\hat{X}_u$). This defines the size-biased version $\hat{X}$ of X.

The significance of size-biasing for the biological model is this. If $(X_1, \ldots, X_r)$ has the density (4.15) for each r, then $X = (X_1, X_2, \ldots)$ describes a population with infinitely many types and mutation rate θ. Now $\hat{X} = (X_{\tau_1}, X_{\tau_2}, \ldots)$, and for a given X, X_{τ_1} is the conditional probability that τ_1 is the oldest type, X_{τ_2} is the conditional probability that τ_2 is the second oldest type, and so on. For us the significance of size-biasing will be that we can use it to show that the densities (4.15) specify the finite-dimensional distributions of the Poisson-Dirichlet distribution on Δ. This will give us the asymptotic distribution of the lengths of the long cycles in a random permutation, and we can also use size-biasing to analyze the distribution of the large prime divisors of a random integer.

Return now to the Z^n of Theorems 4.2 and 4.3. By (4.8) and Dirichlet's formula [M12], $(Z_1^n, \ldots, Z_r^n)$ has at $(z_1, \ldots, z_r)$ the density

$$(4.18)\quad h_r(z_1, \ldots, z_r) = \frac{\Gamma(n\alpha)}{\Gamma^r(\alpha)\Gamma((n-r)\alpha)} z_1^{\alpha-1} \cdots z_r^{\alpha-1}(1 - z_1 - \cdots - z_r)^{(n-r)\alpha-1}$$

if $r < n$, $z_1, \ldots, z_r > 0$, and $z_1 + \cdots + z_r < 1$; and by symmetry, this is also the density for $Z_{i_1}^n, \ldots, Z_{i_r}^n$ for any of the $(n)_r$ permutations $i_1, \ldots, i_r$ of size r from $\{1, 2, \ldots, n\}$. For size-biasing on Z^n,

$$(4.19)\quad \mathsf{P}[\tau_1 = i_1, \ldots, \tau_r = i_r \| Z^n] = Z_{i_1}^n \frac{Z_{i_2}^n}{1 - Z_{i_1}^n} \cdots \frac{Z_{i_r}^n}{1 - Z_{i_1}^n - \cdots - Z_{i_{r-1}}^n} = p_r(Z_{i_1}^n, \ldots, Z_{i_r}^n);$$

the last equality defines p_r and there are no divisions by 0. Since $(\hat{Z}_1^n, \ldots, \hat{Z}_r^n) = (Z_{\tau_1}^n, \ldots, Z_{\tau_r}^n)$,

$$\mathsf{P}[(\hat{Z}_1^n, \ldots, \hat{Z}_r^n) \in H \| Z^n] = \sum p(Z_{i_1}^n, \ldots, Z_{i_r}^n) I_H(Z_{i_1}^n, \ldots, Z_{i_r}^n),$$

where the sum extends over all $(n)_r$ choices for $i_1, \ldots, i_r$. And now it follows by symmetry that

$$(4.20)\quad \mathsf{P}[(\hat{Z}_1^n, \ldots, \hat{Z}_r^n) \in H] = \int_H (n)_r p_r(z_1, \ldots, z_r) h_r(z_1, \ldots, z_r)\, dz_1 \cdots dz_r.$$

Let $[0,1]^\infty$ be the subspace of R^∞ consisting of the points x satisfying $0 \le x_i \le 1$ for all i. Suppose that B^n is a random element of $[0,1]^\infty$ such that the components $B_1^n, B_2^n, \ldots$ are independent and B_i^n has the beta-$(\alpha+1,(n-i)\alpha)$ density

$$h_{ni}(x) = \frac{\Gamma((n-i+1)\alpha+1)}{\Gamma(\alpha+1)\Gamma((n-i)\alpha)} x^\alpha (1-x)^{(n-i)\alpha-1} \tag{4.21}$$

on $(0,1)$. Define G^n by $G_1^n = B_1^n$ and $G_i^n = (1-B_1^n)\cdots(1-B_{i-1}^n)B_i^n$. Consider the map given by $z_1 = x_1$, $z_i = (1-x_1)\cdots(1-x_{i-1})x_i$, $1 < i \le r$. The inverse map is $x_1 = z_1$, $x_i = z_i/(1-z_1-\cdots-z_{i-1})$, $1 < i \le r$, and so $G_1^n, \ldots, G_r^n$ has density

(4.22)
$$h_{n1}(z_1)h_{n2}\Big(\frac{z_2}{1-z_1}\Big)\cdots h_{nr}\Big(\frac{z_r}{1-z_1-\cdots-z_{r-1}}\Big)\frac{\partial(x_1,\ldots,x_r)}{\partial(z_1,\ldots,z_r)}.$$

Since the Jacobian is $\prod_{i=2}^r (1-z_1-\cdots-z_{i-1})^{-1}$, algebra reduces (4.22) to the integrand in (4.20):

$$(\hat{Z}_1^n, \ldots \hat{Z}_r^n) =_d (G_1^n, \ldots, G_r^n), \tag{4.23}$$

in the sense that the two have the same distribution.

As $n \to \infty$ and $n\alpha \to \theta$, (4.21) converges to the beta-$(1,\theta)$ density for $1 \le i \le r$. If $B_1, B_2, \ldots$ are independent, each having this limiting distribution, then of course $(B_1^n, \ldots, B_r^n) \Rightarrow_n (B_1, \ldots, B_r)$. And if $G_1, G_2, \ldots$ are defined in terms of these B_i by (4.4), as before, then the mapping theorem gives $(G_1^n, \ldots, G_r^n) \Rightarrow_n (G_1, \ldots, G_r)$. It follows by (4.23) that $(\hat{Z}_1^n, \ldots, \hat{Z}_r^n) \Rightarrow_n (G_1, \ldots, G_r)$, and since this holds for each r, $\hat{Z}^n \Rightarrow_n G$. And further, since the components of $\hat{Z}^n$ are a (random) permutation of those of Z^n, $\rho\hat{Z}^n$ and ρZ^n coincide, and it follows by Theorem 4.1 that

$$Z_{(\cdot)}^n = \rho Z^n = \rho\hat{Z}^n \Rightarrow_n \rho G = G_{(\cdot)}. \tag{4.24}$$

Since Theorems 4.2 and 4.3 give the limiting densities for the components of $Z_{(\cdot)}^n$, we have a complete description of the Poisson-Dirichlet distribution:

Theorem 4.4. *The random vector $(G_{(1)}, \ldots, G_{(r)})$ has density (4.15), $G_{(r)}$ has density (4.16), and the moments $\mathsf{E}[(G_{(r)})^m]$ are given by (4.17).*

Numerical values for some of these distributional quantities can be calculated. Below is a short table of the distribution of $G_{(1)}$ for the case $\theta = 1$. The median is $e^{-1/2}$, which is about .61, and so there is, for large n, probability about $\frac{1}{2}$ that the length of the longest cycle exceeds $.61 \times n$.

x	.20	.30	.40	.50	.60	.70	.80	.90
$\mathsf{P}[G_{(1)} \le x]$	.00	.02	.13	.31	.49	.65	.78	.90

Moments have also been computed: For $\theta = 1$, the first three components $G_{(1)}$, $G_{(2)}$, $G_{(3)}$ of $G_{(\cdot)}$ have means .62, .21, .01 and standard deviations .19, .11, .07. And here is a graph of $d_1(\,\cdot\,;1)$:

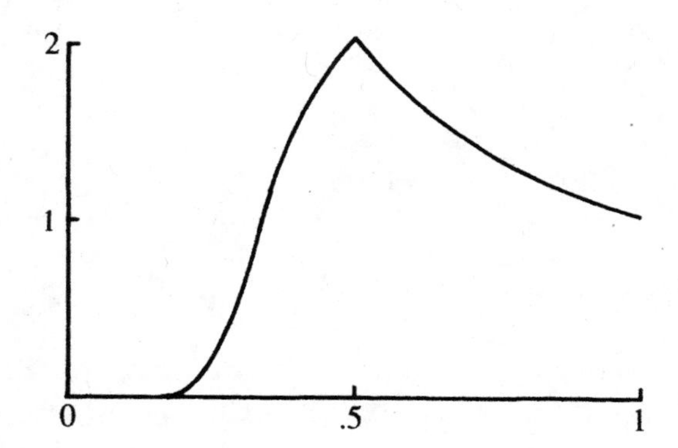

Large Prime Divisors

Let N_n be an integer randomly chosen from among the first n integers: $\mathsf{P}[N_n = m] = 1/n$ for $1 \le m \le n$. Let Q_{ni} be the distinct prime divisors of N_n, ordered by size: $Q_{n1} > Q_{n2} > \cdots$, where $Q_{nv} = 1$ if N_n has fewer than v distinct prime divisors,† and let $T_n = \prod_v Q_{nv}$ be the product of the distinct prime divisors.

Theorem 4.5. *We have*

$$\left(\frac{\log Q_{n1}}{\log T_n}, \frac{\log Q_{n2}}{\log T_n}, \ldots\right) \Rightarrow_n G_{(\cdot)}, \tag{4.25}$$

where $G_{(\cdot)}$ has the Poisson-Dirichlet distribution for $\theta = 1$.

Note that the vector on the left does lie in Δ. We prove below that

$$\frac{\log T_n}{\log N_n} \Rightarrow_n 1, \tag{4.26}$$

† In the limit, the large prime divisors all have multiplicity 1: $\mathsf{P}[Q_{ni}^2 | N_n] \to_n 0$ for each i. See Problem 4.4.

from which it follows that the asymptotic properties of the ratios $\log Q_{ni}/\log N_n$ are the same as those of the ordered relative cycle lengths of a random permutation. For example, the chance is about $\frac{1}{2}$ that the largest prime factor of N_n exceeds $N_n^{.61}$.

PROOF. Since $T_n \le N_n \le n$, (4.26) will follow if we prove that

$$\frac{\log T_n}{\log n} \Rightarrow_n 1. \tag{4.27}$$

We need two facts from number theory:[†]

$$\sum_{p \le x} \frac{\log p}{p} = \log x + \mathrm{O}(1), \qquad \sum_{p \le x} \log p = \mathrm{O}(x) \tag{4.28}$$

as $x \to \infty$. Since

$$\mathsf{E}[\log T_n] = \sum_{p \le n} \frac{1}{n} \left\lfloor \frac{n}{p} \right\rfloor \log p \ge \sum_{p \le n} \frac{\log p}{p} - \frac{1}{n} \sum_{p \le n} \log p,$$

(4.28) implies that $\mathsf{E}[\log n - \log T_n]$ is bounded, and (4.27) follows from this.

Size-bias the vector in (4.25), then multiply each component of the result by $\log T_n$ and apply $\exp(\cdot)$ to it. This gives a sequence $D_{n1}, D_{n2}, \ldots$ of random variables such that, if N_n has exactly r distinct prime factors and $p_1, \ldots, p_r$ is an arbitrary permutation of them, then

$$\begin{aligned} &(4.29)\quad \mathsf{P}[D_{n1} = p_1, \ldots, D_{nr} = p_r \| N_n] \\ &= \frac{\log p_1}{\log T_n} \frac{\log p_2/\log T_n}{1 - (\log p_1/\log T_n)} \cdots \frac{\log p_r/\log T_n}{1 - (\log p_1/\log T_n) - \cdots - (\log p_{r-1}/\log T_n)} \\ &= \frac{\log p_1}{\log T_n} \frac{\log p_2}{\log(T_n/p_1)} \cdots \frac{\log p_r}{\log(T_n/p_1 \cdots p_{r-1})}. \end{aligned}$$

And $D_{ni} = 1$ if $i > r$. Note that the conditional probabilities (4.29) summed over the $r!$ permutations come to 1. Define $V_{n1} = n$ and $V_{ni} = n/D_{n1} \cdots D_{n,i-1}$, and take

$$B_{ni} = \log D_{ni}/\log V_{ni}. \tag{4.30}$$

Finally, define $G_{n1} = B_{n1}$ and $G_{ni} = (1 - B_{n1}) \cdots (1 - B_{n,i-1}) B_{ni}$. If we can prove that

$$(B_{n1}, B_{n2}, \ldots) \Rightarrow_n (B_1, B_2, \ldots), \tag{4.31}$$

[†] Hardy & Wright [37], Theorems 414 and 425, or [PM.85].

where the B_i are independent random variables, each uniformly distriubted over $[0,1]$, then the mapping theorem will imply that $G_n = (G_{n1}, G_{n2}, \ldots) \Rightarrow_n G = (G_1, G_2, \ldots)$, where the G_i are defined by (4.4). Now $G_{ni} = \log D_{ni}/\log n$ (note that $V_{n,i-1}/D_{n,i-1} = V_{ni}$ and that $G_{ni} = 0$ if i exceeds the number of distinct prime divisors of N_n), and if $G'_{ni} = \log D_{ni}/\log T_n$, then it follows by (4.27) that $G'_n \Rightarrow G$. Theorem 4.1 now gives $\rho G'_n \Rightarrow \rho G = G_{(\cdot)}$, and since the D_{ni} are a permutation of the Q_{ni}, this is (4.25).

To understand how (4.31) is proved, consider first an approximate argument for B_{n1}. Suppose that $0 \le a \le 1$. If $t(m)$ is the product of the distinct prime divisors of m ($T_n = t(N_n)$), then

$$\begin{aligned}\mathsf{P}[B_{n1} \le a] = \mathsf{P}[D_{n1} \le n^a] &= \sum_{m=1}^{n} \frac{1}{n} \mathsf{P}[D_{n1} \le n^a | N_n = m] \\ &= \sum_{m=1}^{n} \frac{1}{n} \sum_{p \le n^a, p|m} \frac{\log p}{\log t(m)} \approx \sum_{m=1}^{n} \frac{1}{n} \sum_{p \le n^a, p|m} \frac{\log p}{\log n} \\ &= \sum_{p \le n^a} \frac{\log p}{\log n} \sum_{m \le n, p|m} \frac{1}{n} \approx \frac{1}{\log n} \sum_{p \le n^a} \frac{\log p}{p} \to_n a,\end{aligned}$$

where the first approximation works because $\log x$ increases so slowly, the second works because there are about n/p multiples of p preceding n, and the final step is a consequence of (4.28).

By Theorem 2.5, in order to prove (4.31) we need only show that, for each r,

$$\liminf_n \mathsf{P}[a_i < B_{ni} \le b_i,\ i \le r] \ge \prod_{i=1}^{r} (b_i - a_i) \tag{4.32}$$

if $0 < a_i < b_i < 1$. Note that there are at least r prime divisors if the event in (4.32) occurs. By the definitions,

$$\begin{aligned}\mathsf{P}[a_i < B_{ni} \le b_i, i \le r] &= \mathsf{P}[V_{ni}^{a_i} < D_{ni} \le V_{ni}^{b_i},\ i \le r] \\ &= \sum_{m=1}^{n} \sum \mathsf{P}[N_n = m,\ D_{ni} = p_i,\ i \le r],\end{aligned} \tag{4.33}$$

where the inner sum extends over certain permutations $p_1 \ldots, p_r$ of the distinct prime divisors of m, namely, over those permutations that satisfy

$$\left(\frac{n}{p_1 \cdots p_{i-1}}\right)^{a_i} < p_i \le \left(\frac{n}{p_1 \cdots p_{i-1}}\right)^{b_i}, \quad i \le r. \tag{4.34}$$

By (4.29), each term in the sum satisfies ($t(m)$ defined as before)

$$\mathsf{P}[N_n = m,\ D_{ni} = p_i,\ i \le r] = \frac{1}{n}\prod_{i=1}^{r} \frac{\log p_i}{\log(t(m)/p_1 \cdots p_{i-1})}. \tag{4.35}$$

Decrease the right side by increasing $t(m)$ to n. This gives

$$\mathsf{P}[a_i < B_{ni} \le b_i,\ i \le r] \ge \sum_{m=1}^{n} \frac{1}{n} \sum \prod_{i=1}^{r} \frac{\log p_i}{\log(n/p_1 \cdots p_{i-1})}, \tag{4.36}$$

and we can further decrease the right side by tightening (4.34) to

$$\left(\frac{n}{p_1 \cdots p_{i-1}}\right)^{a_i} < p_i \le \left(\frac{\epsilon n}{p_1 \cdots p_{i-1}}\right)^{b_i}, \quad i \le r \tag{4.37}$$

($0 < \epsilon < 1$). For $i = r$, the right-hand inequality here implies $p_r \le p_r^{1/b_i} \le \epsilon n/p_1 \cdots p_{r-1}$ and hence $p_1 \cdots p_r \le \epsilon n$. But then the number $\lfloor n/p_1 \cdots p_r \rfloor$ of multiples m of $p_1 \cdots p_r$ satisfying $m \le n$ is greater than $(1-\epsilon)n/p_1 \cdots p_r$, and reversing the sums in (4.36) gives

$$\mathsf{P}[a_i < B_{ni} \le b_i,\ i \le r] \ge (1-\epsilon) \sum \prod_{i=1}^{r} \frac{\log p_i}{p_i \log(n/p_1 \cdots p_{i-1})}, \tag{4.38}$$

the sum extending over the permutations satisfying (4.37).

If $n_i = n/p_1 \cdots p_{i-1}$, then (4.37) implies $n_i/n_{i+1} = p_i \le n_i^{b_i}$ and hence $n_{i+1} \ge n_i^{1-b_i}$, and it follows by recursion that $n_i \ge n_r \ge n^\delta$, where $\delta = \prod_{i=1}^{r}(1-b_i) > 0$. By (4.28), there is a u_ϵ such that $u \ge u_\epsilon$ implies

$$\sum_{u^{a_i} < p \le (\epsilon u)^{b_i}} \frac{\log p}{p \log u} \ge (1-\epsilon)(b_i - a_i), \quad i \le r, \tag{4.39}$$

and if $n \ge u_\epsilon^{1/\delta}$, then this holds for each $u = n_i$. Fix $p_1, \ldots p_{r-1}$, sum out p_r, and apply (4.39) for $u = n_r$:

$$\mathsf{P}[a_i < B_{ni} \le b_i,\ i \le r] \ge (1-\epsilon) \sum \prod_{i=1}^{r-1} \frac{\log p_i}{\log n_i} \left[(1-\epsilon)(b_r - a_r) - \frac{r}{\log n^\delta}\right],$$

where here the sum extends over $p_1, \ldots, p_{r-1}$ and the term $-r/\log n^\delta$ is to account for the fact that in summing out p_r we must avoid the values of the earlier p_i. And now successively sum out $p_{r-1}, \ldots, p_1$ in the same way:

$$\mathsf{P}[a_i < B_{n-i} \le b_i,\ i \le r] \ge (1-\epsilon)^r \prod_{i=1}^{r} \left[(1-\epsilon)(b_i - a_i) - \frac{r}{\log n^\delta}\right]$$

for $n \ge u_\epsilon^{1/\delta}$. This proves (4.32). □

Technical Arguments

PROOF OF LEMMA 2. Let

$$m = n - 1 - r, \quad s = \sum_{i=1}^{r} z_i, \quad c = (1-s)/z_r. \tag{4.40}$$

By symmetry, the density of $Z_{(1)}^n, \ldots, Z_{(r)}^n$ at a point $(z_1, \ldots, z_r)$ of M_r is

$$(n)_r \frac{\Gamma(n\alpha)}{\Gamma^n(\alpha)} z_1^{\alpha-1} \cdots z_r^{\alpha-1} \int y_1^{\alpha-1} \cdots y_m^{\alpha-1} \left(1 - s - \sum_{i=1}^{m} y_i\right)^{\alpha-1} dy_1 \cdots dy_m, \tag{4.41}$$

where the domain of integration is given by $0 < y_1, \ldots, y_m < z_r$ and (to account for Z_n^n) $0 < 1 - s - \sum_{i=1}^m y_i < z_r$. The change of variables $y_i = z_r s_i$ reduces the integral here to

$$z_r^{(m+1)\alpha-1} \alpha^{-m} \int \left(\prod_{i=1}^{m} \alpha s_i^{\alpha-1}\right)\left(c - \sum_{i=1}^{m} s_i\right)^{\alpha-1} ds_1 \cdots ds_m, \tag{4.42}$$

where now the variables are constrained by $0 < s_1, \ldots, s_m < 1$ and $c-1 < \sum_{i=1}^m s_i < c$. If S is a (positive) random variable having the density g_m, then the integral in (4.42) is the integral of $(c-S)^{\alpha-1}$ over the event $c - 1 < S < c$ (if $c < 1$, the constraint $c - 1 < S$ is vacuous). Therefore, (4.41) can be written as

$$(n)_r \frac{\Gamma(n\alpha)}{\Gamma^n(\alpha)} \alpha^{-m-1} z_1^{\alpha-1} \cdots z_{r-1}^{\alpha-1} z_r^{(m+2)\alpha-2} \int_{c-1}^{c} \alpha(c-x)^{\alpha-1} g_m(x;\alpha)\, dx.$$

Since the last integral is $g_{m+1}(c;\alpha)$, this proves (4.10). □

PROOF OF LEMMA 3. The Laplace transform $[(1-\alpha\int_0^1(1-e^{-tx})x^{\alpha-1}dx)^{1/\alpha}]^{m\alpha}$ of $g_m(\cdot\,;\alpha)$ converges to the right-hand member of (4.14). This by itself, however, does not imply that the limit transform comes from a density to which $g_m(\cdot\,;\alpha)$ converges; see Example 3.3. Let $A_m(x)$ be the set where $0 < s_1, \ldots, s_m < 1$ and $\sum_{i=1}^m s_i < x$; let $B_k(x)$ be the set where $s_1, \ldots, s_k > 0$ and $\sum_{i=1}^k s_i < x$; and define $C_k(x)$ as in (4.11). First, for $0 < x < m$, we have

$$\int_0^x g_m(u)\, du = \int_{A_m(x)} \alpha^m s_1^{\alpha-1} \cdots s_m^{\alpha-1} ds_1 \cdots ds_m. \tag{4.43}$$

Use the inclusion-exclusion principle, together with symmetry, to express this as the integral over $B_m(x)$ minus an alternating sum of integrals over the regions where $s_1, \ldots, s_k \geq 1$, $s_{k+1}, \ldots, s_m > 0$, and $\sum_{i=1}^m s_i < x$. This reduces (4.43) to

$$\sum_{0 \leq k < x} (-1)^k \binom{m}{k} \alpha^m \int_{C_k(x)} s_1^{\alpha-1} \cdots s_k^{\alpha-1}$$
$$\times \left[\int_{B_{m-k}\left(x - \sum_{i=1}^k s_i\right)} s_{k+1}^{\alpha-1} \cdots s_m^{\alpha-1} ds_{k+1} \cdots ds_m\right] ds_1 \cdots ds_k,$$

where in the term for $k = 0$ the iterated integral is replaced by a single integral over $B_m(x)$. Dirichlet's integral formula [M12] further reduces this to

$$(4.44)\quad \sum_{0 \le k < x} (-1)^k \binom{m}{k} \frac{\Gamma^{m-k}(\alpha)\alpha^m}{(m-k)\alpha\Gamma((m-k)\alpha)} \times \int_{C_k(x)} s_1^{\alpha-1} \cdots s_k^{\alpha-1} \left(x - \sum_{i=1}^k s_i \right)^{(m-k)\alpha} ds_1 \cdots ds_k,$$

where in the term for $k = 0$ the integral is replaced by $x^{m\alpha}$.

The derivative of the term for $k = 0$ in (4.44) matches the corresponding term in (4.12), and we turn to the case $1 \le k < x < m$. Write $\beta = (m-k)\alpha > 0$ and $\sigma = s_1 + \cdots + c_k$. Since (4.43) does have a derivative, we need only find the limit for h decreasing to 0 of the difference quotient

$$(4.45)\quad \frac{1}{h} \int_{C_k(x+h)-C_k(x)} s_1^{\alpha-1} \cdots s_k^{\alpha-1} (x+h-\sigma)^\beta ds_1 \cdots ds_k + \int_{C_k(x)} s_1^{\alpha-1} \cdots s_k^{\alpha-1} \frac{(x+h-\sigma)^\beta - (x-\sigma)^\beta}{h} ds_1 \cdots ds_k.$$

Since $x + h - \sigma \le h$ and $\beta > 0$, the first term here is at most (Dirichlet's formula again)

$$h^\beta \frac{\Gamma^k(\alpha)}{\Gamma(k\alpha)k\alpha} \frac{(x+h)^{k\alpha} - x^{k\alpha}}{h} \sim h^\beta \frac{\Gamma^k(\alpha)}{\Gamma(k\alpha)} x^{k\alpha-1} \to_h 0.$$

The difference quotient in the second term in (4.45) goes to $\beta(x-\sigma)^{\beta-1}$, and so the derivative of the kth term ($k \ge 1$) in (4.44) will agree with the kth term in (4.12) if we can integrate to the limit. But by the mean-value theorem, the difference quotient is bounded by $\beta(2x)^{\beta-1}$ if $\beta \ge 1$ (and $h < x$) and by the limit $\beta(x-\sigma)^{\beta-1}$ if $0 < \beta < 1$, and in either case, the dominated convergence theorem applies. (This argument is like the proof of the chain rule, which does not work here.)

Suppose now that $m \to \infty$ and $m\alpha \to \theta$, and consider a fixed k. Since $\Gamma'(1) = -\gamma$ (for our purposes this can be taken as the definition of γ), we have $\alpha\Gamma(\alpha) = \Gamma(1+\alpha) = 1 - \gamma\alpha + O(\alpha^2)$ and hence $\alpha^m\Gamma^m(\alpha) \to e^{-\gamma\theta}$. Since $(m)_k \sim m^k$ and $m/\Gamma(\alpha) = m\alpha/\Gamma(1+\alpha) \to \theta$, it now follows that the factor in front of the integral in (4.12) converges to the corresponding factor in (4.13). It is now clear that the term for $k = 0$ in (4.12) converges to the term for $k = 0$ in (4.13). Consider the case $1 \le k < x < m$. Let $D_k(x)$ be the set defined by $y_1, \ldots, y_k > 1/x$ and $\sum_{i=1}^k y_i < 1$. Transforming the integrals in (4.11) and (4.12) by $s_i = xy_i$, consider whether

$$x^{m\alpha-1} \int_{D_k(x)} y_1^\alpha \cdots y_k^\alpha \left(1 - \sum_{i=1}^k y_i\right)^{(m-k)\alpha-1} \frac{dy_1 \cdots d_k}{y_1 \cdots y_k} \longrightarrow x^{\theta-1} \int_{D_k(x)} \left(1 - \sum_{i=1}^k y_i\right)^{\theta-1} \frac{dy_1 \cdots dy_k}{y_1 \cdots y_k}$$

holds for each positive x. Of course, $x^{m\alpha-1} \to x^{\theta-1}$. The two integrals themselves differ by at most

$$x^k \int_{B_k(1)} \left| y_1^\alpha \cdots y_k^\alpha \left(1 - \sum_{i=1}^k y_i\right)^{(m-k)\alpha-1} - \left(1 - \sum_{i=1}^k y_i\right)^{\theta-1} \right| dy_1 \cdots dy_k.$$

The integrand here goes to 0 pointwise; if $0 < \theta_0 < \theta$, then $(m-k)\alpha > \theta_0$ for large m, and so the integrand is at most $2(1-\sum_{i=1}^k y_i)^{\theta_0-1}$, which is integrable over $B_k(1)$ (Dirichlet yet again). By the dominated convergence theorem, the integral goes to 0, and we do have $g_m(x;\alpha) \to g_\theta(x)$ pointwise. The limit g_θ is nonnegative, and by Fatou's lemma, it is integrable.

Although the series in (4.13) cannot be integrated term by term, it can be if we multiply through by e^{-tx} for $t > 0$. Reversing the integral in the kth term then gives $\Gamma(\theta)t^{-\theta}(E_1(t))^k$. Putting this into (4.13) and summing over k gives the middle expression in (4.14), and sum and integral here can be reversed because if we work with absolute values (suppress the $(-1)^k$), the resulting sum is finite.

For $t > 0$, the right-hand equality in (4.14) is equivalent to $\gamma + \log t + E_1(t) = \int_0^t (1-e^{-u})u^{-1}du$, which follows by integration by parts on each side, together with the fact that $-\gamma = \Gamma'(1) = \int_0^\infty e^{-u}\log u\, du$. Finally, since the right-hand member of (4.14) converges to 1 as $t \downarrow 0$, g integrates to 1—is a probability density. □

PROOF OF THEOREM 4.2. Recall the definitions (4.40). Since r is fixed, we have $m \to \infty$ and $m\alpha \to \theta$, and so, by Lemma 3 ($m+1$ in place of m), (4.10) converges pointwise to (4.15). By Fatou's lemma, (4.15) integrates to at most 1, and we must show that the integral in fact equals 1; this will be part of the next proof. □

PROOF OF THEOREM 4.3. By (4.15) and (4.13),

$$d_r(x;\theta) = x^{\theta-2}\int \sum_{k=0}^\infty \frac{(-1)^k\theta^{r+k}}{k!} J_k\left(\frac{1-s_1-\cdots-s_{r-1}-x}{x};\theta\right)\frac{ds_1\cdots ds_{r-1}}{s_1\cdots s_{r-1}},$$

where the integration extends over $s_1 > \cdots > s_{r-1} > x$ and $s_1 + \cdots + s_{r-1} + x < 1$. First, restrict the sum by $0 \le k \le x^{-1} - r$ (the other terms are 0) and take it outside the integral; second, lift the constraint $s_1 > \cdots > s_{r-1}$ and divide by $(r-1)!$ to compensate; and third, change variables: $y_i = s_i/x$. This leads to

$$(4.46)\quad d_r(x;\theta) = x^{\theta-2}\sum_{0\le k\le x^{-1}-r}\frac{(-1)^k\theta^{r+k}}{k!(r-1)!} \times \int J_k\left(\frac{1-x}{x} - y_1 - \cdots - y_{r-1};\theta\right)\frac{dy_1\cdots dy_{r-1}}{y_1\cdots y_{r-1}},$$

where here the integration extends over $y_1,\ldots,y_{r-1} > 1$ and $y_1 + \cdots + y_{r-1} < (1-x)/x$. Replace the J_k in (4.46) by its definition (4.11), and intersect the region of integration for the y_i with that for the s_i to arrive at (4.16).

It is probabilistically clear that d_r must be supported by $(0, r^{-1})$. It is convenient to take the support as $(0,1)$, which is consistent because all the summands in (4.16) vanish if $x > r^{-1}$. The mth moment of $d_r(\cdot\,;\theta)$ is

$$(4.47)\quad \int_0^1 x^m d_r(x;\theta)\,dx = \int_1^\infty \frac{1}{y^{m+2}} d_r(y^{-1};\theta)\,dy = \sum_{k=0}^\infty \frac{(-1)^k\theta^{r+k}}{k!(r-1)!}\int_1^\infty \frac{1}{y^{m+\theta}} J_{k+r-1}(y-1;\theta)\,dy.$$

To simplifty the notation, take $v = k + r - 1$; by (4.11), the last integral in (4.47) is

$$\int_1^\infty \frac{1}{y^{m+\theta}} \int_{C_v(y-1)} \left(y - 1 - \sum_{i=1}^v s_i\right)^{\theta-1} \frac{ds_1 \cdots ds_v}{s_1 \cdots s_v} dy$$
$$= \int_{s_i>1} \int_{y>a} \frac{1}{y^{m+1}} (1 - \frac{a}{y})^{\theta-1} dy \frac{ds_1 \cdots ds_v}{s_1 \cdots s_v},$$

where a stands for $1 + s_1 + \cdots + s_v$. The change of variable $s = a/y$ converts the inner integral into [M11]

$$a^{-m} \int_0^1 s^{m-1}(1-s)^{\theta-1} ds = a^{-m} \frac{\Gamma(m)\Gamma(\theta)}{\Gamma(m+\theta)}.$$

Therefore, the last integral in (4.47) is ($v = k + r - 1$)

$$\frac{\Gamma(m)\Gamma(\theta)}{\Gamma(m+\theta)} \int_{s_i>1} (1 + s_1 + \cdots + s_v)^{-m} \frac{ds_1 \cdots ds_v}{s_1 \cdots s_v}. \tag{4.48}$$

Expanding the right-hand exponential in (4.17) in a series gives

$$\sum_{k=0}^\infty \frac{(-1)^k \theta^{r-1+k}}{k!(r-1)!} \frac{\Gamma(\theta+1)}{\Gamma(\theta+m)} \int_0^\infty y^{m-1} e^{-y} (E_1(y))^{r-1+k} dy. \tag{4.49}$$

What (4.17) says is that the right member of (4.47) is identical with (4.49), and this will be true if the two match term for term, for each k. Since the last integral in (4.47) reduces to (4.48), and since $\Gamma(\theta+1) = \theta\Gamma(\theta)$, the question now is whether (take $v = k + r - 1$ again in (4.49))

$$\Gamma(m) \int_{s_i>1} (1 + s_1 + \cdots + s_v)^{-m} \frac{ds_1 \cdots ds_v}{s_1 \cdots s_v} = \int_0^\infty y^{m-1} e^{-y} (E_1(y))^v dy. \tag{4.50}$$

This last integral is (take $s_i = x_i/y$)

$$\int_0^\infty y^{m-1} e^{-y} \int_{x_i>y} e^{-x_1} \cdots e^{-x_v} \frac{dx_1 \cdots dx_v}{x_1 \cdots x_v} dy$$
$$= \int_{s_i>1} \int_0^\infty y^{m-1} e^{-y} \exp\left[-y \sum_{i=1}^v s_i\right] dy \frac{ds_1 \cdots ds_v}{s_1 \cdots s_v}.$$

Since the inner integral on the right here is $(1 + s_1 + \cdots + s_v)^{-m}\Gamma(m)$, we arrive at the integral on the left in (4.50). This proves (4.17).

We have still to prove that (4.15) integrates to 1, which will be true if the marginal density (4.16) does. Since (4.17) holds for $m = 0$, the question is whether

$$\theta \int_0^\infty y^{-1} e^{-y} \frac{(\theta E_1(y))^{r-1}}{(r-1)!} e^{-\theta E_1(y)} dy = 1. \tag{4.51}$$

For $r = 1$, this is

$$\int_0^\infty \theta y^{-1} e^{-y} e^{-\theta E_1(y)} dy = 1, \tag{4.52}$$

and it holds because the integrand is the derivative of $e^{-\theta E_1(y)}$. Multiply each side of (4.52) by θ^{-1} and repeatedly differentiate the resulting identity with respect to θ. This gives a sequence of identities equivalent to (4.51).

The proofs of Lemmas 2 and 3 and Theorems 4.2 and 4.3 are now complete. □

Problems

4.1. Show that

$$J'_k(x;\theta) = \frac{k}{x} J_{k-1}(x-1;\theta), \quad \text{for } x > k \geq 1$$

if $\theta = 1$. Now show that $d_1(x;1)$ is differentiable except at $x = 1/2$.

4.2. Let $\hat{X}$ be the size-biased version of X. Show that $X_n \Rightarrow X$ implies $\hat{X}_n \Rightarrow \hat{X}$.

4.3. Denote the size-biased version of X by σX. Show that $\rho^2 X = \rho X$, $\sigma^2 X =_d \sigma X$, $\rho\sigma X = \rho X$, $\sigma\rho X =_d \sigma X$. Show that, if X has the Poisson-Dirichlet distribution, then $\rho X = X$ and σX has the GEM distribution. Show that, if X has the GEM distribution, then $\sigma X =_d X$ and ρX has the Poisson-Dirichlet distribution.

4.4. **(a)** Let A_b be the set of integers m such that p^2 divides m for some prime p exceeding b. Show that, if $b_n \to \infty$, then $\mathsf{P}[N_n \in A_{b_n}] \to 0$.
(b) Redefine the Q_{ni} to consist of all the prime divisors of N_n, multiplicity accounted for, in nonincreasing order: $Q_{n1} \geq Q_{n2} \geq \cdots$. Show that (4.25) still holds if N_n is substituted for T_n.

4.5. If Z has density (4.13), then the denisty h_θ of $(1+Z)^{-1}$ is related to $d_1(\cdot,\theta)$ by

$$d_1(x;\theta) = \theta\Gamma(\theta)e^{\gamma\theta}x^\theta h_\theta(x), \quad 0 < x < 1.$$

Show that $\mathsf{E}[G_{(1)}^{-\theta}] = \theta\Gamma(\theta)e^{\gamma\theta}$.

SECTION 5. PROHOROV'S THEOREM

Relative Compactness

Let Π be a family of probability measures on $(S, \mathcal{S})$. We call Π *relatively compact* if every sequence of elements of Π contains a weakly convergent subsequence; that is, if for every sequence $\{P_n\}$ in Π there exist a subsequence $\{P_{n_i}\}$ and a probability measure Q (defined on $(S, \mathcal{S})$ but not necessarily an element of Π) such that $P_{n_i} \Rightarrow_i Q$. Even though $P_{n_i} \Rightarrow_i Q$ makes no sense if $QS < 1$, it is to be emphasized that we do require $QS = 1$—we disallow any escape of mass, as discussed below. For the most part we are concerned with the relative compactness of *sequences* $\{P_n\}$; this means that every subsequence $\{P_{n_i}\}$ contains a further subsequence $\{P_{n_{i(m)}}\}$ such that $P_{n_{i(m)}} \Rightarrow_m Q$ for some probability measure Q.

Example 5.1. Suppose we know of probability measures P_n and P on $(C, \mathcal{C})$ that the finite-dimensional distributions of P_n converge weakly to those of P: $P_n\pi_{t_1\cdots t_k}^{-1} \Rightarrow_n P\pi_{t_1\cdots t_k}^{-1}$ for all k and all $t_1, \ldots, t_k$. We have seen (Example 2.5) that P_n need not converge weakly to P. Suppose, however, that we also know that $\{P_n\}$ is relatively compact. Then each $\{P_{n_i}\}$ contains some $\{P_{n_{i(m)}}\}$ converging weakly to some Q.

Since the mapping theorem then gives $P_{n_{i(m)}}\pi^{-1}_{t_1\cdots t_k} \Rightarrow_m Q\pi^{-1}_{t_1\cdots t_k}$, and since $P_n\pi^{-1}_{t_1\cdots t_k} \Rightarrow_n P\pi^{-1}_{t_1\cdots t_k}$ by assumption, we have $Q\pi^{-1}_{t_1\cdots t_k} = P\pi^{-1}_{t_1\cdots t_k}$ for all $t_1, \ldots, t_k$. Thus the finite-dimensional distributions of P and Q are identical, and since the class $\mathcal{C}_f$ of finite-dimensional sets is a separating class (Example 1.3), $P = Q$. Therefore, each subsequence contains a further subsequence converging weakly to P—not to some fortuitous limit, but specifically to P. It follows by Theorem 2.6 that the entire sequence $\{P_n\}$ converges weakly to P. Therefore:

If $\{P_n\}$ is relatively compact and the finite-dimensional distributions of P_n converge weakly to those of P, then $P_n \Rightarrow_n P$.

This idea provides a powerful method for proving weak convergence in C and other function spaces. Note that, if $\{P_n\}$ does converge weakly to P, then it is relatively compact, so that this is not too strong a condition. □

Example 5.2. Now suppose we know that a sequence of measures $\{P_n\}$ on $(C, \mathcal{C})$ is relatively compact and that, for all k and all $t_1, \ldots, t_k$, $P_n\pi^{-1}_{t_1\cdots t_k}$ converges weakly to some probability measure $\mu_{t_1\cdots t_k}$ on $(R^k, \mathcal{R}^k)$—the point being that we do not assume at the outset that the $\mu_{t_1\cdots t_k}$ are the finite-dimensional distributions of a probability measure on $(C, \mathcal{C})$. Some subsequence $\{P_{n_i}\}$ converges weakly to some limit P (sub-subsequences are not relevant here). Since $P_{n_i}\pi^{-1}_{t_1\cdots t_k} \Rightarrow_i P\pi^{-1}_{t_1\cdots t_k}$ (again by the mapping theorem), and since $P_n\pi^{-1}_{t_1\cdots t_k} \Rightarrow_n \mu_{t_1\cdots t_k}$ (again by assumption), $P\pi^{-1}_{t_1\cdots t_k} = \mu_{t_1\cdots t_k}$ for all $t_1, \ldots, t_k$. This time we conclude that there exists a probability measure P having the $\mu_{t_1\cdots t_k}$ as its finite-dimensional distributions. Therefore:

If $\{P_n\}$ is relatively compact, and if $P_n\pi^{-1}_{t_1\cdots t_k} \Rightarrow_n \mu_{t_1\cdots t_k}$ for all $t_1, \ldots t_k$, then some P satisfies $P\pi^{-1}_{t_1\cdots t_k} = \mu_{t_1\cdots t_k}$ for all $t_1, \ldots, t_k$.

This provides a method of proving the existence of measures on $(C, \mathcal{C})$ having prescribed properties. □

Tightness

To use this method requires an effective means of proving relative compactness.

Example 5.3. Suppose P_n are probability measures on the line. How might we try to prove that $\{P_n\}$ is relatively compact? Let F_n be the distribution functions corresponding to the P_n. By the Helly selection theorem [PM.336], every subsequence $\{F_{n_i}\}$ contains a further subsequence $\{F_{n_{i(m)}}\}$ for which there exists a nondecreasing, right-continuous function F such that $F_{n_{i(m)}}(x) \to_m F(x)$ for all continuity points x of F. Of course, $0 \le F(x) \le 1$ for all x, and if F has limits 0 and 1 at $-\infty$ and $+\infty$, then F is the distribution function of a probability measure Q, and $P_{n_{i(m)}} \Rightarrow_m Q$ follows (Example 2.3). One can try in this way to prove $\{P_n\}$ relatively compact.

If $\{P_n\}$ is not, in fact, relatively compact, then of course the effort must fail for certain subsequences. Suppose for example that $P_n = \delta_n$. Then $F(x) \equiv 0$ is the only possibility for the limit function, and accordingly, there are no weak limits at all: If $P_{n_{i(m)}} \Rightarrow_m Q$, then $Q(-k,k) \le \liminf_m P_{n_{i(m)}}(-k,k) = 0$ for all k, and so Q cannot be a probability measure. The reason this sequence is not relatively compact is that mass is "escaping to infinity."

For a second example, let μ_n be the uniform distribution over $[-n,+n]$, and take $P_n = \delta_0$ for even n and $P_n = \frac{1}{3}\delta_0 + \frac{2}{3}\mu_n$ for odd n. Then $\{P_{n_i}\}$ contains a weakly convergent subsequence if n_i is even for infinitely many i, but what if n_i runs through odd integers only? Then the single possible limit $F(x)$ is $\frac{1}{3}$ for $x < 0$ and $\frac{2}{3}$ for $x \ge 0$, and again $P_{n_{i(m)}} \Rightarrow_m Q$ is impossible: It would imply $Q(-k,k) \le \frac{1}{3}$ for all k. In this example, the mass lost (namely $\frac{2}{3}$), rather than heading off in any particular direction, simply evaporates. □

The condition that prevents this escape of mass generalizes the tightness concept of Theorem 1.3. The family Π is *tight* if for every ϵ there exists a compact set K such that $PK > 1 - \epsilon$ for every P in Π.

Theorem 5.1. *If Π is tight, then it is relatively compact.*

This is the direct half of *Prohorov's theorem*, and the main purpose of the section is to prove it. The proof, like that of Helly's theorem, will depend on a diagonal argument. The following corollary extracts the essence of the argument in Example 5.1.

Corollary. *If $\{P_n\}$ is tight, and if each subsequence that converges weakly at all in fact converges weakly to P, then the entire sequence converges weakly to P: $P_n \Rightarrow_n P$.*

PROOF. By the theorem, each subsequence contains a further subsequence converging weakly to some limit, and by the hypothesis, this limit must be P. Apply Theorem 2.6. □

Example 5.4. Return to Example 5.1, and suppose that P_n is a unit mass at the function z_n defined by (1.5). No subsequence can converge weakly (Example 2.7), and the reason for this is that $\{P_n\}$ is not tight: If $P_nK > 1 - \epsilon > 0$ for all n, then K must contain all the z_n and hence cannot be compact. □

Theorem 5.2. *Suppose that S is separable and complete. If Π is relatively compact, then it is tight.*

This is the converse half of Prohorov's theorem. It contains Theorem 1.3, since a Π consisting of a single measure is obviously relatively compact. Although this converse puts things in perspective, the direct half is what is essential to the applications.

PROOF. Consider open sets G_n increasing to S. For each ϵ there is an n such that $PG_n > 1 - \epsilon$ for all P in Π: Otherwise, for each n we have $P_nG_n \le 1 - \epsilon$ for some P_n in Π, and by the assumed relative compactness, $P_{n_i} \Rightarrow_i Q$ for some subsequence and some probability measure Q, which is impossible because then $QG_n \le \liminf_i P_{n_i}G_n \le \liminf_i P_{n_i}G_{n_i} \le 1 - \epsilon$, while $G_n \uparrow S$.

From this it follows that, if $A_{k1}, A_{k2}, \ldots$ is a sequence of open balls of radius $1/k$ covering S (separability), then there is an n_k such that $P(\bigcup_{i \le n_k} A_{ki}) > 1 - \epsilon/2^k$ for all P in Π. If K is the closure of the totally bounded set $\bigcap_{k \ge 1} \bigcup_{i \le n_k} A_{ki}$, then K is compact (completeness), and $PK > 1 - \epsilon$ for all P in Π. □

The Proof

There remains the essential agendum of the section, the proof of Theorem 5.1. Suppose that $\{P_n\}$ is a sequence in the tight family Π. We are to find a subsequence $\{P_{n_i}\}$ and a probability measure P such that $P_{n_i} \Rightarrow_i P$.

CONSTRUCTION. Choose compact sets K_u in such a way that $K_1 \subset K_2 \subset \cdots$ and $P_nK_u > 1 - u^{-1}$ for all u and n. The set $\bigcup_u K_u$ is separable, and hence [M3] there exists a countable class $\mathcal{A}$ of open sets with the property that, if x lies both in $\bigcup_u K_u$ and in G, and if G is open, then $x \in A \subset A^- \subset G$ for some A in $\mathcal{A}$. Let $\mathcal{H}$ consist of $\emptyset$ and the finite unions of sets of the form $A^- \cap K_u$ for $A \in \mathcal{A}$ and $u \ge 1$.

Using the diagonal procedure, choose from the given sequence $\{P_n\}$ a subsequence $\{P_{n_i}\}$ along which the limit

$$\alpha(H) = \lim_i P_{n_i} H \tag{5.1}$$

exists for each H in the countable class $\mathcal{H}$. Our objective is to construct on $\mathcal{S}$ a probability measure P such that

$$PG = \sup_{H \subset G} \alpha(H) \tag{5.2}$$

for all open sets G. If there does exist such a probability measure P, then the proof will be complete: If $H \subset G$, then $\alpha(H) = \lim_i P_{n_i}(H) \le \liminf_i P_{n_i}G$, whence $PG \le \liminf_i P_{n_i}G$ follows via (5.2), and therefore $P_{n_i} \Rightarrow_i P$.

To construct a P satisfying (5.2), note first that $\mathcal{H}$ is closed under the formation of finite unions and that $\alpha(H)$, defined by (5.1) for each H in $\mathcal{H}$, has these three properties:

$$\alpha(H_1) \le \alpha(H_2) \quad \text{if } H_1 \subset H_2; \tag{5.3}$$

$$\alpha(H_1 \cup H_2) = \alpha(H_1) + \alpha(H_2) \quad \text{if } H_1 \cap H_2 = \emptyset; \tag{5.4}$$

$$\alpha(H_1 \cup H_2) \le \alpha(H_1) + \alpha(H_2). \tag{5.5}$$

And clearly $\alpha(\emptyset) = 0$. For open sets G define

$$\beta(G) = \sup_{H \subset G} \alpha(H); \tag{5.6}$$

then β is monotone and $\beta(\emptyset) = \alpha(\emptyset) = 0$. Finally, for arbitrary subsets M of S define

$$\gamma(M) = \inf_{M \subset G} \beta(G); \tag{5.7}$$

it is clear that $\gamma(G) = \beta(G)$ for open G.

Suppose we succeed in proving that γ *is an outer measure*. Recall [PM.165] that M is by definition γ-measurable if $\gamma(L) \ge \gamma(M \cap L) + \gamma(M^c \cap L)$ for all $L \subset S$, that the class $\mathcal{M}$ of γ-measurable sets is a σ-field, and that the restriction of γ to $\mathcal{M}$ is a measure. Suppose also that we are able to prove that *each closed set lies in* $\mathcal{M}$. It will then follow, first, that $\mathcal{S} \subset \mathcal{M}$, and second, that the restriction P of γ to $\mathcal{S}$ is a measure satisfying $PG = \gamma(G) = \beta(G)$, so that (5.2) will hold for open G. But P will be a *probability* measure because (each K_u, having a finite covering by $\mathcal{A}$-sets, lies in $\mathcal{H}$)

$$1 \ge PS = \beta(S) \ge \sup_u \alpha(K_u) \ge \sup_u (1 - u^{-1}) = 1. \tag{5.8}$$

We proceed in steps.

Step 1: If $F \subset G$, where F is closed and G is open, and if $F \subset H$ for some H in $\mathcal{H}$, then $F \subset H_0 \subset G$ for some H_0 in $\mathcal{H}$. To see this, choose, for each x in F, an A_x in the class $\mathcal{A}$ in such a way that $x \in A_x \subset A_x^- \subset G$. These A_x cover F, and since F is compact (being a subset of H), there is a finite subcover $A_{x_1}, \dots A_{x_k}$. Since $F \subset K_u$ for some u, we can take $H_0 = \bigcup_{i=1}^k (A_{x_i}^- \cap K_u)$.

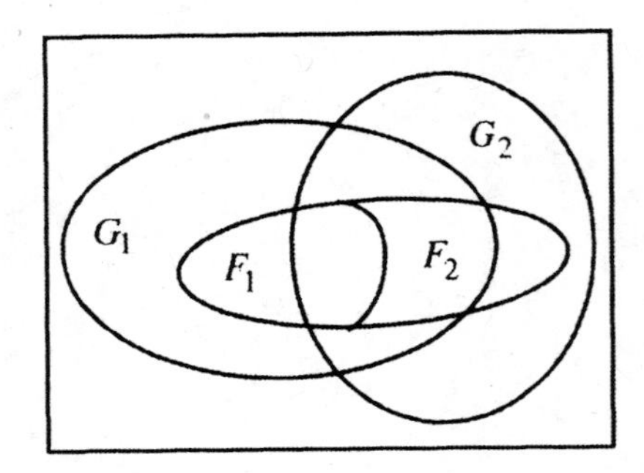

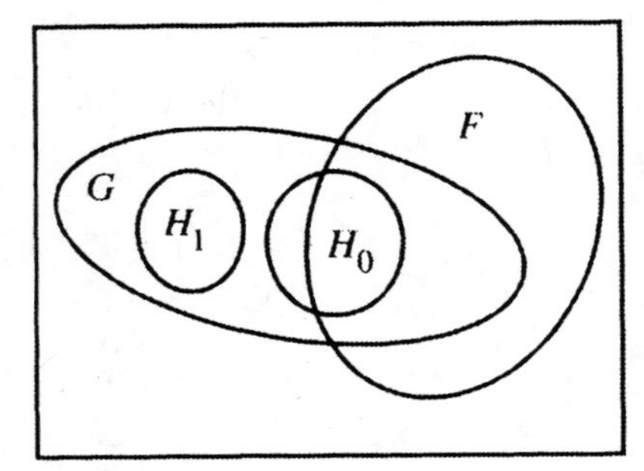

Step 2: β is finitely subadditive (on the open sets). Suppose that $H \subset G_1 \cup G_2$, where $H \in \mathcal{H}$ and G_1 and G_2 are open. Define

$$F_1 = [x \in H: \rho(x, G_1^c) \geq \rho(x, G_2^c)],\ F_2 = [x \in H: \rho(x, G_2^c) \geq \rho(x, G_1^c)].$$

If $x \in F_1$ and $x \notin G_1$, then $x \in G_2$, so that, since G_2^c is closed, $\rho(x, G_1^c) = 0 < \rho(x, G_2^c)$, a contradiction. Thus $F_1 \subset G_1$, and similarly $F_2 \subset G_2$. Since $F_1 \subset H$ and $H \in \mathcal{H}$, it follows by Step 1 that $F_1 \subset H_1 \subset G_1$ for some H_1 in $\mathcal{H}$; similarly, $F_2 \subset H_2 \subset G$ for some H_2 in $\mathcal{H}$. But then $\alpha(H) \leq \alpha(H_1 \cup H_2) \leq \alpha(H_1) + \alpha(H_2) \leq \beta(G_1) + \beta(G_2)$ by (5.3), (5.5), and (5.6). Since we can vary H inside $G_1 \cup G_2$, $\beta(G_1 \cup G_2) \leq \beta(G_1) + \beta(G_2)$ follows.

Step 3: β is countably subadditive (on the open sets). If $H \subset \bigcup_n G_n$, then, since H is compact, $H \subset \bigcup_{n \leq n_0} G_n$ for some n_0, and finite subadditivity implies $\alpha(H) \leq \beta(\bigcup_{n \leq n_0} G_n) \leq \sum_{n \leq n_0} \beta(G_n) \leq \sum_n \beta(G_n)$. Taking the supremum over H contained in $\bigcup_n G_n$ gives $\beta(\bigcup_n G_n) \leq \sum_n \beta(G_n)$.

Step 4: γ is an outer measure. Since γ is clearly monotone and satisfies $\gamma(\emptyset) = 0$, we need only prove that it is countably subadditive. Given a positive ϵ and arbitrary subsets M_n of S, choose open sets G_n such that $M_n \subset G_n$ and $\beta(G_n) < \gamma(M_n) + \epsilon/2^n$. Then, by the countable subadditivity of β, $\gamma(\bigcup_n M_n) \le \beta(\bigcup_n G_n) \le \sum_n \beta(G_n) < \sum_n \gamma(M_n) + \epsilon$; since ϵ was arbitrary, $\gamma(\bigcup_n M_n) \le \sum_n \gamma(M_n)$.

Step 5: $\beta(G) \ge \gamma(F \cap G) + \gamma(F^c \cap G)$ for F closed and G open. Choose, for given ϵ, an H_1 in $\mathcal{H}$ for which $H_1 \subset F^c \cap G$ and $\alpha(H_1) > \beta(F^c \cap G) - \epsilon$. Now choose an H_0 in $\mathcal{H}$ for which $H_0 \subset H_1^c \cap G$ and $\alpha(H_0) > \beta(H_1^c \cap G) - \epsilon$. Since H_0 and H_1 are disjoint and are contained in G, it follows by (5.6), (5.4), and (5.7) that $\beta(G) \ge \alpha(H_0 \cup H_1) = \alpha(H_0) + \alpha(H_1) > \beta(H_1^c \cap G) + \beta(F^c \cap G) - 2\epsilon \ge \gamma(F \cap G) + \gamma(F^c \cap G) - 2\epsilon$. But ϵ was arbitrary.

Step 6: $F \in \mathcal{M}$ if F is closed. By the inequality of Step 5, $\beta(G) \ge \gamma(F \cap L) + \gamma(F^c \cap L)$ if G is open and $G \supset L$; taking the infimum over these G shows that F is γ-measurable. This completes the construction. □

Prohorov's theorem (the direct half) will be used many times in what follows. Here is a standard application to classical analysis:

Example 5.5. Let P be a probability measure on the Borel subsets of the half-line $S = [0, \infty)$. The Laplace transform of P is the function defined for nonnegative t by $L(t) = \int_{x \ge 0} e^{-tx} P(dx)$. There is a uniqueness theorem [PM.286] according to which L completely determines P. Suppose we also have a sequence of measures P_n with transforms L_n. If $P_n \Rightarrow P$, then obviously $L_n(t) \to L(t)$ for each t.

Let us prove the converse. The argument depends on the inequality

$$\begin{aligned} (5.9)\quad \frac{1}{u}\int_0^u (1 - L_n(t))\,dt &= \frac{1}{u}\int_{x \ge 0}\int_0^u (1 - e^{-tx})\,dt\,P_n(dx) \\ &\ge \frac{1}{u}\int_{x > u^{-1}}\int_0^u (1 - e^{-t/u})\,dt\,P_n(dx) = e^{-1}P_n(u^{-1}, \infty). \end{aligned}$$

Since $L(t)$ is continuous and $L(0) = 1$, there is for given ϵ a u such that $u^{-1}\int_0^u (1 - L(t))\,dt < \epsilon/e$. But then, if $L_n(t) \to L(t)$ for all t, $u^{-1}\int_0^u (1 - L_n(t))\,dt < \epsilon/e$ for n beyond some n_0. And if $a = u^{-1}$, then (5.9) gives $P_n[0, a] \ge 1 - \epsilon$ for $n > n_0$. Therefore, $\{P_n\}$ is tight (increase a to take care of $P_1, \ldots, P_{n_0}$), and by the corollary to Theorem 5.1 it is enough to show that if a subsequence converges weakly at all then it must converge to P. But $P_{n_i} \Rightarrow_i Q$ implies that Q has transform $\lim_i L_{n_i}(t) = L(t)$, so that, by uniqueness, P must be the limit. □

The arguments in Examples 5.1 and 5.5 prove very different propositions but have the same structure. Each argument rests on three things: (i) the concept of tightness, (ii) a previously established converse proposition, and (iii) a previously established proposition concerning uniqueness.

As to Example 5.1, on conditions under which $P_n\pi_{t_1\cdots t_k}^{-1} \Rightarrow P\pi_{t_1\cdots t_k}^{-1}$ for all $t_1, \ldots, t_k$ implies $P_n \Rightarrow P$ on C:

(i) Methods for proving tightness will be developed in Chapter 2.
(ii) The previously established converse says that $P_n \Rightarrow P$ implies $P_n\pi_{t_1\cdots t_k}^{-1} \Rightarrow P\pi_{t_1\cdots t_k}^{-1}$ for all $t_1, \ldots, t_k$.
(iii) The previously established uniqueness proposition says that the finite-dimensional distributions $P\pi_{t_1\cdots t_k}^{-1}$ uniquely determine P.

As to Example 5.5, on the fact that pointwise convergence of the corresponding Laplace transforms ($L_n(t) \to L(t)$ for $t \geq 0$) implies $P_n \Rightarrow P$ on $[0, \infty)$:

(i) Tightness was proved by means of (5.9).
(ii) The previously established converse says that $P_n \Rightarrow P$ implies $L_n(t) \to L(t)$ for all t.
(iii) The previously established uniqueness proposition says that the transform L uniquely determines P.

Although these two arguments have the same structure, there is an essential difference: If the Laplace transforms in Example 5.5 converge pointwise, then, because of (5.9), $\{P_n\}$ is necessarily tight. On the other hand, if the finite-dimensional distributions in Example 5.1 converge weakly, then $\{P_n\}$ may or may not be tight—and the essential task in Chapter 2 is to sort out the cases.

Problems

5.1. In connection with Example 5.2, assume that $P_n\pi_{t_1\cdots t_k}^{-1} \Rightarrow \mu_{t_1\cdots t_k}$ for all $t_1 \cdots t_k$ and prove directly that the $\mu_{t_1\cdots t_k}$ are consistent in the sense of Kolmogorov's existence theorem.

5.2. If S is compact, every separating class is a limit-determining class.

5.3. Let Π consist of the δ_x for x in A. Show directly from the definition that Π is relatively compact if and only if A^- is compact in S.

5.4. A weak limit of a tight sequence is tight.

5.5. If Π is tight, then its elements have a common σ-compact support, but the converse is false.

5.6. A sequence of probability measures on the line is tight if and only if, for the corresponding distribution functions, we have $\lim_{x\to\infty} F_n(x) = 1$ and $\lim_{x\to-\infty} F_n(x) = 0$ uniformly in n.

5.7. A class of normal distributions on the line is tight if and only if the means and variances are bounded (a normal distribution with variance 0 being a point mass).

5.8. A sequence of distributions of random variables X_n is tight if it is uniformly integrable.

5.9. Probability measures on $S' \times S''$ are tight if and only if the two sets of marginal distributions are tight on S' and S''.

5.10. As in Problem 1.10, suppose that S is separable and locally compact. Assume that $P_n f \to Pf$ for all continuous f with compact support; show first that $\{P_n\}$ is tight and second that $P_n \Rightarrow P$.

5.11. Consider the following analogues of the concepts of this section: Points on the line play the role of probability measures on S, ordinary convergence on the line plays the role of weak convergence, relative compactness has the same definition (every sequence has a convergent subsequence), and boundedness plays the role of tightness. For each of the following facts, find the analogue in the theory of weak convergence.

(a) If each subsequence of $\{x_n\}$ contains a further subsequence converging to x, then $x_n \to x$.

(b) If $\{x_n\}$ is bounded, then it is relatively compact (and conversely).

(c) If $\{x_n\}$ is relatively compact and every subsequence that converges at all converges to x, then $x_n \to x$.

(d) The same with "bounded" in place of "relatively compact."

(e) If $\{x_n\}$ is bounded, $\sin x_n \to \sin x$, and $\sin \pi x_n \to \sin \pi x$, then $x_n \to x$.

5.12. The example in Problem 1.13 shows that Theorem 1.3 fails without the assumption of completeness, and hence so does Theorem 5.2. Here is an example of a separable S on which is defined a relatively compact (even compact), non-tight set of probability measures the individual elements of which are tight.

(a) Let Q be the closed unit square and put $L_x = [(x, y): 0 \le y \le 1]$ for $0 \le x \le 1$. Let $\mathcal{K}$ be the class of compact K in Q that meet each L_x. From the fact that $\mathcal{K}$ contains each $[(x, y): 0 \le x \le 1]$, $0 \le y \le 1$, conclude that $\mathcal{K}$ has the power of the continuum.

(b) Let $x \leftrightarrow K_x$ be a one-to-one correspondence between $[0, 1]$ and $\mathcal{K}$. For each x choose a point p_x in $K_x \cap L_x$, and let $A = [p_x: 0 \le x \le 1]$. Show that (i) A meets each L_x and (ii) A meets each K in $\mathcal{K}$.

(c) Let $S = Q - A$. Show that (i′) S misses exactly one point on each L_x and (ii′) if K is a compact subset of S, then K misses some L_x.

(d) Let P_x be Lebesgue measure on $L_x \cap S$ (just one point on L_x is missing). Show that $x_n \to x$ implies $P_{x_n} \Rightarrow P_x$. Show that each P_x is tight but that $[P_x: 0 \le x \le 1]$ is not tight.

SECTION 6. A MISCELLANY*

Of the four topics in this section, the first, on the ball σ-field, is used only in Section 15, and the second, on Skorohod's representation theorem, is not used at all. The last two, on the Prohorov metric and a coupling theorem, are used only in Section 21.

The Ball σ-Field

Applications to nonseparable function spaces require a theory of weak convergence for probability measures on the ball σ-field $\mathcal{S}_0$, the one generated by the open balls. Although $\mathcal{S}_0$ coincides with the full Borel σ-field $\mathcal{S}$ in the separable case, it may be smaller if S is not separable. In this case, a continuous real function on S may not be measurable $\mathcal{S}_0$, which creates problems: If S is discrete and uncountable (Example 1.4) and A is neither countable nor cocountable, then I_A is continuous but not measurable $\mathcal{S}_0$.

Lemma 1. (i) *If M is separable, then $\rho(\cdot, M)$ is $\mathcal{S}_0$-measurable as well as uniformly continuous.*
(ii) *If M is separable, then $M^\delta \in \mathcal{S}_0$ for positive δ.*
(iii) *If M is closed and separable, in particular if it is compact, then $M \in \mathcal{S}_0$.*

PROOF. Since $[x: \rho(x,y) < u] = B(y,u) \in \mathcal{S}_0$, it follows that $\rho(\cdot, y)$ is $\mathcal{S}_0$-measurable for each y. If D is countable, then $\rho(\cdot, D)$ is $\mathcal{S}_0$-measurable; and if D is dense in M, then $\rho(\cdot, D) = \rho(\cdot, M)$. Hence (i), and (ii) follows. Finally, (iii) follows from (ii) and the fact that $M = \bigcap_k M^{1/k}$ for closed M. □

The theory of weak convergence for $\mathcal{S}_0$ is closely analogous to that for $\mathcal{S}$.

Theorem 6.1. *If P is a probability measure on $(S, \mathcal{S}_0)$, A is an $\mathcal{S}_0$-set, and ϵ is positive, then there exist a closed $\mathcal{S}_0$-set F and an open $\mathcal{S}_0$-set G such that $F \subset A \subset G$ and $P(G - F) < \epsilon$.*

PROOF. The proof is analogous to that of Theorem 1.1. Let $\mathcal{G}$ be the class of $\mathcal{S}_0$-sets with the asserted property. If $A = [y : \rho(x,y) \le r]$ is a closed ball, take $F = A$ and $G = [y : \rho(x,y) < r + \delta]$ for small δ: $A \in \mathcal{G}$. It is now enough to show that $\mathcal{G}$ is a σ-field, and the argument in Section 1 carries over without change. □

Call a probability measure on $\mathcal{S}_0$ *separable* if it has a separable support (in $\mathcal{S}_0$). By Lemma 1(iii), the support can be assumed closed.

Theorem 6.2. *Suppose probability measures P and Q on $\mathcal{S}_0$ are separable. If $Pf = Qf$ for all bounded, uniformly continuous, $\mathcal{S}_0$-measurable functions f, then $P = Q$.*

PROOF. Let M be a common closed, separable support for P and Q. By part (i) of Lemma 1, the function

$$f(x) = 1 - (1 - \rho(x, (F^\epsilon)^c \cap M)/\epsilon)^+ \tag{6.1}$$

is $\mathcal{S}_0$-measurable as well as uniformly continuous, and by part (iii), the set $A_\epsilon = (F^\epsilon)^c \cap M$ lies in $\mathcal{S}_0$. If $x \in F$, then $\rho(x, A_\epsilon) \geq \epsilon$, and hence $f(x) = 1$. And f is supported by $A_\epsilon^c = F^\epsilon \cup M^c$. Thus $I_F \leq f \leq I_{A_\epsilon^c}$. If $F \in \mathcal{S}_0$, then $PF \leq Pf = Qf \leq QA_\epsilon^c = Q(F^\epsilon \cup M^c) = Q(F^\epsilon)$. Suppose that F is closed and let $\epsilon \downarrow 0$: $PF \leq QF$. Since, by symmetry, $QF \leq PF$ as well, P and Q agree for closed $\mathcal{S}_0$-sets and hence, by Theorem 6.1, agree on $\mathcal{S}_0$. □

For probability measures P_n and P on $\mathcal{S}_0$, if $P_n f \to Pf$ for every bounded, continuous, $\mathcal{S}_0$–measurable real function f on S, we say that P_n converges weak°ly to P and write $P_n \Rightarrow^\circ P$. By Theorem 6.2, a sequence cannot have two distinct, separable weak° limits.

Theorem 6.3. *If P has a separable support, then these five conditions are equivalent:*

(i) $P_n \Rightarrow^\circ P$.
(ii) $P_n f \to Pf$ *for all bounded, $\mathcal{S}_0$–measurable, uniformly continuous f.*
(iii) $\limsup_n P_n F \leq PF$ *for all closed F in $\mathcal{S}_0$.*
(iv) $\liminf_n P_n G \geq PG$ *for all open G in $\mathcal{S}_0$.*
(v) $P_n A \to PA$ *for every $\mathcal{S}_0$-set A for which $\mathcal{S}_0$ contains an open G and a closed F such that $G \subset A \subset F$ and $P(F - G) = 0$.*

PROOF. As in Theorem 2.1, the implication (i) → (ii) is trivial. To prove (ii) → (iii) we can use (6.1) and the set A_ϵ again, where M is a closed, separable support for P. Since $\limsup_n P_n F \leq \limsup_n P_n f = Pf \leq PA_\epsilon^c = P(F^\epsilon \cup M^c) = P(F^\epsilon)$, we can let $\epsilon \downarrow 0$, as before.

The equivalence of (iii) and (iv) is again simple. To prove that (iii) and (iv) together imply (v), replace A° by G and A^- by F in (2.3).

Finally, suppose that (v) holds and that f is bounded, continuous, and measurable with respect to $\mathcal{S}_0$. Let $G_t = A_t = [f > t]$ and $F_t = [f \geq t]$. Then the $\mathcal{S}_0$-sets G_t and F_t are open and closed, respectively, and $F_t - G_t \subset [f = t]$. Except for countably many t, therefore, we have $P(F_t - G_t) = 0$ and hence $P_n[f > t] \to P[f > t]$, and so the old argument goes through. □

There is also a mapping theorem. Let S' be a second metric space, with ball σ-field $\mathcal{S}_0'$. Let h map S into S'.

Theorem 6.4. *Suppose that M (in $\mathcal{S}_0$) is a separable support for P and that h is measurable $\mathcal{S}_0/\mathcal{S}_0'$ and continuous at each point of M. If $P_n \Rightarrow^\circ P$ (in S), then $P_n h^{-1} \Rightarrow^\circ Ph^{-1}$ (in S').*

PROOF. Let G' be an open $\mathcal{S}'_0$-set, and put $A = h^{-1}G'$. Since h is continuous at each point of M, $A \cap M \subset A^\circ$. Since $A \cap M$ is separable, there is a countable union G of open balls which satisfies $A \cap M \subset G \subset A^\circ \subset A$. But since $\mathcal{S}_0$ contains $A \cap M$ and the open set G, $\liminf_n P_n h^{-1}(G') = \liminf_n P_n A \geq \liminf_n P_n G \geq PG \geq P(A \cap M) = PA = Ph^{-1}(G')$. □

Define a sequence $\{P_n\}$ of probability measures on $\mathcal{S}_0$ to be tight° if for every ϵ there is a compact set K such that, for each δ,†

$$\liminf_n P_n K^\delta > 1 - \epsilon. \tag{6.2}$$

By Lemma 1, K and K^δ are in $\mathcal{S}_0$, and so is $K_\delta = [x: \rho(x, K) \leq \delta]$. If $\{P_n\}$ is tight° and $P_n \Rightarrow^\circ P$, then, since K_δ is closed and contains K^δ, it follows by Theorem 6.3, together with (6.2), that $PK_\delta \geq 1 - \epsilon$. Let $\delta \downarrow 0$: $PK \geq 1 - \epsilon$, and so P is tight in the old sense.

The next result is the analogue of Prohorov's theorem.

Theorem 6.5. *If $\{P_n\}$ is tight°, then each subsequence $\{P_{n_i}\}$ contains a further subsequence $\{P_{n_{i(m)}}\}$ such that $P_{n_{i(m)}} \Rightarrow^\circ_m P$ for some separable probability measure P on $\mathcal{S}_0$.*

PROOF. Since a subsequence of a tight° sequence is itself tight°, it is enough to construct a weakly convergent subsequence of $\{P_n\}$ itself. Choose compact sets K_u in such a way that $K_1 \subset K_2 \subset \cdots$ and

$$\liminf_n P_n K_u^\delta > 1 - u^{-1} \tag{6.3}$$

for all u and δ. Now follow the proof of Theorem 5.1.

Define the classes $\mathcal{A}$ and $\mathcal{H}$ just as before. By the diagonal method, choose a subsequence $\{P_{n_i}\}$ along which the limit $\alpha_0(H^\delta) = \lim_i P_{n_i} H^\delta$ exists for each of the countably many sets H^δ for H in $\mathcal{H}$ and δ a positive rational. Now put $\alpha(H) = \lim_\delta \alpha_0(H^\delta)$, where δ decreases to 0 through the rationals.

Clearly, α satisfies (5.3). Since $(H_1 \cup H_2)^\delta = H_1^\delta \cup H_2^\delta$, it also satisfies (5.5). And if H_1 and H_2 are disjoint, then (compactness) H_1^δ and H_2^δ are also disjoint for small δ, and so α satisfies (5.4). Define $\beta(G)$ for open G by (5.6) and $\gamma(M)$ for arbitrary M by (5.7). Then Steps 1 through 6 of the previous proof go through exactly as before.

† Problem 6.1 shows why the δ and the limit inferior are needed here, even though they are not needed for the theory based on $\mathcal{S}$.

Let P be the restriction of γ to $\mathcal{S}_0$. By (6.3), $\alpha_0(K_u^\delta) \geq 1 - u^{-1}$ $(K_u \in \mathcal{H})$, and so $\alpha(K_u) \geq 1 - u^{-1}$. Therefore, (5.8) holds again, and P is a probability measure on $\mathcal{S}_0$. Finally, if $H \subset G$, where $H \in \mathcal{H}$, G is open, and $G \in \mathcal{S}_0$, then (compactness again) $H^\delta \subset G$ for small δ, and so $\alpha(H) \leq \alpha_0(H^\delta) \leq \liminf_i P_{n_i}G$. Therefore, $PG = \beta(G) \leq \liminf_i P_{n_i}G$, and $P_{n_i} \Rightarrow_i^\circ P$.

The measure P, as the weak° limit of the tight° sequence $\{P_n\}$, is tight in the old sense, and so it is certainly separable. □

Suppose that ρ and ρ' are two metrics on S, and let $\mathcal{S}_0$, $\mathcal{S}$, $\mathcal{S}_0'$, $\mathcal{S}'$ be the corresponding ball and Borel σ-fields. If ρ' is finer [M2] than ρ, then each ρ-open set is also ρ'-open, and hence $\mathcal{S} \subset \mathcal{S}'$. In Section 15 there arise a function space S and a pair of metrics for which

$$\mathcal{S} = \mathcal{S}_0' \subset \mathcal{S}'. \tag{6.4}$$

If P_n and P are probability measures on $\mathcal{S} = \mathcal{S}_0'$, one can ask whether

$$P_n \Rightarrow P \quad \text{re}\ \rho \tag{6.5}$$

("re" for "with respect to") and whether

$$P_n \Rightarrow^\circ P \quad \text{re}\ \rho'. \tag{6.6}$$

Suppose P has a separable support M (in $\mathcal{S} = \mathcal{S}_0'$). If (6.6) holds, then $\limsup_n P_nF \leq PF$ for each ρ'-closed F in $\mathcal{S}_0'$ and hence holds for each ρ-closed F in $\mathcal{S}$. Thus (6.6) implies (6.5). Now assume also that ρ-convergence to a limit in M implies ρ'-convergence. If the $\mathcal{S}_0'$-set F is ρ'-closed, and if F_1 is its ρ-closure, then $M \cap F_1 \subset F$, and (6.5) implies $\limsup_n P_nF \leq \limsup_n P_nF_1 \leq PF_1 = P(M \cap F_1) \leq PF$, from which (6.6) follows.

Theorem 6.6. *Suppose that ρ' is finer than ρ, that ρ-convergence to a limit in M implies ρ'-convergence, that* (6.4) *holds, and that M (in $\mathcal{S} = \mathcal{S}_0'$) is separable and supports P. Then* (6.5) *and* (6.6) *are equivalent.*

Even though (6.5) and (6.6) are equivalent here, (6.6) can give more information: If h is a real $\mathcal{S}_0$-measurable function on S, then $P_nh^{-1} \Rightarrow Ph^{-1}$ follows from (6.5) if h is ρ-continuous and from (6.6) if h is ρ'-continuous. And if ρ' is strictly finer than ρ, then there are more ρ'-continuous functions than there are ρ-continuous functions.†

† Problem 6.4.

Skorohod's Representation Theorem

If $P_n \Rightarrow_n P$, then there are (see (3.4)) random elements X_n and X such that X_n has distribution P_n, X has distribution P, and $X_n \Rightarrow X$. According to the following theorem of Skorohod, if P has a separable support, then it is possible to constuct the X and all the X_n on a common probability space and to do it in such a way that $X_n(\omega) \to_n X(\omega)$ for each ω. This leads to some simple proofs. For example, if $P_n \Rightarrow_n P$ (and P has separable support), and if $h: S \to S'$ is continuous, construct X_n and X as just described; then $h(X_n(\omega)) \to_n h(X(\omega))$ for all ω, and hence (see the corollary to Theorem 3.1) $hX_n \Rightarrow_n hX$, which is equivalent to $P_n h^{-1} \Rightarrow_n Ph^{-1}$. This gives an alternative approach to the mapping theorem.

Theorem 6.7. *Suppose that $P_n \Rightarrow_n P$ and P has a separable support. Then there exist random elements X_n and X, defined on a common probability space $(\Omega, \mathcal{F}, \mathsf{P})$, such that $\mathcal{L}(X_n) = P_n$, $\mathcal{L}(X) = P$, and $X_n(\omega) \to_n X(\omega)$ for every ω.*

PROOF. We first show that, for each ϵ, there is a finite $\mathcal{S}$-partition $B_0, B_1, \ldots, B_k$ of S such that

$$\begin{cases} PB_0 < \epsilon, \\ P(\partial B_i) = 0, \quad i = 0, 1, \ldots, k, \\ \text{diam } B_i < \epsilon, \quad i = 1, \ldots, k. \end{cases} \tag{6.7}$$

Let M be a separable $\mathcal{S}$-set for which $PM = 1$. For each x in M, choose r_x so that $0 < r_x < \epsilon/2$ and $P(\partial B(x, r_x)) = 0$. Since M is separable, it can be covered by a countable subcollection $A_1, A_2, \ldots$ of the $B(x, r_x)$. As $k \uparrow \infty$, $P(\bigcup_{i=1}^k A_i) \uparrow P(\bigcup_{i=1}^\infty A_i) \geq PM = 1$, and we can choose k so that $P(\bigcup_{k=1}^k A_i) > 1 - \epsilon$. Let $B_1 = A_1$ and $B_i = A_1^c \cap \cdots \cap A_{i-1}^c \cap A_i$, $i = 2, \ldots, k$, and take $B_0 = (\bigcup_{i=1}^k A_i)^c$. Then the middle relation in (6.4) holds because $\partial B_i \subset \bigcup_{j=1}^k \partial A_j$ for $i = 0, 1, \ldots, k$, and the other two are obvious.

Take $\epsilon_m = 1/2^m$; there are $\mathcal{S}$-partitions $B_0^m, B_1^m, \ldots, B_{k_m}^m$ such that

$$\begin{cases} P(B_0^m) < \epsilon_m, \\ P(\partial B_i^m) = 0, \quad i = 0, 1, \ldots, k_m, \\ \text{diam } B_i^m < \epsilon_m, \quad i = 1, \ldots, k_m. \end{cases} \tag{6.8}$$

Amalgamate with B_0^m (which need not have small diameter) all the B_i^m for which $P(B_i^m) = 0$, so that $P(\cdot | B_i^m)$ is well defined for $i \geq 1$.

By the middle relation in (6.8) and the assumption that $P_n \Rightarrow P$, there is for each m an n_m such that $n \geq n_m$ implies

$$P_n(B_i^m) \geq (1-\epsilon_m)P(B_i^m), \quad i = 0, 1, \ldots, k_m. \tag{6.9}$$

We can assume $n_1 < n_2 < \cdots$.

There is a probability space supporting a random element with any given distribution on $(S, \mathcal{S})$, and by passing to the appropriate infinite product space† we can find an $(\Omega, \mathcal{F}, \mathsf{P})$ that supports random elements X, Y_n $(n \geq 1)$, Y_{ni} $(n, i \geq 1)$, Z_n $(n \geq 1)$ of S and a random variable ξ, all independent of one another, having these three properties: First, $\mathcal{L}(X) = P$ and $\mathcal{L}(Y_n) = P_n$. Second, if $n_m \leq n < n_{m+1}$, then $\mathcal{L}(Y_{ni}) = P_n(\,\cdot\,|B_i^m)$. Third, if $n_m \leq n < n_{m+1}$, then $\mathcal{L}(Z_n) = \mu_n$, where

$$\mu_n(A) = \epsilon_m^{-1} \sum_{i=0}^{k_m} P_n(A|B_i^m)[P_n(B_i^m) - (1-\epsilon_m)P(B_i^m)];$$

by (6.9), μ_n is a probability measure on $(S, \mathcal{S})$. Finally, ξ is uniformly distributed over $[0, 1]$.

For $n < n_1$ (if $n_1 > 1$), take $X_n = Y_n$. For $n_m \leq n < m_{m+1}$, define

$$X_n = \sum_{i=0}^{k_m} I_{[\xi \leq 1-\epsilon_m,\, X \in B_i^m]} Y_{ni} + I_{[\xi > 1-\epsilon_m]} Z_n.$$

If $n_m \leq n < n_{m+1}$, then, by independence and the definitions,

$$\begin{aligned}
&\mathsf{P}[X_n \in A] \\
&= \sum_{i=0}^{k_m} \mathsf{P}[\xi \leq 1-\epsilon_m,\, X \in B_i^m,\, Y_{ni} \in A] + \mathsf{P}[\xi > 1-\epsilon_m,\, Z_m \in A] \\
&= (1-\epsilon_m) \sum_{i=0}^{k_m} \mathsf{P}[X \in B_i^m] P(A|B_i^m) + \epsilon_m \mu_n(A) = P_n A.
\end{aligned}$$

Thus $\mathcal{L}(X_n) = P_n$ for each n. Let $E_m = [X \notin B_0^m,\, \xi \leq 1 - \epsilon_m]$ and $E = \liminf_m E_m$. If $n_m \leq n < n_{m+1}$, then on the set E_m, X_n and X lie in the same B_i^m, which has diameter less than ϵ_m. It follows that $X_n \to_n X$ on E, and since $\mathsf{P}(E_m^c) < 2\epsilon_m$, we have $\mathsf{P}E = 1$ by the Borel–Cantelli lemma. Redefine X_n as X outside E. This does not change its distribution, and now $X_n(\omega) \to_n X(\omega)$ for all ω. □

---

† See Halmos [36], p. 157.

The Prohorov Metric

Let $\mathcal{P}$ be the space of probability measures on the Borel σ-field $\mathcal{S}$ of a metric space S. If one topologizes $\mathcal{P}$ by taking as the general basic neighborhood of P the set of Q such that $|Pf_i - Qf_i| < \epsilon$ for $i = 1, \ldots, k$, where the f_i are bounded and continuous on S, then weak convergence is convergence in this topology. Here we study the case where $\mathcal{P}$ can be metrized.

The *Prohorov distance* $\pi(P, Q)$ between elements P and Q of $\mathcal{P}$ is defined as the infimum of those positive ϵ for which the two inequalities

$$PA \le QA^\epsilon + \epsilon, \quad QA \le PA^\epsilon + \epsilon \tag{6.10}$$

hold for all Borel sets A. We first prove a sequence of facts connecting this distance with weak convergence and then, at the end, state the most important of them as a theorem.

(i) *The Prohorov distance is a metric on* $\mathcal{P}$. Obviously $\pi(P,Q) = \pi(Q,P)$ and $\pi(P,P) = 0$. If $\pi(P,Q) = 0$, then for positive ϵ, $PF \le QF^\epsilon + \epsilon$, and if F is closed, letting $\epsilon \downarrow 0$ gives $PF \le QF$; the symmetric inequality and Theorem 1.1 now imply $P = Q$. If $\pi(P,Q) < \epsilon_1$ and $\pi(Q,R) < \epsilon_2$, then $PA \le QA^{\epsilon_1} + \epsilon_1 \le R(A^{\epsilon_1})^{\epsilon_2} + \epsilon_1 + \epsilon_2 \le R(A^{\epsilon_1+\epsilon_2}) + \epsilon_1 + \epsilon_2$. The triangle inequality follows from this and the symmetric relation.

(ii) *If* $PA \le QA^\epsilon + \epsilon$ *for all Borel sets* A, *then* $\pi(P,Q) \le \epsilon$. In other words, if the first inequality in (6.10) holds for all A, then so does the second. Note that $A \subset S - B^\epsilon$ and $B \subset S - A^\epsilon$ are equivalent because each is equivalent to the condition that $\rho(x,y) \ge \epsilon$ for all x in A and y in B. Given A, take $B = S - A^\epsilon$; if the first inequality in (6.10) holds for B, then $PA^\epsilon = 1 - PB \ge 1 - QB^\epsilon - \epsilon = Q(S - B^\epsilon) - \epsilon \ge QA - \epsilon$.

(iii) *If* $\pi(P_n, P) \to 0$, *then* $P_n \Rightarrow P$. Suppose that $\pi(P_n, P) < \epsilon_n \to 0$. If F is closed, then $\limsup_n P_nF \le \limsup_n (PF^{\epsilon_n} + \epsilon_n) = PF$.

(iv) *If* S *is separable and* $P_n \Rightarrow P$, *then* $\pi(P_n, P) \to 0$. For given ϵ, let $\{A_i\}$ be an $\mathcal{S}$-partition of S into sets of diameter less than ϵ. Choose k so that $P(\bigcup_{i>k} A_i) < \epsilon$ and let $\mathcal{G}$ be the finite class of open sets $(A_{i_1} \cup \cdots \cup A_{i_m})^\epsilon$ for $1 \le i_1 < \cdots < i_m \le k$. If $P_n \Rightarrow P$, then there exists an n_0 such that $n \ge n_0$ implies that $P_nG > PG - \epsilon$ for each G in $\mathcal{G}$. For a given A, let A_0 be the union of those sets among $A_1, \ldots, A_k$ that meet A. Then $A_0^\epsilon \in \mathcal{G}$, and $n \ge n_0$ implies $PA \le PA_0 + P(\bigcup_{i>k} A_i) \le PA_0 + \epsilon < P_nA_0^\epsilon + 2\epsilon \le P_nA^{2\epsilon} + 2\epsilon$. Use (ii).

(v) *If S is separable, then a set in $\mathcal{P}$ is relatively compact in the sense of Section 5 if and only if its π-closure is π-compact.* In the separable case, weak convergence and π-convergence are the same thing, and so this is just a matter of translating the terms.

(vi) *If S is separable, then $\mathcal{P}$ is separable.* Given ϵ, choose an $\mathcal{S}$-partition $\{A_i\}$ as in the proof of (iv). In each nonempty A_i choose a point x_i, and let Π_ϵ be the countable set of probability measures having the form $\sum_{i\le k} r_i \delta_{x_i}$ for some k and some set of rationals r_i. Given $P \in \mathcal{P}$, choose k so that $P(\bigcup_{i>k} A_i) < \epsilon$, then choose rationals $r_1, \ldots, r_k$ in such a way that $\sum_{i\le k} r_i = 1$ and $\sum_{i\le k} |r_i - P(A_i)| < \epsilon$, and put $Q = \sum_{i\le k} r_i \delta_{x_i}$. Given a set A, let I be the set of indices i $(i \le k)$ for which A_i meets A. If $A_0 = \bigcup_{i\in I} A_i$, then $PA \le PA_0 + \epsilon = \sum_{i\in I} P(A_i) + \epsilon \le \sum_{i\in I} r_i + 2\epsilon = QA_0 + 2\epsilon \le QA^\epsilon + 2\epsilon$. By (ii), Π_ϵ is a countable 2ϵ-net for $\mathcal{P}$.

(vii) *If S is separable and complete, then $\mathcal{P}$ is complete.* Suppose that $\{P_n\}$ is π-fundamental. It will be enough to show that the sequence is tight, since then it must (by Theorem 5.1 and (iv) above) contain a π-convergent subsequence. The proof can be completed by the argument in the proof of Theorem 5.2 if we show that for all ϵ and δ there exist finitely many δ-balls C_i such that $P_n(C_1 \cup \cdots \cup C_m) > 1 - \epsilon$ for all n. First, choose η so that $0 < 2\eta < \epsilon \wedge \delta$. Second, choose n_0 so that $n \ge n_0$ implies $\pi(P_{n_0}, P_n) < \eta$. Third, cover S by balls $A_i = B(x_i, \eta)$ and choose m so that $P_n(A_1 \cup \cdots \cup A_m) > 1 - \eta$ for $n \le n_0$. Third, let $B_i = B(x_i, 2\eta)$. If $n \ge n_0$, then $P_n(B_1 \cup \cdots \cup B_m) \ge P_n((A_1 \cup \cdots \cup A_m)^\eta) \ge P_{n_0}(A_1 \cup \cdots \cup A_m) - \eta \ge 1 - 2\eta$. If $n \le n_0$, then $P_n(B_1 \cup \cdots \cup B_m) \ge P_n(A_1 \cup \cdots \cup A_m) \ge 1 - \eta$. Take $C_i = B(x_i, \delta)$, $i \le m$.

Theorem 6.8. *Suppose that S is separable and complete. Then weak convergence is equivalent to π-convergence, $\mathcal{P}$ is separable and complete, and a set in $\mathcal{P}$ is relatively compact if and only if its π-closure is π-compact.*

A Coupling Theorem

Suppose M is a probability measure on $S \times S$ that has marginal measures P and Q on S and satisfies

$$M[(s,t)\colon \rho(s,t) > \alpha] < \alpha, \tag{6.11}$$

where ρ is the metric on S. Then, for positive η, $QA = M(S \times A) \le M[(s,t)\colon \rho(s,t) > \alpha] + M((A^\alpha)^- \times S) < \alpha + P(A^\alpha)^- \le \alpha + P(A^{\alpha+\eta})$.

This and the symmetric relation imply that the Prohorov distance satisfies $\pi(P,Q) \le \alpha$. The following converse, the Strassen-Dudley theorem, is used in proving the approximation theorem in Section 21.

Theorem 6.9. *If S is separable and $\pi(P,Q) < \alpha$, then there is a probability measure M on $S \times S$ that has marginals P and Q and satisfies* (6.11).

To begin the proof, choose β so that $\pi(P,Q) < \beta < \alpha$, and then choose ϵ so that $\beta + 2\epsilon < \alpha$ and $\epsilon < \beta$. As in the proof of Theorem 6.7, use separability to find a partition $B_0, B_1, \ldots, B_k$ such that $PB_0 < \epsilon$ and $\operatorname{diam} B_i < \epsilon$ for $i > 1$. Find a similar partition for Q and replace $\{B_i\}$ by the common refinement, so that $PB_0 < \epsilon$, $QB_0 < \epsilon$, and $\operatorname{diam} B_i < \epsilon < \beta$ for $i = 1, \ldots, k$.

Let $p_i = PB_i$ and $q_j = QB_j$ for $0 \le i, j \le k$, and let Δ be the set of pairs (i,j) such that $1 \le i, j \le k$ and $\operatorname{dist}(B_i, B_j) < \beta$. It will be shown later that there exists a $(k+1) \times (k+1)$ array of nonnegative numbers p_{ij} such that

$$p_i = \sum\nolimits_j p_{ij}, \quad q_j = \sum\nolimits_i p_{ij}, \quad \sum\nolimits_{(i,j)\in\Delta} p_{ij} \ge 1 - \beta - 2\epsilon. \tag{6.12}$$

Assuming for the moment that there exists such an array, we can complete the proof this way: Let M_{ij} be the product of the conditional probability measures $P(\cdot|B_i)$ and $Q(\cdot|B_j)$; if PB_i or QB_j is 0, take $M_{ij}(S \times S) = 0$. Then M_{ij} is a measure on $S \times S$ supported by $B_i \times B_j$. Let M be the probability measure $\sum_{ij} p_{ij} M_{ij}$. Then (sum over the (i,j) for which $p_{ij} > 0$) $M(A \times S) = \sum_{ij} p_{ij} M_{ij}(A \times S) = \sum_{ij} p_{ij} P(A|B_i) Q(S|B_j) = \sum_{ij} p_{ij} P(A \cap B_i)/p_i = \sum_i P(A \cap B_i) = PA$. Thus P is the first marginal measure for M, and similarly Q is the second.

To prove (6.11), observe first that, if $(i,j) \in \Delta$, then B_i and B_j have diameters less than ϵ (since $i, j \ge 1$) and $\operatorname{dist}(B_i, B_j) < \beta$. Choose s_{ij} in B_i and t_{ij} in B_j in such a way that $\rho(s_{ij}, t_{ij}) < \beta$. If $s \in B_i$ and $t \in B_j$, then $\rho(s,t) \le \rho(s, s_{ij}) + \rho(s_{ij}, t_{ij}) + \rho(t_{ij}, t) < \beta + 2\epsilon < \alpha$. It follows by (6.12) that

$$\begin{aligned} M[(s,t)\colon \rho(s,t) < \alpha] &\ge M\Big(\bigcup\nolimits_{(i,j)\in\Delta} (B_i \times B_j)\Big) \\ &= \sum\nolimits_{(i,j)\in\Delta} p_{ij} \ge 1 - \beta - 2\epsilon > 1 - \alpha. \end{aligned}$$

Hence (6.11).

There remains the (considerable) task of constructing an array $\{p_{ij}\}$ satisfying (6.12). We start with finite sets X, Y, and $R \subset X \times Y$. If (x, y) lies in R, write xRy: x and y stand in the relation (represented by) R. For $A \subset X$, let A^R be the set of y in Y such that xRy for some x in A. Indicate cardinality by bars.

Lemma 2. *Suppose that $|A^R| \geq |A|$ for each subset A of X. Then there is a one-to-one map φ of X into Y such that $xR\varphi(x)$ for every x in X.*

This can be understood through its interpretation as "the marriage lemma": X is a set of women, Y is a set of men, and xRy means that x and y are mutually compatible. The hypothesis is that for each group of women there is a group of men, at least as large, each of whom is compatible with at least one woman in the group. The conclusion is that each woman in X can be married off to a compatible man. Some men may be left celibate, but not if the total numbers of men and women are the same; in the notation of the lemma, if $|X| = |Y|$, then the map φ of the conclusion carries X *onto* Y.

PROOF. For $U \subset X$ and $V \subset Y$, call the pair (U, V) an R-couple if $|A^R \cap V| \geq |A|$ for every $A \subset U$. This is just the hypothesis of the theorem formulated for the relation $R \cap (U \times V)$ on $U \times V$. Call f an R-pairing for (U, V) if it is a one-to-one map of U into V and $uRf(u)$ for all $u \in U$. We want to find an R-pairing φ for the R-couple (X, Y).

The proof of the lemma goes by induction on $n = |X|$. It obviously holds for $n = 1$. Assume that it holds for sets of size less than n. The induction hypothesis implies this:

(T) *If U is a proper subset of X and (U, V) is an R-couple, then there is an R-pairing for (U, V).*

Fix an x_0 in X; by the hypothesis, $x_0 R y_0$ for some y_0 in Y. There are two possibilities: (a) $(X - \{x_0\}, Y - \{y_0\})$ is an R-couple, and (b) it is not. Assume (a). It then follows by (T) that there is an R-pairing f for $(X - \{x_0\}, Y - \{y_0\})$. Extend f to an R-pairing φ for (X, Y) by taking $\varphi(x_0) = y_0$, and we are done.

So assume (b). Then $(X - \{x_0\}, Y - \{y_0\})$ is not an R-couple, and there is an $A \subset X - \{x_0\}$ such that $|A^R \backslash \{y_0\}| < |A|$. But since $|A^R| \geq |A|$ by the hypothesis of the lemma, we have $|A^R| \geq |A| > |A^R \backslash \{y_0\}|$. This means that ($y_0 \in A^R$ and) $|A^R| = |A|$. Since (by the hypothesis of the lemma) $|B^R \cap A^R| \geq |B|$ for $B \subset A$, (A, A^R) is an R-couple, and by (T) there is an R-pairing ψ for (A, A^R). Again consider two possibilities: (a°) $(X - A, Y - A^R)$ is an R-couple, and (b°) it is not. If (a°) holds, then, by (T), there is an R-pairing χ for $(X - A, Y - A^R)$. But this χ can be combined with ψ to give an R-pairing φ for (X, Y).

It remains only to rule out (b°) as a possibility. If (b°) does hold, then there is a $C \subset X - A$ such that $|C^R \backslash A^R| < |C|$. But then (recall that $|A^R| = |A|$) we have $|(A \cup C)^R| = |A^R| + |C^R \backslash A^R| < |A| + |C| = |A \cup C|$, which contradicts the hypothesis of the lemma. □

Next consider finite sets I, J, and a subset D of $I \times J$. Consider also a rectangular array of cells corresponding to the (i, j) in $I \times J$, together with nonnegative numbers c_i and r_j. Think of the c_i in a row below the array (i running from left to right) and the r_j in a column to the left (j running from bottom to top). For $E \subset I$, let E^D be the set of j such that $(i, j) \in D$ for some $i \in E$.

Lemma 3. *Suppose that, for each $E \subset I$,*

$$\sum_{j \in E^D} r_j \geq \sum_{i \in E} c_i. \tag{6.13}$$

Then there are nonnegative numbers m_{ij} such that $m_{ij} > 0$ only for $(i, j) \in D$, and

$$\sum_{j \in J} m_{ij} = c_i, \quad \sum_{i \in I} m_{ij} \leq r_j. \tag{6.14}$$

If $\sum_{i \in I} c_i = \sum_{j \in J} r_j$, then there is equality on the right in (6.14).

PROOF. Suppose first that the c_i and r_j are rational. By multiplying through by a common multiple of the denominators, we can ((6.13) and (6.14) are homogeneous) arrange that the c_i and r_j are nonnegative integers. Take A_i to be disjoint sets with $|A_i| = c_i$, take B_j to be disjoint sets with $|B_j| = r_j$, and take $X = \bigcup_i A_i$, $Y = \bigcup_j B_j$. Now define $R \subset X \times Y$ by putting into R those (x, y) such that the indices of the A_i and B_j containing them ($x \in A_i$, $y \in B_j$) satisfy $(i, j) \in D$.

There is an interpretation of this setup in terms of wholesale marriage. Suppose the women are divided into clans and the men are divided into hordes. Certain clans and hordes are compatible, and a woman-man pair is compatible if and only if the clan and horde they belong to are compatible.

We can verify the hypothesis of Lemma 2. Suppose that $A \subset X$, and take $E = [i: A \cap A_i \neq \emptyset]$. If $i \notin E$, then $(A \cap A_i)^R = \emptyset$. If $i \in E$, then $(A \cap A_i)^R = \bigcup_{j:(i,j) \in D} B_j$. Therefore, $A^R = \bigcup_{i \in E} \bigcup_{j:(i,j) \in D} B_j = \bigcup_{j \in E^D} B_j$. By (6.13), $|A^R| = \sum_{j \in E^D} |B_j| \geq \sum_{i \in E} |A_i| = |A|$. The hypothesis of Lemma 2 is thus satisfied, and there is a one-to-one map φ

of X into Y such that, for each x in X, $xR\varphi(x)$. Let m_{ij} be the number of points in A_i that are mapped into B_j by φ. Then $\sum_j m_{ij} = c_i$ (every point of A_i is mapped into some B_j), and $\sum_i m_{ij} \le |B_j| = r_j$ (φ is one-to-one). Hence (6.14). If $m_{ij} > 0$, then there are points x and y of A_i and B_j such that xRy, which implies $(i,j) \in D$.

To treat the general case, choose rational $c_i(n)$ and $r_j(n)$ such that $0 \le c_j(n) \uparrow c_j$ and $r_j(n) \downarrow r_j$ (D does not change). Let $\{m_{ij}(n)\}$ be a solution for these marginals, and pass to a sequence along which each $m_{ij}(n)$ converges to some m_{ij}. This gives a solution to (6.14).

If $\sum_i c_i = \sum_j r_j$, then $\sum_j \sum_i m_{ij} = \sum_j r_j$, and none of the inequalities in (6.14) can be strict. $\square$

COMPLETION OF THE PROOF OF THEOREM 6.9. We can use Lemma 3 to finish the proof of Theorem 6.9. We must construct a $(k+1) \times (k+1)$ array $\{p_{ij}\}$ satisfying (6.12). Let $I = \{0, 1, \ldots, k\}$, $J = \{0, 1, \ldots, k, \infty\}$, $c_i = p_i = PB_i$ for $0 \le i \le k$, $r_j = q_j = QB_j$ for $0 \le j \le k$, and $r_\infty = \beta + 2\epsilon$. Use for D the elements of the set Δ (defined just above (6.12)) together with the pairs (i, ∞) for $i = 0, 1, \ldots, k$. If E is a nonempty subset of I, then, since β was chosen so that $\pi(P, Q) < \beta$,

$$\sum_{i \in E} c_i \le PB_0 + P\Big(\bigcup_{i \in E \setminus \{0\}} B_i\Big) < \epsilon + Q\Big(\bigcup_{i \in E \setminus \{0\}} B_i\Big)^\beta + \beta$$
$$\le \epsilon + Q\Big(B_0 \cup \bigcup_{i \in E \setminus \{0\}} (B_i^\beta \cap B_0^c)\Big) + \beta.$$

Suppose that $x \in B_i^\beta$, where $1 \le i \le k$, and $x \notin B_0$; then $x \in B_j$ for some j, $1 \le j \le k$, and $\operatorname{dist}(B_i, B_j) < \beta$, which implies $(i,j) \in \Delta$, which in turn implies $j \in E^D \setminus \{\infty\}$. Therefore, $\bigcup_{i \in E \setminus \{0\}} (B_i^\beta \cap B_0^c) \subset \bigcup_{j \in E^D \setminus \{\infty\}} B_j$, and it follows that

$$\sum_{i \in E} c_i \le \sum_{j \in E^D \setminus \{\infty\}} q_j + 2\epsilon + \beta = \sum_{j \in E^D} r_j.$$

Thus (6.13) holds, and there exist m_{ij} for $(i,j) \in I \times J$, positive only for $(i,j) \in D$, such that

$$(6.15) \qquad \begin{cases} \text{(a)} & \sum_{j \in J} m_{ij} = p_i, & \text{for } 0 \le i \le k, \\ \text{(b)} & \sum_{i \in I} m_{ij} \le q_j, & \text{for } 0 \le j \le k, \\ \text{(c)} & \sum_{i \in I} m_{i\infty} \le 2\epsilon + \beta. \end{cases}$$

Reduce J to $J' = \{0, 1, \ldots, k\}$, let $D' = I \times J'$, and apply Lemma 3 a second time, with $c_i' = m_{i\infty}$ and $r_j' = q_j - \sum_{i \in I} m_{ij}$ (all nonnegative).

If $E \subset I$ and $E \neq \emptyset$, then $E^{D'} = J'$, and by (6.15a) and the fact that $\sum_{i \in I} p_i = \sum_{j \in J'} q_j = 1$, we have

$$(6.16) \quad \sum_{j \in E^{D'}} r'_j = \sum_{j \in J'} (q_j - \sum_{i \in I} m_{ij})$$
$$= 1 - \sum_{i \in I} \Big(\sum_{j \in J} m_{ij} - m_{i\infty} \Big) = \sum_{i \in I} m_{i\infty} \geq \sum_{i \in E} m_{i\infty}.$$

The hypothesis of Lemma 3 is therefore satisfied. Since (6.16) also implies $\sum_{i \in I} c'_i = \sum_{j \in J'} r'_j$, there exist m'_{ij} such that

$$(6.17) \qquad \begin{cases} \text{(a)} & \sum_{j \in J'} m'_{ij} = c'_i \quad \text{for } i \in I, \\ \text{(b)} & \sum_{i \in I} m'_{ij} = r'_j \quad \text{for } j \in J'. \end{cases}$$

Take $p_{ij} = m_{ij} + m'_{ij}$ for $(i,j) \in D'$. From (6.15a) and (6.17a) follows $\sum_{j \in J'} p_{ij} = p_i$; from (6.17.b) follows $\sum_{i \in I} p_{ij} = q_j$. And finally, since $m_{ij} > 0$ only if $(i,j) \in \Delta$, it follows by (6.15a) and (6.15c) that

$$\sum_{(i,j) \in \Delta} p_{ij} \geq \sum_{(i,j) \in \Delta} m_{ij} = \sum_{(i,j) \in I \times J} m_{ij} - \sum_{i \in I} m_{i\infty}$$
$$= \sum_{i \in I} p_i - \sum_{i \in I} m_{i\infty} \geq 1 - 2\epsilon - \beta.$$

Therefore, (6.12) is satisfied. □

Problems

6.1. Metrize $S = [0,1)$ by $\rho(x,y) = x + y$ for $x \neq y$. Describe the balls in this space; show that $\mathcal{S}_0$ consists of the ordinary Borel sets and that $\mathcal{S} = 2^{[0,1)}$. Show that each point of $(0,1)$ is isolated. An infinite compact set must consist of 0 and a sequence converging to 0 in the ordinary sense. Define P_n on $\mathcal{S}_0$ as the uniform distribution over $[2^{-n-1}, 2^{-n}]$. If $K = \{0\}$, then $P_n K^\delta = 1$ for large n, and so $\{P_n\}$ is tight°. On the other hand, if K is compact, then, for each n, $P_n K^\delta = 0$ for small enough δ.

6.2. Does $\mathcal{S}_0$ consist of the $\mathcal{S}$-sets that are either separable or coseparable?

6.3. Prove in steps that a separable probability measure on $\mathcal{S}_0$ has a unique extension to $\mathcal{S}$. Let P be the measure and M the separable support; let $\mathcal{B}$ and $\mathcal{O}$ be the classes of open balls and open sets in S.

(a) Show [PM.159] that the ball and the Borel σ-fields in M (with the relative topology) are $M \cap \sigma(\mathcal{B}) = M \cap \mathcal{S}_0$ and $M \cap \sigma(\mathcal{O}) = M \cap \mathcal{S}$ and that they coincide.

(b) Define a probability measure P' on $M \cap \mathcal{S}_0 = M \cap \mathcal{S}$ by $P'(M \cap B) = PB$ for $B \in \mathcal{S}_0$. Define a probability measure P'' on $\mathcal{S}$ by setting $P''A = P'(M \cap A)$ for $A \in \mathcal{S}$.

(c) Now show that $P''B = PB$ for $B \in \mathcal{S}_0 \subset \mathcal{S}$.

(d) Suppose that Q'' is a second extension. If $A \in \mathcal{S}$, then $M \cap A = M \cap B$ for some $B \in \mathcal{S}_0$; use this fact to prove that P'' and Q'' are identical (on $\mathcal{S}$).

6.4. Suppose that every ρ'-continuous function is also ρ-continuous; show that ρ is finer than ρ'.

6.5. There is another way to define weak convergence for $\mathcal{S}_0$. If μ is a probability measure on $\mathcal{S}_0$ and f is bounded, define the upper [lower] integral $\int^* f\,d\mu$ [$\int_* f\,d\mu$] as the infimum [supremum] of $\int g\,d\mu$ for bounded, $\mathcal{S}_0$-measurable functions g satisfying $g \ge f$ [$g \le f$]. Let P_n be probability measures on $\mathcal{S}_0$, let P be a probability measure on $\mathcal{S}$, and take P_0 to be the restriction of P to $\mathcal{S}_0$. Define $P_n \to P$ (weak*) to mean that

$$\lim_n \int^* f\,dP_n = \lim_n \int_* f\,dP_n = \int f\,dP$$

for all bounded, continuous f. Show that this is equivalent to $P_n \Rightarrow^\circ P_0$.

6.6. Let h and h_n be maps from S to S', each measurable $\mathcal{S}/\mathcal{S}'$. Let E be the set of x such that $h_n x_n \not\to hx$ for some sequence $\{x_n\}$ converging to x. Suppose that $P_n \Rightarrow P$, P has separable support, $E \in \mathcal{S}$, and $PE = 0$. Use Theorem 6.7 to show that $P_n h_n^{-1} \Rightarrow Ph^{-1}$.

6.7. Suppose that S is separable. For probability measures P and Q, take $\mathcal{M}(P,Q)$ to be the set of probability measures on $S \times S$ having marginal measures P and Q. Let $\pi'(P,Q) = \inf\inf[\alpha\colon M[(s,t)\colon \rho(s,t) > \alpha] < \alpha]$, where the outer infimum is over the M in $\mathcal{M}(P,Q)$. Show that $\pi'(P,Q) = \pi(P,Q)$.

6.8. Show that $\pi(P,Q)$ is unchanged if we require (6.10) only for closed sets A.

6.9. For measures on the line, show that, if $Q(A) = P(A + x)$ and $|x| < \epsilon$, then $\pi(P,Q) < \epsilon$.

6.10. For distribution functions F on the line, consider the completed graphs $\Gamma_F = [(x,y)\colon F(x-) \le y \le F(x)]$. Each line $y = a - x$ meets each of Γ_F and Γ_G at a single point; let $L(F,G)$ be the supremum over a of the distance between these two points. Show that L, the *Lévy* metric, is indeed a metric. Show that this metric carried over to $\mathcal{P}$ in the natural way is equivalent to the topology of weak convergence and to the Prohorov metric for the case of the line.

CHAPTER 2

THE SPACE C

SECTION 7. WEAK CONVERGENCE AND TIGHTNESS IN C

The Introduction shows by example some of the reasons for studying weak convergence in the space $C = C[0,1]$ of continuous functions on the unit interval, where we give to C the uniform topology, defining the distance between two points x and y (two continuous functions $x(\cdot)$ and $y(\cdot)$ on $[0,1]$) as

$$\rho(x,y) = \|x - y\| = \sup_t |x(t) - y(t)|.$$

Although weak convergence in C need not follow from the weak convergence of the finite-dimensional distributions alone (Example 2.5), it does in the presence of relative compactness (Example 5.1). And since tightness implies relative compactness by Prohorov's Theorem 5.1, we have this basic result:

Theorem 7.1. *Let P_n, P be probability measures on $(C, \mathcal{C})$. If the finite-dimensional distributions of P_n converge weakly to those of P, and if $\{P_n\}$ is tight, then $P_n \Rightarrow P$.*

Tightness and Compactness in C

In order to use Theorem 7.1 to prove weak convergence in C, we need an exact understanding of tightness and hence of compactness in this space. The *modulus of continuity* of an arbitrary function $x(\cdot)$ on $[0,1]$ is defined by

$$w_x(\delta) = w(x,\delta) = \sup_{|s-t| \le \delta} |x(s) - x(t)|, \qquad 0 < \delta \le 1. \tag{7.1}$$

A necessary and sufficient condition for x to be uniformly continuous over $[0,1]$ is

$$\lim_{\delta \to 0} w_x(\delta) = 0. \tag{7.2}$$

And an x in C satisfies (7.2). Note that, since $|w_x(\delta) - w_y(\delta)| \leq 2\rho(x,y)$, $w(x,\delta)$ is, for fixed positive δ, continuous in x.

Recall that A is relatively compact if A^- is compact, which is equivalent to the condition that each sequence in A contains a convergent subsequence (the limit of which may not lie in A).† The *Arzelà–Ascoli theorem* completely characterizes relative compactness in C:

Theorem 7.2. *The set A is relatively compact if and only if*

$$\sup_{x \in A} |x(0)| < \infty \tag{7.3}$$

and

$$\lim_{\delta \to 0} \sup_{x \in A} w_x(\delta) = 0. \tag{7.4}$$

The functions in A are by definition *equicontinuous at t_0* if, as $t \to t_0$, $\sup_{x \in A} |x(t) - x(t_0)| \to 0$; and (7.4) defines *uniform* equicontinuity (over $[0,1]$) of the functions in A.

If A consists of the functions z_n defined by (1.5), then A is not relatively compact; since (7.3) holds, (7.4) must fail, and in fact $w(z_n, \delta) = 1$ for $n \geq \delta^{-1}$.

PROOF. If A^- is compact, (7.3) follows easily. Since $w(x, n^{-1})$ is continuous in x and nonincreasing in n, (7.2) holds uniformly on A if A^- is compact [M8], and (7.4) follows.

Suppose now that (7.3) and (7.4) hold. Choose k large enough that $\sup_{x \in A} w_x(k^{-1})$ is finite. Since

$$|x(t)| \leq |x(0)| + \sum_{i=1}^{k} |x(it/k) - x((i-1)t/k)|,$$

it follows that

$$\sup_t \sup_{x \in A} |x(t)| < \infty. \tag{7.5}$$

The idea now is to use (7.4) and (7.5) to prove that A is totally bounded; since C is complete, it will follow that A^- is compact.

---

† Although this is analogous to the relative compactness in Prohorov's theorem, it is technically different because in Section 5 the space of probability measures on $(S, \mathcal{S})$ is not assumed metrizable. But see Theorem 6.8.

Let α be the supremum in (7.5). Given ϵ, choose a finite ϵ-net H in the interval $[-\alpha, \alpha]$ on the line, and choose k large enough that $w_x(1/k) < \epsilon$ for all x in A. Take B to be the finite set consisting of the (polygonal) functions in C that are linear on each interval $I_{ki} = [(i-1)/k, i/k]$, $1 \leq i \leq k$, and take values in H at the endpoints. If $x \in A$, then $|x(i/k)| \leq \alpha$, and therefore there is a point y in B such that $|x(i/k) - y(i/k)| < \epsilon$ for $i = 0, 1, \ldots, k$. Now $y(i/k)$ is within 2ϵ of $x(t)$ for $t \in I_{ki}$, and similarly for $y((i-1))/k$. Since $y(t)$ is a convex combination of $y((i-1)/k)$ and $y(i/k)$, it too is within 2ϵ of $x(t)$: $\rho(x, y) < 2\epsilon$. Thus B is a finite 2ϵ-net for A. □

Let P_n be probability measures on $(C, \mathcal{C})$.

Theorem 7.3. *The sequence $\{P_n\}$ is tight if and only if these two conditions hold:*

(i) *For each positive η, there exist an a and an n_0 such that*

$$P_n[x: |x(0)| \geq a] \leq \eta, \quad n \geq n_0. \tag{7.6}$$

(ii) *For each positive ϵ and η, there exist a δ, $0 < \delta < 1$, and an n_0 such that*

$$P_n[x: w_x(\delta) \geq \epsilon] \leq \eta, \quad n \geq n_0. \tag{7.7}$$

In connection with (7.7), note that $w(\cdot, \delta)$ is measurable because it is continuous. Condition (ii) can be put in a more compact form: For each positive ϵ,

$$\lim_{\delta \to 0} \limsup_{n \to \infty} P_n[x: w_x(\delta) \geq \epsilon] = 0. \tag{7.8}$$

PROOF. Suppose $\{P_n\}$ is tight. Given η, choose a compact K such that $P_n K > 1 - \eta$ for all n. By the Arzelà–Ascoli theorem, we have $K \subset [x: |x(0)| \leq a]$ for large enough a and $K \subset [x: w_x(\delta) \leq \epsilon]$ for small enough δ, and so (i) and (ii) hold, with $n_0 = 1$ in each case. Hence the necessity.

Since C is separable and complete, a single measure P is tight (Theorem 1.3), and so by the necessity of (i) and (ii), for a given η there is an a such that $P[x: |x(0)| \geq a] \leq \eta$, and for given ϵ and η there is a δ such that $P[x: w_x(\delta) > \epsilon] \leq \eta$. If $\{P_n\}$ satisfies (i) and(ii), therefore, we may ensure that the inequalities in (7.6)and (7.7) hold for the finitely many n preceding n_0 by increasing a and decreasing δ if necessary: In proving sufficiency, we may assume that the n_0 is always 1.

Assume then that $\{P_n\}$ satisfies (i) and (ii), with $n_0 \equiv 1$. Given η, choose a so that, if $B = [x: |x(0)| \le a]$, then $P_n B \ge 1-\eta$ for all n; then choose δ_k so that, if $B_k = [x: w_x(\delta_k) < 1/k]$, then $P_n B_k \ge 1 - \eta/2^k$ for all n. If K is the closure of $A = B \cap \bigcap_k B_k$, then $P_n K \ge 1 - 2\eta$ for all n. Since A satisfies (7.3) and (7.4), K is compact. Therefore, $\{P_n\}$ is tight. □

Theorem 7.3 transforms the concept of tightness in C simply by substituting for relative compactness its Arzelà–Ascoli characterization. The next theorem and its corollary go only a small step beyond this but, even so, fill our present needs.

Theorem 7.4. *Suppose that* $0 = t_0 < t_1 < \cdots < t_v = 1$ *and*

$$\min_{1<i<v} (t_i - t_{i-1}) \ge \delta. \tag{7.9}$$

Then, for arbitrary x,

$$w_x(\delta) \le 3 \max_{1 \le i \le v} \sup_{t_{i-1} \le s \le t_i} |x(s) - x(t_{i-1})|, \tag{7.10}$$

and, for arbitrary P,

$$P[x: w_x(\delta) \ge 3\epsilon] \le \sum_{i=1}^{v} P\Big[x: \sup_{t_{i-1} \le s \le t_i} |x(s) - x(t_{i-1})| \ge \epsilon\Big]. \tag{7.11}$$

Note that (7.9) does not require $t_i - t_{i-1} \ge \delta$ for $i = 1$ or $i = v$.

PROOF. Let m be the maximum in (7.10). If s and t lie in the same interval $I_i = [t_{i-1}, t_i]$, then $|x(s) - x(t)| \le |x(s) - x(t_{i-1})| + |x(t) - x(t_{i-1})| \le 2m$. If s and t lie in adjacent intervals I_i and I_{i+1}, then $|x(s) - x(t)| \le |x(s) - x(t_{i-1})| + |x(t_{i-1}) - x(t_i)| + |x(t_i) - x(t)| \le 3m$. If $|s - t| \le \delta$, then s and t must lie in the same I_i or in adjacent ones, which proves (7.10). And now (7.11) follows by subadditivity. □

Corollary. *Condition* (ii) *of Theorem* 7.3 *holds if, for each positive* ϵ *and* η, *there exist a* δ, $0 < \delta < 1$, *and an integer* n_0 *such that*

$$\frac{1}{\delta} P_n[x: \sup_{t \le s \le t+\delta} |x(s) - x(t)| \ge \epsilon] \le \eta, \quad n \ge n_0, \tag{7.12}$$

for every t in $[0,1]$.

If $t > 1-\delta$, restrict s in the supremum in (7.12) to $t \le s \le 1$. Note that (7.12) is formally satisfied if $\delta > 1/\eta$; but we require $\delta < 1$.

PROOF. Take $t_i = i\delta$ for $i < v = \lfloor 1/\delta \rfloor$. If (7.12) holds, then, by (7.11), (7.7) holds as well (3ϵ in place of ϵ). □

Random Functions

Let $(\Omega, \mathcal{F}, \mathsf{P})$ be a probability space, and let X map Ω into C: $X(\omega)$ is an element of C with value $X_t(\omega) = X(t,\omega)$ at t. For fixed t, let $X_t = X(t)$ denote the real function on Ω with value $X_t(\omega)$ at ω; X_t is the composition $\pi_t X$, where π_t is the natural projection defined in Example 1.3. And let $(X_{t_1}, \ldots, X_{t_k})$ denote the map carrying ω to the point $(X_{t_1}(\omega), \ldots, X_{t_k}(\omega)) = \pi_{t_1 \ldots t_k}(X(\omega))$ in R^k.

If X is a random function—that is, measurable $\mathcal{F}/\mathcal{C}$—then the composition $\pi_{t_1 \ldots t_k} X$ is measurable $\mathcal{F}/\mathcal{R}^k$, so that $(X_{t_1}, \ldots, X_{t_k})$ is a random vector. But the argument can be reversed: The general finite-dimensional set has the form $A = \pi_{t_1 \ldots t_k}^{-1} H$, $H \in \mathcal{R}^k$, and if $\pi_{t_1 \ldots t_k} X$ is measurable $\mathcal{F}/\mathcal{R}^k$, then $X^{-1}A = (\pi_{t_1 \ldots t_k} X)^{-1} H \in \mathcal{F}$; but since the class $\mathcal{C}_f$ of finite-dimensional sets generates $\mathcal{C}$, it follows that X is measurable $\mathcal{F}/\mathcal{C}$. Therefore, X is a random function if and only if each $(X_{t_1}, \ldots, X_{t_k})$ is a random vector, and of course this holds if and only if each X_t is a random variable. Let $P = \mathsf{P}X^{-1}$ be the distribution of X; see (3.1). Then $\mathsf{P}[(X_{t_1}, \ldots, X_{t_k}) \in H] = P\pi_{t_1 \ldots t_k}^{-1} H$, and so the finite-dimensional distributions of P (Example 1.3 again) are the distributions of the various random vectors $(X_{t_1}, \ldots, X_{t_k})$.

Suppose that $X, X^1, X^2, \ldots$ are random functions.

Theorem 7.5. *If*

$$(X_{t_1}^n, \ldots, X_{t_k}^n) \Rightarrow_n (X_{t_1}, \ldots, X_{t_k}) \tag{7.13}$$

holds for all $t_1, \ldots, t_k$, and if

$$\lim_{\delta \to 0} \limsup_{n \to \infty} \mathsf{P}[w(X^n, \delta) \ge \epsilon] = 0 \tag{7.14}$$

for each positive ϵ, then $X^n \Rightarrow_n X$.

FIRST PROOF. Let P and P_n be the distributions of X and the X_n. Since (7.13) is the same thing as $P_n \pi_{t_1 \ldots t_k}^{-1} \Rightarrow_n P\pi_{t_1 \ldots t_k}^{-1}$ and $X^n \Rightarrow_n X$ is the same thing as $P_n \Rightarrow_n P$, the result will follow by Theorem 7.1 if we show that $\{X^n\}$ is tight in the sense that $\{P_n\}$ is tight. Now $X_0^n \Rightarrow_n X_0$ implies that $\{P_n \pi_0^{-1}\}$ is tight, which in turn implies condition (i) of Theorem 7.3. Since (7.14) translates into (7.8), condition (ii) of the theorem also holds, and $\{P_n\}$ is indeed tight. □

There is a second proof, very different, based on Theorem 3.2.

SECOND PROOF. For $u = 1, 2, \ldots$, define a map M_u of C into itself this way: Let $M_u x$ be the polygonal function in C that agrees with x at the points i/u, $0 \le i \le u$, and is defined by linear interpolation between these points:

$$(M_u x)(t) = (i - ut)x\left(\frac{i-1}{u}\right) + (ut - (i-1))x\left(\frac{i}{u}\right), \quad \frac{i-1}{u} \le t \le \frac{i}{u}.$$

Since $M_u x$ agrees with x at the endpoints of each $[(i-1)/n, i/n]$, it is clear that $\rho(M_u x, x) \le 2w_x(1/u)$. And now define a map $L_u: R^{u+1} \to C$ this way: For $\alpha = (\alpha_0, \ldots, \alpha_u)$, take $(L_u \alpha)(t) = (i - ut)\alpha_{i-1} + (ut - (i-1))\alpha_i$ for t in $[(i-1)/u, i/u]$; this is another polygonal function, and its values are α_i at the corner points. Clearly, $\rho(L_u\alpha, L_u\beta) = \max_i |\alpha_i - \beta_i|$, so that L_u is continuous; and $M_u = L_u \pi_{t_0 \ldots t_u}$ if $t_i = i/u$.

By (7.13), $\pi_{t_0 \ldots t_u} X^n \Rightarrow_n \pi_{t_0 \ldots t_u} X$ ($t_i = i/u$ again), and since L_u is continuous, the mapping theorem gives $M_u X^n \Rightarrow_n M_u X$. Since $\rho(M_u X, X) \le 2w(X, 1/u)$ and X is an element of C, $\rho(M_u X, X)$ goes to 0 everywhere and hence goes to 0 in probability, so that $M_u X \Rightarrow_u X$ by the corollary to Theorem 3.1. Finally, from (7.14) and the inequality $\rho(M_u X^n, X^n) \le 2w(X^n, 1/u)$, it follows that

$$\lim_u \limsup_n \mathsf{P}[\rho(M_u X^n, X^n) \ge \epsilon] = 0.$$

Combine this with $M_u X^n \Rightarrow_n M_u X \Rightarrow_u X$ and apply Theorem 3.2. □

This second proof does not require the concepts of relative compactness and tightness, and it makes no use of Prohorov's theorem. See the discussion following the proof of Donsker's theorem in the next section.

Coordinate Variables

The projection π_t, with value $\pi_t(x) = x(t)$ at x, is a random variable on $(C, \mathcal{C})$. We denote it by x_t: For fixed t, x_t has value $x(t)$ at x. If P is a probability measure on $(C, \mathcal{C})$ and t is thought of as a time parameter, then $[x_t: 0 \le t \le 1]$ becomes a stochastic process, and the

x_t are commonly called the *coordinate* varables or functions. We speak of the distribution of x_t *under* P and often write $P[x_t \in H]$ in place of $P[x: x_t \in H]$ and $\int x_t \, dP$ in place of $\int_C x_t \, P(dx)$. Finally, when t is a complicated expression, we sometimes revert back to $x(t)$, still intended as a coordinate variable.

Problems

7.1. Let $X(t) = t\xi$, where ξ is a random variable satisfying $\mathsf{P}[|\xi| \geq \alpha] \sim \alpha^{-1/2}$; for every n, let P_n be the distribution of X. Then $\{P_n\}$ is tight but does not satisfy (7.12).

7.2. Show that, if the functions in A are equicontinuous at each point of $[0, 1]$, then A is uniformly equicontinuous.

SECTION 8. WIENER MEASURE AND DONSKER'S THEOREM

Wiener Measure

Wiener measure, denoted here by W, is a probability measure on $(C, \mathcal{C})$ having the following two properties. First, each x_t is normally distributed under W with mean 0 and variance t:

$$W[x_t \leq \alpha] = \frac{1}{\sqrt{2\pi t}} \int_{-\infty}^{\alpha} e^{-u^2/2t} \, du. \tag{8.1}$$

For $t = 0$ this is interpreted to mean that $W[x_0 = 0] = 1$. Second, the stochastic process $[x_t: 0 \leq t \leq 1]$ has independent increments under W: If

$$0 \leq t_0 \leq t_1 \leq \cdots \leq t_k = 1, \tag{8.2}$$

then the random variables

$$x_{t_1} - x_{t_0}, \, x_{t_2} - x_{t_1}, \, \ldots, \, x_{t_k} - x_{t_{k-1}} \tag{8.3}$$

are independent under W. We must prove the existence of such a measure.

If W has these two properties, and if $s \leq t$, then x_t (normal with mean 0 and variance t) is the sum of the independent random variables x_s (normal with mean 0 and variance s) and $x_t - x_s$, so that $x_t - x_s$

must be normal with mean 0 and variance $t-s$, as follows from dividing the characteristic functions. Therefore, when (8.2) holds,

$$(8.4)\quad W[x_{t_i} - x_{t_{i-1}} \le \alpha_i,\ i = 1, \ldots, k] = \prod_{i=1}^{k} \frac{1}{\sqrt{2\pi(t_i - t_{i-1})}} \int_{-\infty}^{\alpha_i} e^{-u^2/2(t_i - t_{i-1})}\, du.$$

In particular, the increments are stationary (the distribution of $x_t - x_s$ under W depends only on the difference $t - s$) as well as independent.

Since $x_s x_t = x_s^2 + x_s(x_t - x_s)$, it follows from the independence of the increments that the covariance of x_s and x_t under W is s if $s \le t$. By a linear change of variables, the joint distribution of $(x_{t_1}, \ldots, x_{t_k})$ can therefore be writen down explicitly; it is the centered normal distribution for which x_{t_i} and x_{t_j} have covariance $t_i \wedge t_j$. But (8.4) is the clearest way to specify the finite-dimensional distributions.

Wiener measure gives a model for Brownian motion. In proving its existence, we face the problem of proving the existence on $(C, \mathcal{C})$ of a probability measure having specified finite-dimensional distributions. There can for an arbitrary specification be at most one such measure (Example 1.3), and for some specifications there can be none at all (there is, for example, no P under which the distribution of x_t is a unit mass at 0 for $t < \frac{1}{2}$ and at 1 for $t \ge \frac{1}{2}$).

Construction of Wiener Measure

We start with a sequence $\xi_1, \xi_2, \ldots$ of independent and identically distributed random variables (on some probability space) having mean 0 and finite, positive variance σ^2. Let $S_n = \xi_1 + \cdots + \xi_n$ ($S_0 = 0$), and let $X^n(\omega)$ be the element of C having the value

$$(8.5)\qquad X_t^n(\omega) = \frac{1}{\sigma\sqrt{n}} S_{\lfloor nt \rfloor}(\omega) + (nt - \lfloor nt \rfloor) \frac{1}{\sigma\sqrt{n}} \xi_{\lfloor nt \rfloor + 1}(\omega)$$

at t. Thus $X^n(\omega)$ is the function defined by linear interpolation between its values $X_{i/n}^n(\omega) = S_i(\omega)/\sigma\sqrt{n}$ at the points i/n. Since (8.5) defines a random variable for each t, X^n is a random function, the one discussed in the Introduction. If ξ_i takes the values ± 1 with probability $\frac{1}{2}$ each, then $\sigma = 1$ and X^n is the path corresponding to a symmetric random walk.

If $\psi_{n,t}$ is the rightmost term in (8.5), then $\psi_{n,t} \Rightarrow_n 0$ by Chebyshev's inequality. Therefore, by the Lindeberg–Lévy central limit theorem (together with Theorem 3.1, Example 3.2, and the fact that

$\lfloor nt \rfloor / n \to t$), $X_t^n \Rightarrow_n \sqrt{t}N$, where, as always, N has the standard normal distribution. Similarly, if $s \le t$, then

$$(8.6) \qquad (X_s^n, X_t^n - X_s^n) = \frac{1}{\sigma\sqrt{n}}(S_{\lfloor ns \rfloor}, S_{\lfloor nt \rfloor} - S_{\lfloor ns \rfloor}) + (\psi_{n,s}, \psi_{n,t} - \psi_{n,s}) \Rightarrow_n (N_1, N_2),$$

where N_1 and N_2 are independent and normal with variances s and $t-s$. And by the mapping theorem, $(X_s^n, X_t^n) \Rightarrow_n (N_1, N_1 + N_2)$. The obvious extension shows that the limiting distributions of the random vectors $(X_{t_1}^n, \ldots X_{t_k}^n)$ are exactly those specified as the finite-dimensional distributions of the measure W we are to construct. To put it another way, if P_n is the distribution on C of X^n, then for each $t_1, \ldots, t_k$, $P_n \pi_{t_1 \ldots t_k}^{-1}$ converges weakly to what we want $W \pi_{t_1 \ldots t_k}^{-1}$ to be.

Suppose we can show that $\{P_n\}$ is tight. It will then follow by the direct half of Prohorov's theorem that some subsequence $\{P_{n_i}\}$ converges weakly to a limit we can call W. But then $P_{n_i} \pi_{t_1 \ldots t_k}^{-1} \Rightarrow_i W \pi_{t_1 \ldots t_k}^{-1}$, and therefore, by what we *have* in fact just proved, $W \pi_{t_1 \ldots t_k}^{-1}$ will be the probability measure on R^k we want. See Example 5.2.

The argument for the tightness of $\{X^n\}$ will be clearer if we first consider the more general case in which $\{\xi_n\}$ is stationary (the distribution of $(\xi_k, \ldots, \xi_{k+j})$ is the same for all k).

Lemma. *Suppose that X^n is defined by* (8.5), *that $\{\xi_n\}$ is stationary, and that*

$$(8.7) \qquad \lim_{\lambda \to \infty} \limsup_{n \to \infty} \lambda^2 \mathsf{P}\Big[\max_{k \le n} |S_k| \ge \lambda\sigma\sqrt{n}\Big] = 0.$$

Then $\{X^n\}$ is tight.†

PROOF. Use Theorem 7.3. Since $X_0^n = 0$, $\{P_n\}$ obviously satisfies condition (i) of the theorem. Condition (ii) we prove by (7.8), which translates into the requirement that

$$(8.8) \qquad \lim_{\delta \to 0} \limsup_{n \to \infty} \mathsf{P}[w(X^n, \delta) \ge \epsilon] = 0$$

for each ϵ. And to this we can apply Theorem 7.4. If (7.9) holds, then so do (7.10) and (7.11):

$$(8.9) \qquad \mathsf{P}[w(X^n, \delta) \ge 3\epsilon] \le \sum_{i=1}^{v} \mathsf{P}\Big[\sup_{t_{i-1} \le s \le t_i} |X_s^n - X_{t_{i-1}}^n| \ge \epsilon\Big], \quad \text{if } \min_{1<i<v} (t_i - t_{i-1}) \ge \delta.$$

† The condition (8.7) is in fact necessary for $X^n \Rightarrow W$. See Problem 8.3.

This becomes easier to analyze if we take $t_i = m_i/n$ for integers m_i satisfying $0 = m_0 < m_1 < \cdots < m_v = n$. The point is that, because of the polygonal character of the random function X^n, if the t_i have this special form, then the supremum in (8.9) becomes a maximum of differences $|S_k - S_{m_{i-1}}|/\sigma\sqrt{n}$:

$$\text{(8.10)} \qquad \mathsf{P}[w(X^n,\delta) \geq 3\epsilon] \leq \sum_{i=1}^{v} \mathsf{P}\left[\max_{m_{i-1} \leq k \leq m_i} \frac{|S_k - S_{m_{i-1}}|}{\sigma\sqrt{n}} \geq \epsilon\right]$$

$$= \sum_{i=1}^{v} \mathsf{P}\left[\max_{k \leq m_i - m_{i-1}} |S_k| \geq \epsilon\sigma\sqrt{n}\right],$$

where the equality is a consequence of the assumed stationarity. The inequality holds if the condition on the right in (8.9) does, which requires that

$$\text{(8.11)} \qquad \frac{m_i}{n} - \frac{m_{i-1}}{n} \geq \delta, \quad 1 < i < v.$$

For a further simplification, take $m_i = im$ for $0 \leq i < v$ (and $m_v = n$), where m is an integer (a function of n and δ) chosen according to these criteria: Since we need $m_i - m_{i-1} = m \geq n\delta$ for $i < v$, take $m = \lceil n\delta \rceil$; since we also need $(v-1)m < n \leq vm$, take $v = \lceil n/m \rceil$. Then

$$m_v - m_{v-1} \leq m, \qquad v = \left\lceil \frac{n}{m} \right\rceil \to_n \frac{1}{\delta} < \frac{2}{\delta}, \qquad \frac{n}{m} \to_n \frac{1}{\delta} > \frac{1}{2\delta},$$

and it follows by (8.10) that, for large n,

$$\text{(8.12)} \qquad \mathsf{P}[w(X^n,\delta) \geq 3\epsilon] \leq v \cdot \mathsf{P}\left[\max_{k \leq m} |S_k| \geq \epsilon\sigma\sqrt{n}\right]$$

$$\leq \frac{2}{\delta} \cdot \mathsf{P}\left[\max_{k \leq m} |S_k| \geq \frac{\epsilon}{\sqrt{2\delta}}\sigma\sqrt{m}\right].$$

Take λ and δ to be functions of one another: $\lambda = \epsilon/\sqrt{2\delta}$. Expressed in terms of λ, (8.12) is

$$\mathsf{P}[w(X^n,\delta) \geq 3\epsilon] \leq \frac{4\lambda^2}{\epsilon^2} \cdot \mathsf{P}\left[\max_{k \leq m} |S_k| \geq \lambda\sigma\sqrt{m}\right].$$

For given positive ϵ and η, there is, by (8.7), a λ such that

$$\frac{4\lambda^2}{\epsilon^2} \limsup_m \mathsf{P}\left[\max_{k \leq m} |S_k| \geq \lambda\sigma\sqrt{m}\right] < \eta.$$

Once λ and δ are fixed, m goes to infinity along with n, and so (8.8) follows. □

We can complete the construction of Wiener measure by using the independence of the random variables ξ_n in the definition (8.5). Because of this independence, Etemadi's inequality [M19] implies

$$\mathsf{P}[\max_{u \le m} |S_u| \ge \alpha] \le 3 \max_{u \le m} \mathsf{P}[|S_u| \ge \alpha/3]. \tag{8.13}$$

Therefore, (8.7) will follow if

$$\lim_{\lambda \to \infty} \limsup_{n \to \infty} \lambda^2 \max_{k \le n} \mathsf{P}[|S_k| \ge \lambda\sigma\sqrt{n}] = 0. \tag{8.14}$$

For the construction of Wiener measure, we can use any sequence $\{\xi_i\}$ that is convenient. Suppose the ξ_i are independent and each has the standard normal distribution ($\sigma = 1$), in which case $S_k/\sqrt{k}$ also has the standard normal distribution. Since

$$\mathsf{P}[|N| \ge \lambda] < \mathsf{E}N^4 \cdot \lambda^{-4} = 3\lambda^{-4}, \tag{8.15}$$

we have $\mathsf{P}[|S_k| \ge \lambda\sigma\sqrt{n}] = \mathsf{P}[\sqrt{k}|N| \ge \lambda\sigma\sqrt{n}] < 3/\lambda^4\sigma^4$ for $k \le n$, which implies (8.14). This proves the existence of Wiener measure:

Theorem 8.1. *There exists on* $(C, \mathcal{C})$ *a probability measure* W *having the finite-dimensional distributions specified by* (8.4).

Denote by W not only Wiener measure, but also a random function having Wiener measure as its distribution over C. It is possible to reverse the approach taken above, constructing the random function first and the corresponding measure second. There are a number of ways to do this. For one, use Komogorov's existence theorem to show that there exists a stochastic process $[W(t): 0 \le t \le 1]$ having the finite-dimensional distributions specified by the right-hand side of (8.4). By a further argument involving Etemadi's inequality (or something similar), modify the process so as to ensure that the sample path $W(\cdot, \omega)$ is continuous for each ω [PM.503]. None of this involves the space C at all. But once we have this $W(\omega)$, we can regard it as a random element of $(C, \mathcal{C})$, and then we can define Wiener measure as its distribution.

Donsker's Theorem

Donsker's theorem is the one discussed in the Introduction:

Theorem 8.2. *If* $\xi_1, \xi_2, \ldots$ *are independent and identically distributed with mean* 0 *and variance* σ^2, *and if* X^n *is the random function defined by* (8.5), *then* $X^n \Rightarrow_n W$.

PROOF. The proof depends on Theorem 7.5. Now that the existence of Wiener measure and the corresponding random function W have been established, we can write (8.6) as $(X_s^n, X_t^n - X_s^n) \Rightarrow_n (W_s, W_t - W_s)$, and this implies $(X_s^n, X_t^n) \Rightarrow_n (W_s, W_t)$. An easy extension gives (7.13) with W in the role of X.

We have already proved tightness by means of (8.14), but under the additional assumption that the ξ_i are normal. In place of this, we can use the central limit theorem, provided we consider separately the small and the large values of k in the maximum in (8.14). By the central limit theorem, if k_λ is large enough and $k_\lambda \le k \le n$, then (use (8.15)) $\mathsf{P}[|S_k| \ge \lambda\sigma\sqrt{n}] \le \mathsf{P}[|S_k| \ge \lambda\sigma\sqrt{k}] < 3/\lambda^4$. For $k \le k_\lambda$ we can use Chebyshev's inequality: $\mathsf{P}[|S_k| \ge \lambda\sigma\sqrt{n}] \le k_\lambda/\lambda^2 n$. The maximum in (8.14) is therefore dominated by $(3/\lambda^4) \vee (k_\lambda/\lambda^2 n)$. □

This argument is based on Theorem 7.5, the second proof of which makes no use of Section 5. But Theorem 8.2 assumes the prior existence of W, and our proof that W does exist depended heavily on the theory of Section 5 (compare Examples 5.1 and 5.2). On the other hand, as pointed out above, the existence of W can be proved in an entirely different way, and so one *can* do Donsker's theorem without reference to tightness and Prohorov's theorem. On the *other* other hand, without the concept of tightness, the condition (7.14) is somewhat artificial, an ad hoc device.

An Application

Donsker's theorem has this qualitative interpretation: $X^n \Rightarrow W$ says that, if τ is small, then a particle subject to independent displacements $\xi_1, \xi_2, \ldots$ at successive times $\tau, 2\tau \ldots$ will, viewed from far off, appear to perform a Brownian motion.

Specifically, we can use the theorem to derive limit laws for various functions of the partial sums S_n. The Introduction indicates how to use the relation $X^n \Rightarrow W$ to derive the limiting distribution of

$$M_n = \max_{0 \le i \le n} S_i. \tag{8.16}$$

We are now in a position to carry through the details.

Since $h(x) = \sup_t x(t)$ is a continuous function on C, it follows from $X^n \Rightarrow W$ and the mapping theorem that $\sup_t X_t^n \Rightarrow \sup_t W_t$. Obviously, $\sup_t X_t^n = M_n/\sigma\sqrt{n}$, and so

$$\frac{M_n}{\sigma\sqrt{n}} \Rightarrow \sup_t W_t. \tag{8.17}$$

Thus we would have the limiting distribution of $M_n/\sigma\sqrt{n}$ (under the hypothesis of Theorem 8.2) if we knew the distribution of $\sup_t W_t$. The idea now is to find the latter distribution by calculating the limiting distribution of $M_n/\sigma\sqrt{n}$ in an easy special case.

For the easy special case, assume that the independent ξ_i take the values ± 1 with probability $\frac{1}{2}$ each, so that $S_0, S_1, \ldots$ are the successive positions in a symmetric random walk starting from the origin. We first show that for each nonnegative integer a,

$$\mathsf{P}[M_n \geq a] = 2\mathsf{P}[S_n > a] + \mathsf{P}[S_n = a]. \tag{8.18}$$

The case $a = 0$ being easy ($M_n \geq S_0 = 0$), assume $a > 0$. Since

$$\mathsf{P}[M_n \geq a] - \mathsf{P}[S_n = a] = \mathsf{P}[M_n \geq a,\ S_n < a] + \mathsf{P}[M_n \geq a,\ S_n > a],$$

and since the second term on the right is just $\mathsf{P}[S_n > a]$, (8.18) will follow if

$$\mathsf{P}[M_n \geq a,\ S_n < a] = \mathsf{P}[M_n \geq a,\ S_n > a]. \tag{8.19}$$

Now all 2^n possible paths $(S_1, \ldots, S_n)$ have the same probability, and so (8.19) will follow provided the number of paths contributing to the left-hand event is the same as the number of paths contributing to the right-hand event. Given a path contributing to the left-hand event in (8.19), match it with the path obtained by reflecting through a all the partial sums after the first one that achieves the height a (replace S_k by $a-(S_k-a)$).

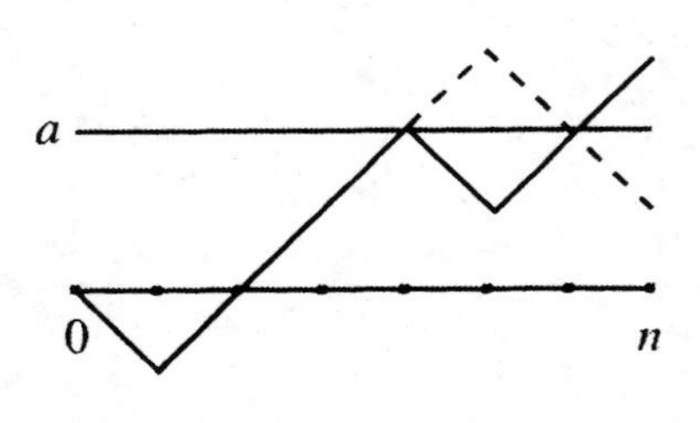

Since this describes a one-to-one correspondence, (8.19) and (8.18) follow. This is an example of the *reflection principle.*

Assume $\alpha \geq 0$ and put $a_n = \lceil \alpha n^{1/2} \rceil$. From (8.18) it follows that $\mathsf{P}[M_n/\sqrt{n} \geq \alpha] = 2\mathsf{P}[S_n > a_n] + \mathsf{P}[S_n = a_n]$. The second term here goes to 0, and $\mathsf{P}[S_n > a_n] \to \mathsf{P}[N > \alpha]$ by the central limit theorem, and so $\mathsf{P}[M_n/\sqrt{n}] \to 2\mathsf{P}[N > \alpha]$ for $\alpha \geq 0$. Combine this with (8.17):

$$\mathsf{P}[\sup_t W_t \leq \alpha] = \frac{2}{\sqrt{2\pi}} \int_0^\alpha e^{u^2/2}\, du, \quad \alpha \geq 0 \tag{8.20}$$

(the left side is 0 if $\alpha < 0$). We have derived a fact about Brownian motion by combining Donsker's theorem with a computation involving random walk, a computation which is simple because it reduces to enumeration and because a random walk cannot pass above an integer without passing through it.

Under the hypotheses of Donsker's theorem (drop the assumption $\xi_i = \pm 1$), (8.17) holds, and from (8.20) it now follows that

$$\mathsf{P}\left[\frac{M_n}{\sigma\sqrt{n}} \le \alpha\right] \to \frac{2}{\sqrt{2\pi}} \int_0^\alpha e^{u^2/2}\, du, \quad \alpha \ge 0. \tag{8.21}$$

Under the hypotheses of the Lindeberg–Lévy theorem, the limiting distribution of M_n (normalized) is the folded normal law.

The next section has further examples that conform to this pattern. If h is continuous on C—or continuous outside a set of Wiener measure 0—then $X^n \Rightarrow W$ implies $h(X^n) \Rightarrow h(W)$. We can find the limiting distribution of $h(X^n)$ if we can find the distribution of $h(W)$, and we can in many cases find the distribution of $h(W)$ by finding the limiting distribution of $h(X^n)$ in some simple special case and then using $h(X^n) \Rightarrow h(W)$ in the other direction.

Therefore, if the ξ_i are independent and identically distributed with mean 0 and variance σ^2, then the limiting distribution of $h(X^n)$ does not depend on any further properties of the ξ_i. For this reason, Donsker's theorem is often called the (or an) *invariance principle.* It is perhaps better called a *functional limit theorem*, or since the limit is Brownian motion in this case, a functional *central* limit theorem.

The Brownian Bridge

A random element X of $(C, \mathcal{C})$ is *Gaussian* if all its finite-dimensional distributions are normal. The distribution over C of a Gaussian X is completely specified by the means $\mathsf{E}[X_t]$ and the product moments $\mathsf{E}[X_s X_t]$, because these determine the finite-dimensional distributions. For W, the moments are $\mathsf{E}[W_t] = 0$ and $\mathsf{E}[W_s W_t] = s \wedge t$.

A second important random element of C is the *Brownian bridge*, a Gaussian random function W° specified by the requirements $\mathsf{E}[W_t^\circ] = 0$ and $\mathsf{E}[W_s^\circ W_t^\circ] = s(1-t)$ for $s \le t$. The simplest way to show that there is such a random function W° is to construct it from W by setting $W_t^\circ = W_t - tW_1$ for $0 \le t \le 1$. Obviously, W°, thus defined, is a Gaussian random element of C, and a calculation shows that it has the required moments.

We also use W° to denote the distribution on C of the random function W°. If $h: C \to C$ carries x to the function with the value $x(t) - tx(1)$ at t, then the measures W and W° are related by $W^\circ = Wh^{-1}$.

Problems

8.1. Show that W has no locally compact support. See Problem 1.17.

8.2. Define $V_t = (1+t)W^\circ_{t/(1+t)}$. The process $[V_t: t \ge 0]$ has continuous sample paths, the finite-dimensional distributions are Gaussian, and the moments are $\mathsf{E}V_t = 0$ and $\mathsf{E}[V_s V_t] = s \wedge t$. The process thus represents a Brownian motion over the time interval $[0, \infty)$.

8.3. Assume that X^n is defined by (8.5) and that $X^n \Rightarrow W$. Use (8.20) and symmetry to show that $\sup_t |W_t|$ has a fourth moment. Conclude that (8.7) holds.

8.4. Let $\xi_{n1}, \ldots, \xi_{nk_n}$ be independent random variables with mean 0 and variances σ^2_{ni}; put $S_{ni} = \xi_{n1} + \cdots + \xi_{ni}$, $s^2_{ni} = \sigma^2_{n1} + \cdots + \sigma^2_{ni}$, and $s^2_n = s^2_{nk_n}$. Let X^n be the random function that is linear on each interval $[s^2_{n,i-1}/s^2_n, s^2_{ni}/s^2_n]$ and has values $X^n(s^2_{ni}/s^2_n) = S_{ni}/s_n$ at the points of division. Assume the Lindeberg condition holds and generalize Donsker's theorem by showing that $X^n \Rightarrow W$.

SECTION 9. FUNCTIONS OF BROWNIAN MOTION PATHS

The technique used in the preceding section to find the distribution of $\sup_t W_t$ and the limiting distribution of $M_n/\sigma\sqrt{n}$ we here apply to other functions of Brownian motion paths and partial sums. We also compute some distributions associated with the Brownian bridge. In subsequent sections, convergence in distribution to W will be proved for a great variety of random functions, and in each case the calculations carried out here apply.

Although the theory of weak convergence of probability measures on function spaces had its origin in problems of the kind considered here, and although these problems and their solutions are indeed interesting, it is possible to understand the general theory of the subsequent sections without studying this one.

In this section, the Lindeberg-Lévy case will mean the one in which the S_n are partial sums of independent, identically distributed random variables ξ_i with mean 0 and finite, positive variance σ^2, and X^n will always denote the random function defined by (8.5). The random walk case will be that in which each ξ_i takes the values ± 1 with probabilities $\frac{1}{2}$ each ($\sigma^2 = 1$).

Maximum and Minimum

Let $m = \inf_t W_t$ and $M = \sup_t W_t$, and let

$$m_n = \min_{0 \le i \le n} S_i, \quad M_n = \max_{0 \le i \le n} S_i \tag{9.1}$$

be the corresponding quantities for the partial sums. The mapping carying the point x of C to the point $(\inf_t x(t), \sup_t x(t), x(1))$ of R^3 is everywhere continuous, and so by Donsker's theorem and the mapping theorem,

$$\frac{1}{\sigma\sqrt{n}}(m_n, M_n, S_n) \Rightarrow (m, M, W_1) \tag{9.2}$$

in the Lindeberg-Lévy case.

We first find an explicit formula for

$$p_n(a, b, v) = \mathsf{P}[a < m_n \le M_n < b,\ S_n = v] \tag{9.3}$$

in the random walk case. We show that if

$$p_n(j) = \mathsf{P}[S_n = j], \tag{9.4}$$

then

$$p_n(a, b, v) = \sum_{k=-\infty}^{\infty} p_n(v+2k(b-a)) - \sum_{k=-\infty}^{\infty} p_n(2b-v+2k(b-a)) \tag{9.5}$$

for integers a, b, and v satisfying

$$a \le 0 \le b, \quad a < b, \quad a \le v \le b. \tag{9.6}$$

Since $a < b$, the series in (9.5) are really finite sums. Notice that both sides of (9.5) vanish if n and v have opposite parity or if v is either a or b.

For particular values of n, a, b, v, denote the equation (9.5) by $[n, a, b, v]$. Then $[n, a, b, v]$ is valid if (9.6) holds, and the proof of this goes by induction on n. For $n = 1$, this follows by a straightforward examination of cases. Assume as induction hypothesis that $[n-1, a, b, v]$ holds for a, b, v satisfying (9.6). If $a = 0$, then (9.3) vanishes (i starts at 0 in the minimum in (9.1), and $S_0 = 0$), and the sums on the right in (9.5) cancel because $p_n(j) = p_n(-j)$. Thus $[n, a, b, v]$ is valid if (9.6) holds and $a = 0$; we may dispose of the case $b = 0$ in the same way. To complete the induction step, we must verify $[n, a, b, v]$ under the assumption that $a < 0 < b$ and $a \le v \le b$. But in this case, $a+1 \le 0$ and $b-1 \ge 0$, so that $[n-1, a-1, b-1, v-1]$ and $[n-1, a+1, b+1, v+1]$ both come under the induction hypothesis and hence both hold. And now $[n, a, b, v]$ follows by the probabilistically obvious recursions (condition on the direction of the first step of the random walk)

$$p_n(j) = \frac{1}{2}p_{n-1}(j-1) + \frac{1}{2}p_{n-1}(j+1)$$

and

$$p_n(a,b,v) = \frac{1}{2}p_{n-1}(a-1,b-1,v-1) + \frac{1}{2}p_{n-1}(a+1,b+1,v+1).$$

This proves (9.5), and it follows by summation over v that, if

$$a \le 0 \le b, \quad a \le u < v \le b, \tag{9.7}$$

then

$$\begin{aligned} \mathsf{P}[a < m_n \le M_n < b,\ u < S_n < v] &= \sum_{k=-\infty}^{\infty} \mathsf{P}[u + 2k(b-a) < S_n < v + 2k(b-a)] \\ &\quad - \sum_{k=-\infty}^{\infty} \mathsf{P}[2b - v + 2k(b-a) < S_n < 2b - u + 2k(b-a)]. \end{aligned} \tag{9.8}$$

Taking $a = -n-1$ in this formula leads to

$$\begin{aligned} \mathsf{P}[M_n < b,\ u < S_n < v] = \mathsf{P}[u < S_n < v] \\ - \mathsf{P}[2b - v < S_n < 2b - u], \end{aligned} \tag{9.9}$$

valid for $-n-1 \le u < v \le b$, $b \ge 0$. From (9.9) it is possible to derive (8.18) again.

Now (9.8) holds in the random walk case, and because of (9.2), we can find the distribution of (m, M, W_1) by passing to the limit. If a, b, u, and v are real numbers satisfying (9.7), replace them in (9.8) by the integers $\lfloor a\sqrt{n} \rfloor$, $\lceil b\sqrt{n} \rceil$, $\lfloor u\sqrt{n} \rfloor$, and $\lceil v\sqrt{n} \rceil$, respectively. Because of the central limit theorem and the continuity of the normal distribution, a termwise passage to the limit in (9.8) yields

$$\begin{aligned} \mathsf{P}[a < m \le M < b,\ u < W_1 < v] &= \sum_{k=-\infty}^{\infty} \mathsf{P}[u + 2k(b-a) < N < v + 2k(b-a)] \\ &\quad - \sum_{k=-\infty}^{\infty} \mathsf{P}[2b - v + 2k(b-a) < N < 2b - u + 2k(b-a)]. \end{aligned} \tag{9.10}$$

The interchange of the limit with the summation over k can be justified by the series form of Scheffé's theorem [PM.215].

The joint distribution of M and W_1 alone could be obtained by letting a tend to $-\infty$ in (9.10), but it is simpler to return to the random walk case and pass to the limit in (9.9), which gives

$$\text{(9.11)} \qquad \mathsf{P}[M < b,\ u < W_1 < v] = \mathsf{P}[u < N < v] - \mathsf{P}[2b - v < N < 2b - u],$$

valid for $u < v \leq b$, $b \geq 0$. Taking $v = b$ and letting $u \to -\infty$ leads to

$$\text{(9.12)} \qquad \mathsf{P}[0 \leq M < b] = 2\mathsf{P}[0 \leq N < b];$$

this is the same thing as (8.20).

From (9.10) for $u = a$ and $v = b$ we get

$$\text{(9.13)} \qquad \mathsf{P}[a < m \leq M < b] = \sum_{k=-\infty}^{\infty} (-1)^k \mathsf{P}[a + k(b-a) < N < b + k(b-a)],$$

valid for $a \leq 0 \leq b$. And this result with $a = -b$ gives

$$\text{(9.14)} \qquad \mathsf{P}[\sup_t |W_t| < b] = \sum_{k=-\infty}^{\infty} (-1)^k \mathsf{P}[(2k-1)b < N < (2k+1)b]$$

for $b \geq 0$. By continuity, the strict inequalities in all these formulas can be relaxed to allow equality. And the right sides can all be written out as sums of normal integrals.

Although we derived (9.10) through (9.14) by passing to the limit in the random walk case, we have the limiting distributions for (m_n, M_n, S_n), (M_n, S_n), M_n, (m_n, M_n), and $\max_{i \leq n} |S_n|$ (all normalized by $\sigma\sqrt{n}$) in the more general Lindeberg-Lévy case because (9.2) holds there.

The Arc Sine Law

For x in C, let $h_1(x)$ be the supremum of those t in $[0, 1]$ for which $x(t) = 0$; let $h_2(x)$ be the Lebesgue measure of those t in $[0, 1]$ for which $x(t) > 0$; and let $h_3(x)$ be the Lebesgue measure of those t in $[0, h_1(x)]$ for which $x(t) > 0$. Then $T = h_1(W)$ is the time at which W last passes through 0, $U = h_2(W)$ is the total amount of time W

spends above 0, and $V = h_3(W)$ is the amount of time W spends above 0 in the interval $[0, T]$. The object is to find the joint distribution of (T, U, V, W_1).

It can be shown [M15] that each of the functions h_1, h_2, and h_3 is measurable and is continuous outside a set of Wiener measure 0. Therefore,

$$(h_1(X^n), h_2(X^n), h_3(x^n), X_1^n) \Rightarrow_n (T, U, V, W_1) \tag{9.15}$$

in the Lindeberg-Lévy case. In the random walk case, the vector on the left has a simple interpretation: $T_n = nh_1(X^n)$ is the maximum i, $1 \leq i \leq n$, for which $S_i = 0$; $U_n = nh_2(X^n)$ is the number of i, $1 \leq i \leq n$, for which S_{i-1} and S_i are both nonnegative; $V_n = nh_3(X^n)$ is the number of i, $1 \leq i \leq T_n$, for which S_{i-1} and S_i are both nonnegative; and, of course, $X_1^n = S_n/\sqrt{n}$.

With these definitions we therefore have

$$\Big(\frac{1}{n}T_n, \frac{1}{n}U_n, \frac{1}{n}V_n, \frac{1}{\sqrt{n}}S_n\Big) \Rightarrow_n (T, U, V, W_1) \tag{9.16}$$

in the random walk case; the distribution of (T, U, V, W_1) will be found by a passage to the limit. In the general Lindeberg-Lévy case, the left side of (9.15) is a somewhat more complicated function of the partial sums S_i, but it will still be possible to derive limit theorems associated in a natural way with these partial sums.

Since the random vector (T, U, V, W_1) is constrained by

$$U = \begin{cases} 1 - T + V & \text{if } W_1 \geq 0, \\ V & \text{if } W_1 \leq 0, \end{cases} \tag{9.17}$$

it suffices to consider (T, V, W_1) and the related vector (T_n, V_n, S_n). The distribution of the latter vector in the random walk case will be derived from three facts which admit of elementary proofs we do not carry through here.

First, we need the local central limit theorem for random walk: If m tends to infinity and j varies with m in such a way that j and m have the same parity and $j/\sqrt{m} \to y$, then[†]

$$\frac{\sqrt{m}}{2} p_m(j) \to \frac{1}{\sqrt{2\pi}} e^{y^2/2}. \tag{9.18}$$

[†] Feller [28], p. 184, has the local limit theorem for the binomial distribution, and (9.18) follows because $\frac{1}{2}(S_m + m)$ is binomially distributed.

Second, we need the fact that

$$\text{(9.19)} \qquad \mathsf{P}[S_1 > 0, \ldots, S_{m-1} > 0, S_m = j] = \frac{j}{m} p_m(j)$$

for j positive.† If $S_{2m} = 0$, then $U_{2m} = V_{2m}$ assumes one of the values $0, 2, \ldots, 2m$; the third fact we need‡ is that these $m+1$ values all have the same conditional probability:

$$\text{(9.20)} \qquad \mathsf{P}[V_{2m} = 2i \mid S_{2m} = 0] = \frac{1}{m+1}, \quad i = 0, 1, \ldots, m.$$

To compute the probability that $T_n = 2k$, $V_n = 2i$, and $S_n = j$, condition on the event $S_{2k} = 0$. Conditionally on this event, $(S_0, \ldots, S_{2k})$ and $(S_{2k+1}, \ldots, S_n)$ are independent, V_n depends only on the first sequence, and $T_n = 2k$ and $S_n = j$ both hold if and only if the elements of the second sequence are nonzero and the last one is j. By (9.19) and (9.20) we conclude that

$$\text{(9.21)} \quad \mathsf{P}[T_n = 2k,\ V_n = 2i,\ S_n = j] = p_{2k}(0) \frac{1}{k+1} \frac{j}{n-2k} p_{n-2k}(j)$$

if

$$\text{(9.22)} \qquad 0 \le 2i \le 2k < n, \quad j > 0.$$

Both sides of (9.21) vanish if n and j have opposite parity. For j negative, the same formula holds with $|j|$ in place of j on the right.

We apply Theorem 3.3 to the lattice of points $(2k/n, 2i/n, j/\sqrt{n})$ for which j and n have the same parity. Suppose that k, i, and j tend to infinity with n in such a way that

$$\frac{2k}{n} \to t, \quad \frac{2i}{n} \to v, \quad \frac{j}{\sqrt{n}} \to x,$$

where $0 < v < t < 1$ and $x > 0$. Then (9.22) holds for large n, and it follows by (9.21) and (9.18) that

$$\left(\frac{2}{n} \cdot \frac{2}{n} \cdot \frac{2}{\sqrt{n}}\right)^{-1} \mathsf{P}[T_n = 2k,\ V_n = 2i,\ S_n = j] \to g(t, x),$$

† See Feller [28], p. 73.

‡ See Feller [28], p. 94.

where

$$g(t,x) = \frac{1}{2\pi}\frac{|x|}{[t(1-t)]^{3/2}}e^{-x^2/2(1-t)}, \quad 0 < t < 1. \tag{9.23}$$

The same result holds for negative x by symmetry. Since local limit theorems imply global ones (Theorem 3.3),

$$\left(\frac{1}{n}T_n, \frac{1}{n}V_n, \frac{1}{\sqrt{n}}S_n\right) \tag{9.24}$$

has (in the random walk case) the limiting distribution in R^3 specified by the density

$$f(t,v,x) = \begin{cases} g(t,x) & \text{if } 0 < v < t < 1, \\ 0 & \text{otherwise.} \end{cases} \tag{9.25}$$

By (9.16), (T, V, W_1) has this density. Because of (9.17), the distribution of (T, U, V, W_1) can be written out explicitly as well.

From (9.25) it follows that the conditional distribution of V given T and W_1 is uniform on $[0, T]$; this corresponds to (9.20). By (9.17), if $T = t$ and $W_1 = x$, then U is distributed uniformly over $[1-t, 1]$ for $x > 0$ and uniformly over $[0, t]$ for $x < 0$. Using (9.25) to account for the possible values of t and x, we find the density of U alone:

$$\iint_{\substack{x>0 \\ 1-u<t}} g(t,x)\,dt\,dx + \iint_{\substack{x<0 \\ u<t}} g(t,x)\,dt\,dx. \tag{9.26}$$

Now the integral of $g(t,x)$ over the range $x > 0$ is $1/[2\pi t^{3/2}(1-t)^{1/2}]$, which is the derivative of $-\pi^{-1}\left((1-t)/t\right)^{1/2}$, and hence (9.26) reduces to $1/[\pi u^{1/2}(1-u)^{1/2}]$. Therefore

$$\mathsf{P}[U \le u] = \int_0^u \frac{ds}{\sqrt{s(1-s)}} = \frac{2}{\pi}\arcsin\sqrt{u}, \quad 0 < u < 1. \tag{9.27}$$

This is Paul Lévy's arc sine distribution. A similar computation shows that T also follows the arc sine law:

$$\mathsf{P}[T \le t] = \frac{2}{\pi}\arcsin\sqrt{t}, \quad 0 < t < 1. \tag{9.28}$$

Let us now combine (9.15) with the facts just derived to obtain a limit theorem for the general Lindeberg-Lévy case. Let us agree to say that a zero-crossing takes place at i if the event

$$E_i = [S_i = 0] \cup [S_{i-1} > 0 > S_i] \cup [S_{i-1} < 0 < S_i] \tag{9.29}$$

occurs (which in the random walk case is to say that $S_i = 0$). Let T'_n be the largest i, $1 \le i \le n$, for which a zero-crossing takes place at i; let U'_n be the number of i, $1 \le i \le n$, for which $S_i > 0$; and let V'_n be the number of i, $1 \le i \le T'_n$, for which $S_i > 0$. It will follow that

$$\left(\frac{1}{n}T'_n, \frac{1}{n}U'_n, \frac{1}{n}V'_n, \frac{1}{\sigma\sqrt{n}}S_n\right) \Rightarrow_n (T, U, V, W_1) \tag{9.30}$$

if we can show that the left side here approximates the left side of (9.15).

Clearly, T'_n/n is within $1/n$ of $h_1(X^n)$. If γ_n is the number of i, $1 \le i \le n$, for which E_i occurs—the number of zero-crossings—then U'_n/n and V'_n/n are within γ_n/n of $h_2(X^n)$ and $h_3(X^n)$, respectively. Therefore, (9.30) will follow from (9.15) and Theorem 3.1 if we prove that $\gamma_n/n \Rightarrow_n 0$, and for this it is enough to show that

$$\mathsf{E}[\gamma_n/n] = \frac{1}{n}\sum_{i=1}^{n} \mathsf{P}E_i \to 0. \tag{9.31}$$

But

$$\mathsf{P}E_i \le \mathsf{P}\big[|\xi_i| \ge \epsilon\sqrt{i}\big] + \mathsf{P}\big[|S_{i-1}| \le \epsilon\sqrt{i}\big]$$

for each positive ϵ, and hence, by the central limit theorem, $\mathsf{P}E_i \to 0$. And now (9.31) is a consequence of the theorem on Cesàro means.

From (9.30) we may conclude for example that U'_n/n and T'_n/n have arc sine distributions in the limit.

The Brownian Bridge

The Brownian bridge W° behaves like a Wiener path W conditioned by the requirement $W_1 = 0$. With an appropriate passage to the limit to take account of the fact that $[W_1 = 0]$ is an event of probability 0, this observation can be used to derive distributions associated with W°.

Let P_ϵ be the probability measure on $(C, \mathcal{C})$ defined by

$$P_\epsilon A = \mathsf{P}[W \in A \mid 0 \le W_1 \le \epsilon], \quad A \in \mathcal{C}.$$

The first step is to prove that

$$P_\epsilon \Rightarrow_\epsilon W^\circ \tag{9.32}$$

as ϵ tends to 0. Take W as a random function defined on some probability space, and take W° to be defined on the same space by $W_t^\circ = W_t - tW_1$. If we prove that

$$\limsup_{\epsilon \to 0} \mathsf{P}[W \in F \mid 0 \le W_1 \le \epsilon] \le \mathsf{P}[W^\circ \in F] \tag{9.33}$$

for every closed F in C, then (9.32) will follow by Theorem 2.1.

From the normality of the finite-dimensional distributions it follows that W_1 is independent of each $(W_{t_1}^\circ, \ldots, W_{t_k}^\circ)$ because it is uncorrelated of each component. Therefore,

$$\mathsf{P}[W^\circ \in A,\ W_1 \in B] = \mathsf{P}[W^\circ \in A]\mathsf{P}[W_1 \in B] \tag{9.34}$$

if A is a finite-dimensional set in C and B lies in R^1. But for B fixed the set of A in $\mathcal{C}$ that satisfy (9.34) is a monotone class and hence [PM.43] coincides with $\mathcal{C}$. Therefore,

$$\mathsf{P}[W^\circ \in A \mid 0 \le W_1 \le \epsilon] = \mathsf{P}[W^\circ \in A].$$

Since $\rho(W, W^\circ) = |W_1|$, where ρ is the metric on C, $|W_1| \le \delta$ and $W \in F$ imply $W^\circ \in F_\delta = [x: \rho(x, F) \le \delta]$. Therefore, if $\epsilon < \delta$,

$$\mathsf{P}[W \in F \mid 0 \le W_1 \le \epsilon] \le \mathsf{P}[W^\circ \in F_\delta \mid 0 \le W_1 \le \epsilon] = \mathsf{P}[W^\circ \in F_\delta].$$

The limit superior in (9.33) is thus at most $\mathsf{P}[W^\circ \in F_\delta]$, which decreases to $\mathsf{P}[W^\circ \in F]$ as $\delta \downarrow 0$ if F is closed. This proves (9.33) and hence (9.32).†

Suppose now that h is a measurable mapping from C to R^k and that the set D_h of its discontinuities satisfies $\mathsf{P}[W^\circ \in D_h] = 0$. It follows by (9.32) and the mapping theorem that

$$\mathsf{P}[h(W^\circ) \le \alpha] = \lim_{\epsilon \to 0} \mathsf{P}[h(W) \le \alpha \mid 0 \le W_1 \le \epsilon] \tag{9.35}$$

holds for each α at which the left side is continuous (as a funtion of α ranging over R^k). From (9.35) we can find explicit forms for some distributions connected with W°. Sometimes an alternative form of (9.35) is more convenient:

$$\mathsf{P}[h(W^\circ) \le \alpha] = \lim_{\epsilon \to 0} \mathsf{P}[h(W) \le \alpha \mid -\epsilon \le W_1 \le 0]. \tag{9.36}$$

† This part of the argument in effect repeats the proof of Theorem 3.1.

The proof is the same (we could use any subset of $[-\epsilon, \epsilon]$ of positive Lebesgue measure).

Define

$$m^\circ = \inf_t W_t^\circ, \quad M^\circ = \sup_t W_t^\circ.$$

Suppose that $a < 0 < b$ and $0 < \epsilon < b$; by (9.10) we have, if $c = b - a$,

$$\text{(9.37)} \quad \begin{aligned} \mathsf{P}[a < m \le M < b,\ 0 < W_1 < \epsilon] &= \sum_{k=-\infty}^{\infty} \mathsf{P}[2kc < N < 2kc + \epsilon] \\ &- \sum_{k=-\infty}^{\infty} \mathsf{P}[2kc + 2b - \epsilon < N < 2kc + 2b]. \end{aligned}$$

Since

$$\text{(9.38)} \quad \lim_{\epsilon \to 0} \frac{1}{\epsilon} \mathsf{P}[x < N < x + \epsilon] = \frac{1}{\sqrt{2\pi}} e^{-x^2/2},$$

and since the series in (9.37) converge uniformly in ϵ, we can take the limit ($\epsilon \to 0$) inside the sums, and it follows by (9.35) that

$$\text{(9.39)} \quad \mathsf{P}[a < m^\circ \le M^\circ \le b] = \sum_{k=-\infty}^{\infty} e^{-2(kc)^2} - \sum_{k=-\infty}^{\infty} e^{-2(b+kc)^2}.$$

Thus we have the distribution of (m°, M°). Taking $-a = b$ gives

$$\text{(9.40)} \quad \mathsf{P}[\sup_t |W_t^\circ| \le b] = 1 + 2 \sum_{k=-\infty}^{\infty} (-1)^k e^{-2k^2 b^2}, \quad b > 0.$$

By an entirely similar analysis applied to (9.11),

$$\text{(9.41)} \quad \mathsf{P}[m^\circ < b] = 1 - e^{-2b^2}, \quad b > 0.$$

Let U° be the Lebesgue measure of those t in $[0, 1]$ for which $W_t^\circ > 0$. It will follow that U° is uniformly distributed over $[0, 1]$ if we prove

$$\text{(9.42)} \quad \lim_{\epsilon \to 0} \mathsf{P}[U \le \alpha \mid -\epsilon \le W_1 \le 0] = \alpha, \quad 0 < \alpha < 1.$$

Because of (9.17), the conditional probability here is $\mathsf{P}[V \le \alpha \mid -\epsilon \le W_1 \le 0]$. From the form of the density (9.25) we saw that the distribution of V for given T and W_1 is uniform on $(0, T)$. In other words, $L = V/T$ is uniformly distributed on $(0, 1)$ and is independent of (T, W_1). Therefore, the conditional probability in (9.42) is

$$\mathsf{P}[TL \le \alpha \mid -\epsilon \le W_1 \le 0] = \int_0^1 \mathsf{P}[T \le \alpha/s \mid -\epsilon < W_1 \le 0]\, ds,$$

and (9.42) will follow by the bounded convergence theorem if we prove the intuitively obvious relation

$$\mathsf{P}[T \le \theta \mid -\epsilon \le W_1 \le 0] \to 0, \quad 0 < \theta < 1.$$

But this follows by (9.38) and the form of the density (9.25). Therefore,

$$\mathsf{P}[U^\circ \le \alpha] = \alpha, \quad 0 < \alpha < 1. \tag{9.43}$$

Problems

9.1. Show by reflection in the random walk case that

$$\mathsf{P}[M_n \ge b, > S_n = v] = \begin{cases} p_n(v) & \text{if } v \ge b, \\ p_n(2b - v) & \text{if } v \le \mathrm{b} \end{cases} \tag{9.44}$$

for $b \ge 0$. Derive (9.9) from this.

9.2. For nonnegative integers c_i, let $\pi(c_1, \ldots, c_k; v)$ be the probability that an n-step random walk (n fixed) meets c_1 (one or more times), then meets $-c_2$, then meets $c_3, \ldots,$ then meets $(-1)^{k+1}c_k$, and ends at v. Use (9.44) and induction on k to show that

$$\pi(c_1, \ldots, c_k; v) = \begin{cases} p_n(2c_1 + \cdots + 2c_{k-1} + (-1)^{k+1}v) & \text{if } (-1)^{k+1}v \ge c_k, \\ p_n(2c_1 + \cdots + 2c_k - (-1)^{k+1}v) & \text{if } (-1)^{k+1}v \le c_k. \end{cases}$$

(Reflect through $(-1)^k c_{k-1}$ the part of the path to the right of the first passage through that point following successive passages through $c_1, -c_2, \ldots, (-1)^{k-1}c_{k-2}$.) Derive (9.5) by showing that $p_n(a, b, v)$ is

$$\begin{aligned} p_n(v) - \pi(b; v) + \pi(b, a; v) + \pi(b, a, b, ; v) + \cdots \\ - \pi(a; v) + \pi(a, b; v) - \pi(a, b, a; v) + \cdots . \end{aligned}$$

9.3. For x in C let $h(x)$ be the smallest t for which $x(t) = \sup_s x(s)$. Show that h is measurable and continuous on a set of W-measure 1. Let τ_n be the smallest k for which $S_k = \max_{i \le n} S_i$ and prove

$$\mathsf{P}\left[\frac{\tau_n}{n} \le \alpha\right] \to \frac{2}{\pi} \arcsin \sqrt{\alpha}, \quad 0 < \alpha < 1,$$

in the Lindeberg-Lévy case. (See Feller [28], p. 94, for the random walk case.)

9.4. Derive the joint limiting distribution of the maximum and minimum of $S_i - in^{-1}S_n$, $0 < i < n$, in the Lindeberg-Lévy case. (Consider $Y_t^n = X_t^n - tX_1^n$ with X^n defined by (8.5).)

SECTION 10. MAXIMAL INEQUALITIES

To prove tightness in Section 8, we used Etemadi's inequality (8.13), which requires the assumption of independence. Since we are also interested in functional limit theorems for dependent sequences of random variables, we want variants of (8.13) itself. This means that we need usable upper bounds for probabilities of the form

$$\mathsf{P}[\max_{k\le n}|S_k| \ge \lambda],$$

that is, *maximal inequalities*. The inequalities derived here are useful in probability theory itself and also in applications of probability to analysis and number theory.

Maxima of Partial Sums

Let $\xi_1, \ldots, \xi_n$ be random variables (stationary or not, independent or not), let $S_k = \xi_1 + \cdots + \xi_k$ ($S_0 = 0$), and put

$$M_n = \max_{k\le n}|S_k|. \tag{10.1}$$

(Because of the absolute values, the notation differs from (8.16) and (9.1).) We derive upper bounds for $\mathsf{P}[M_n \ge \lambda]$ by an indirect approach. Let

$$m_{ijk} = |S_j - S_i| \wedge |S_k - S_j|, \tag{10.2}$$

and put

$$L_n = \max_{0\le i\le j\le k\le n} m_{ijk}. \tag{10.3}$$

From $|S_k| \le |S_n - S_k| + |S_n|$ and $|S_k| \le |S_k| + |S_n|$ follows $|S_k| \le m_{0kn} + |S_n|$, which gives the useful inequality

$$M_n \le L_n + |S_n|. \tag{10.4}$$

There is a useful companion inequality. If $|S_n| = 0$, then, trivially,

$$|S_n| \le 2L_n + \max_{k\le n}|\xi_k|. \tag{10.5}$$

But this also holds if $|S_n| > 0$. For in that case, $|S_k| \ge |S_n - S_k|$ holds for $k = n$ but not for $k = 0$, and hence there is a k, $1 \le k \le n$, for which $|S_k| \ge |S_n - S_k|$ but $|S_{k-1}| < |S_n - S_{k-1}|$; for this k, $|S_n - S_k| = m_{0kn} \le L_n$ and $|S_{k-1}| = m_{0,k-1,n} \le L_n$, which implies that $|S_n| \le |S_{k-1}| + |\xi_k| + |S_n - S_k| \le 2L_n + |\xi_k|$: (10.5) again holds. Finally, (10.4) and (10.5) combine to give

$$M_n \le 3L_n + \max_{k \le n} |\xi_k|. \tag{10.6}$$

If we have a bound on L_n—that is to say, an upper bound on the right tail of its distibution—as well as a bound either on $|S_n|$ or on $\max_k |\xi_k|$, then we can use (10.4) or (10.6) to get a bound on M_n, as required to establish tightness. And in the next chapter, bounds on L_n itself will play an essential role. We can derive useful bounds on L_n by assuming bounds on the m_{ijk}.

Theorem 10.1. *Suppose that $\alpha > \frac{1}{2}$ and $\beta \ge 0$ and that $u_1, \ldots, u_n$ are nonnegative numbers such that*

$$\mathsf{P}[m_{ijk} \ge \lambda] \le \frac{1}{\lambda^{4\beta}} \Big(\sum_{i < l \le k} u_l \Big)^{2\alpha}, \quad 0 \le i \le j \le k \le n, \tag{10.7}$$

for $\lambda > 0$. Then

$$\mathsf{P}[L_n \ge \lambda] \le \frac{K}{\lambda^{4\beta}} \Big(\sum_{0 < l \le n} u_l \Big)^{2\alpha} \tag{10.8}$$

for $\lambda > 0$, where $K = K_{\alpha,\beta}$ depends only on α and β.

Before turning to the proof, consider the independent case as a first illustration.

Example 10.1. Take $\alpha = \beta = 1$ and suppose that the ξ_k are independent and identically distributed with mean 0 and variance 1. By Chebyshev's inequality and the inequality $xy \le (x+y)^2$,

$$\begin{aligned} \mathsf{P}[m_{ijk} \ge \lambda] &= \mathsf{P}[|S_j - S_i| \ge \lambda] \mathsf{P}[|S_k - S_j| \ge \lambda] \\ &\le \frac{(j-i)}{\lambda^2} \frac{(k-j)}{\lambda^2} \le \frac{(k-i)^2}{\lambda^4}. \end{aligned}$$

Thus (10.7) holds with $u_l \equiv 1$, and by (10.8), $\mathsf{P}[L_n \ge \lambda] \le Kn^2/\lambda^4$ and hence $\mathsf{P}[L_n/\sqrt{n} \ge \lambda] \le K/\lambda^4$ (normalize as usual by $\sqrt{n}$). By stationarity,

$$\mathsf{P}\left[\max_{k\le n}\frac{|\xi_k|}{\sqrt{n}}\ge\lambda\right]\le n\mathsf{P}\left[\frac{|\xi_1|}{\sqrt{n}}\ge\lambda\right]$$
$$\le\frac{1}{\lambda^2}\int_{|\xi_1|\ge\lambda\sqrt{n}}\xi_1^2\,d\mathsf{P}\le\frac{1}{\lambda^2}\int_{|\xi_1|\ge\lambda}\xi_1^2\,d\mathsf{P}.$$

Write $\psi(\lambda)$ for this last integal. By (10.6),

$$\mathsf{P}\left[\frac{M_n}{\sqrt{n}}\ge\lambda\right]\le\mathsf{P}\left[\frac{3L_n}{\sqrt{n}}\ge\frac{1}{2}\lambda\right]+\mathsf{P}\left[\max_{k\le n}\frac{|\xi_k|}{\sqrt{n}}\ge\frac{1}{2}\lambda\right]\le\frac{K}{(\lambda/6)^4}+\frac{\psi(\frac{1}{2}\lambda)}{(\lambda/2)^2}.$$

Since $\psi(\frac{1}{2}\lambda)\to 0$ as $\lambda\to\infty$, we can use this inequality to prove (8.7) and hence tightness. Thus we have a new proof of Donsker's theorem: We have proved tightness by Theorem 10.1 instead of by the central limit theorem and Etemadi's inequality. □

This alternative proof still uses independence, however, and the real point of Theorem 10.1 is that it does not require independence. See, for example, the proof of Theorem 11.1, on trigonometric series, and the proof of Theorem 17.2, on prime divisors. Postponing again the proof of Theorem 10.1, we turn to the following variant of it, where conditions are put on the individual $|S_j - S_i|$ rather than on the minima m_{ijk}.

Theorem 10.2. *Suppose that $\alpha > \frac{1}{2}$ and $\beta \ge 0$ and that $u_1, \ldots, u_n$ are nonnegative numbers such that*

$$\mathsf{P}[|S_j - S_i|\ge\lambda]\le\frac{1}{\lambda^{4\beta}}\Big(\sum_{i<l\le j}u_l\Big)^{2\alpha},\quad 0\le i\le j\le n, \tag{10.9}$$

for $\lambda > 0$. Then

$$\mathsf{P}[M_n\ge\lambda]\le\frac{K'}{\lambda^{4\beta}}\Big(\sum_{0<l\le n}u_l\Big)^{2\alpha} \tag{10.10}$$

for $\lambda > 0$, where $K' = K'_{\alpha,\beta}$ depends only on α and β.

PROOF. By Schwarz's inequality, $\mathsf{P}(A\cap B)\le \mathsf{P}^{1/2}(A)\mathsf{P}^{1/2}(B)$, and therefore, by (10.9) $(xy\le(x+y)^2)$,

$$\mathsf{P}[m_{ijk}\ge\lambda]\le\frac{1}{\lambda^{2\beta}}\Big(\sum_{i<l\le j}u_l\Big)^{\alpha}\frac{1}{\lambda^{2\beta}}\Big(\sum_{j<l\le k}u_l\Big)^{\alpha}\le\frac{1}{\lambda^{4\beta}}\Big(\sum_{i<l\le k}u_l\Big)^{2\alpha}.$$

If $s=\sum_{l\le n}u_l$, then $\mathsf{P}[L_n\ge\lambda]\le Ks^{2\alpha}/\lambda^{4\beta}$ by Theorem 10.1. But $\mathsf{P}[|S_n|\ge\lambda]\le s^{2\alpha}/\lambda^{4\beta}$ by (10.9), and so it follows by (10.4) that $\mathsf{P}[M_n\ge\lambda]\le(K+1)2^{4\beta}s^{2\alpha}/\lambda^{4\beta}$. □

By Markov's inequality, (10.9) holds if

$$\mathsf{E}[|S_j - S_i|^{4\beta}] \le \Big(\sum_{i<l\le j} u_l\Big)^{2\alpha} \tag{10.11}$$

does.

Example 10.2. Return to Example 10.1, where $\alpha = \beta = 1$. If the ξ_k have fourth moment τ^4, then it is a simple matter to show [PM.85] that $\mathsf{E}S_k^4 = k\tau^4 + 3k(k-1) \le 3k\tau^4$ $(\tau \ge \sigma = 1)$. In this case, (10.11) holds for $u_l \equiv \sqrt{3}\tau^2$. By applying Theorem 10.2 we can now conclude that $\mathsf{P}[M_n \ge \lambda] \le 3K'n^2\tau^4/\lambda^4$, or $\mathsf{P}[M_n/\sqrt{n} \ge \lambda] \le 3K'\tau^4/\lambda^4$. This is enough to establish tightness in Theorems 8.1 and 8.2. And as before, the real point of Theorem 10.2 is that it does not require independence. □

Just as (10.11) implies (10.9), there is a moment inequality that implies (10.7), namely

$$\mathsf{E}[|S_j - S_i|^{2\beta}|S_k - S_j|^{2\beta}] \le \Big(\sum_{i<l\le k} u_l\Big)^{2\alpha}. \tag{10.12}$$

A More General Inequality

If we generalize Theorem 10.1, the proof becomes simpler. Let T be a Borel subset of $[0,1]$ and suppose that $\gamma = [\gamma_t : t \in T]$ is a stochastic process with time running through T. We suppose that the paths of the process are right-continuous in the sense that if points s of T converge from the right to a point t of T, then $\gamma_s(\omega) \to \gamma_t(\omega)$ at all sample points ω (if T is finite, this imposes no restriction). Let

$$m_{rst} = |\gamma_s - \gamma_r| \wedge |\gamma_t - \gamma_s|, \tag{10.13}$$

and define

$$L(\gamma) = \sup_{\substack{r\le s\le t\\ r,s,t\in T}} m_{rst}. \tag{10.14}$$

Theorem 10.3. *Suppose that $\alpha > \frac{1}{2}$ and $\beta \ge 0$ and that μ is a finite measure on T such that*

$$\mathsf{P}[m_{rst} \ge \lambda] \le \frac{1}{\lambda^{4\beta}}\mu^{2\alpha}(T \cap (r,t]), \quad r \le s \le t, \tag{10.15}$$

for $\lambda > 0$ and $r, s, t \in T$. Then

$$\mathsf{P}[L(\gamma) \geq \lambda] \leq \frac{K}{\lambda^{4\beta}} \mu^{2\alpha}(T) \tag{10.16}$$

for $\lambda > 0$, where $K = K_{\alpha,\beta}$ depends only on α and β.

To deduce Theorem 10.1 from Theorem 10.3, we need only take $T = [i/n: 0 \leq i \leq n]$ and $\gamma(i/n) = S_i$, $0 \leq i \leq n$, and let μ have mass u_i at i/n, $1 \leq i \leq n$.

PROOF. Write $m(r, s, t)$ for $m_{r,s,t}$. The argument goes by cases.

Case 1: Suppose first that $T = [0, 1]$ and that μ is Lebesgue measure. Let D_k be the set of dyadic rationals $i/2^k$, $0 \leq i \leq 2^k$. Let B_k be the maximum of $m(t_1, t_2, t_3)$ over triples in D_k satisfying $t_1 \leq t_2 \leq t_3$. Let A_k be the same maximum but with the further constraint that t_1, t_2, t_3 are adjacent: $t_2 - t_1 = t_3 - t_2 = 2^{-k}$. For t in D_k, define a point t' in D_{k-1} by

$$t' = \begin{cases} t & \text{if } t \in D_{k-1} \\ t - 2^{-k} & \text{if } t \notin D_{k-1} \text{ and } |\gamma(t) - \gamma(t - 2^{-k})| \\ & \qquad \leq |\gamma(t) - \gamma(t + 2^{-k})|, \\ t + 2^{-k} & \text{if } t \notin D_{k-1} \text{ and } |\gamma(t) - \gamma(t - 2^{-k})| \\ & \qquad > |\gamma(t) - \gamma(t + 2^{-k})|. \end{cases}$$

Then $|\gamma(t) - \gamma(t')| \leq A_k$ for t in D_k, and therefore, for t_1, t_2, t_3 in D_k,

$$\begin{aligned} |\gamma(t_2) - \gamma(t_1)| &\leq |\gamma(t_2) - \gamma(t'_2)| + |\gamma(t'_2) - \gamma(t'_1)| + |\gamma(t'_1) - \gamma(t_1)| \\ &\leq |\gamma(t'_2) - \gamma(t'_1)| + 2A_k \end{aligned}$$

and

$$\begin{aligned} |\gamma(t_2) - \gamma(t_3)| &\leq |\gamma(t_2) - \gamma(t'_2)| + |\gamma(t'_2) - \gamma(t'_3)| + |\gamma(t'_3) - \gamma(t_3)| \\ &\leq |\gamma(t'_2) - \gamma(t'_3)| + 2A_k. \end{aligned}$$

If $t_1 < t_2 < t_3$, then $t'_1 \leq t'_2 \leq t'_3$, and since t'_1, t'_2, t'_3 lie in D_{k-1}, it follows that $m(t_1, t_2, t_3) \leq B_{k-1} + 2A_k$, and therefore, $B_k \leq B_{k-1} + 2A_k$. Since $A_0 = B_0 = 0$, it follows by induction that $B_k \leq 2(A_1 + \cdots + A_k)$ for $k \geq 1$, and it follows further by the right-continuity of the paths that $L(\gamma) \leq 2\sum_{k=1}^{\infty} A_k$.

We need to control $\sum_k A_k$. Suppose that $0 < \theta < 1$ and choose C so that $C \sum_{k=1}^{\infty} \theta^k = \frac{1}{2}$. Then

$$\mathsf{P}[L(\gamma) \geq \lambda] \leq \mathsf{P}\Big[2 \sum_{k=1}^{\infty} A_k \geq \lambda\Big] \leq \sum_{k=1}^{\infty} \mathsf{P}[A_k \geq C\lambda\theta^k]$$

$$\leq \sum_{k=1}^{\infty} \sum_{i=1}^{2^k-1} \mathsf{P}\Big[m\Big(\frac{i-1}{2^k}, \frac{i}{2^k}, \frac{i+1}{2^k}\Big) \geq C\lambda\theta^k\Big].$$

Since μ is assumed to be Lebesgue measure, (10.15) implies

$$\mathsf{P}[L(\gamma) \geq \lambda] \leq \sum_{k=1}^{\infty} 2^k \frac{1}{(C\lambda\theta^k)^{4\beta}} \Big(\frac{2}{2^k}\Big)^{2\alpha} = \frac{2^{2\alpha}}{C^{4\beta}\lambda^{4\beta}} \sum_{k=1}^{\infty} \Big(\frac{1}{\theta^{4\beta} 2^{2\alpha-1}}\Big)^k.$$

Since $4\beta \geq 0$ and $2\alpha - 1 > 0$, there is a θ in (0,1) for which the series here converges. This shows how to define K and disposes of Case 1.

Case 2: Suppose that $T = [0,1]$ and μ has no atoms, so that $F(t) = \mu[0,t]$ is continuous. If F is strictly increasing and $F(1) = c$, take $a = c^{-\alpha/2\beta}$, so that $a^{4\beta}c^{2\alpha} = 1$, and define a new process ζ by

$$\zeta(t) = a\gamma(F^{-1}(ct)).$$

Then ζ comes under Case 1, and the theorem holds for η because $L(\gamma) = aL(\zeta)$. If F is continuous but not strictly increasing, consider first the measure having distribution function $F(t) + \epsilon t$, and then let ϵ go to 0.

Case 3: Suppose that T is finite (which in fact suffices for Theorem 10.1). If $0 \notin T$, let $\gamma(0) = \gamma(t_1)$, where t_1 is the first point of T, and take $\mu\{0\} = 0$. If $1 \notin T$, take $\gamma(1) = \gamma(t_{v-1})$, where t_{v-1} is the last point of T, and take $\mu\{1\} = 0$. Then the new process γ and measure μ satisfy the hypotheses, and so we may assume that T consists of the points

$$0 = t_0 < t_1 < \cdots < t_v = 1.$$

Define a process γ' by

$$\gamma'(t) = \begin{cases} \gamma(t_i) & \text{if } t_i \leq t < t_{i+1}, \quad 0 \leq i < v, \\ \gamma(1) & \text{if } t = 1. \end{cases}$$

If m'_{rst} denotes (10.13) for the process γ', then m'_{rst} vanishes unless r, s, t all lie in different subintervals $[t_i, t_{i+1})$. Suppose that

$$(10.17) \quad r \in [t_i, t_{i+1}), \quad s \in [t_j, t_{j+1}), \quad t \in [t_k, t_{k+1}), \qquad i < j < k.$$

Then $m'_{rst} = m(t_i, t_j, t_k)$, and hence, by the hypothesis of the theorem for the process γ,

$$\mathsf{P}[m'_{rst} \geq \lambda] \leq \frac{1}{\lambda^{4\beta}} \mu^{2\alpha}((t_i, t_k] \cap T). \tag{10.18}$$

Now let ν be the measure that corresponds to a uniform distribution of mass $\mu\{t_{l-1}\} + \mu\{t_l\}$ over the interval $[t_{l-1}, t_l]$, for $1 \leq l \leq v$. Then

$$\mu((t_i, t_k] \cap T) \leq \nu[t_{i+1}, t_k] \leq \nu(r, t],$$

and so (10.18) implies

$$\mathsf{P}[m'_{rst} \geq \lambda] \leq \frac{1}{\lambda^{4\beta}} \nu^{2\alpha}(r, t]. \tag{10.19}$$

Although (10.17) requires $t < 1$, (10.19) follows for $t = 1$ by a small modification of the argument.

Thus (10.19) holds for $0 \leq r \leq s \leq t \leq 1$, and Case 2 applies to the process γ':

$$\mathsf{P}[L(\gamma') \geq \lambda] \leq \frac{K}{\lambda^{4\beta}} \mu^{2\alpha}(0, 1] \leq \frac{K}{\lambda^{4\beta}} (2\mu(T))^{2\alpha}.$$

Since $L(\gamma') = L(\gamma)$, if we replace the K that works in Cases 1 and 2 by $2^{2\alpha}K$, then the new K works in Cases 1, 2, and 3.

Case 4: For the general T and μ, consider finite sets

$$T_n : 0 \leq t_{n0} < t_{n1} < \cdots < t_{nv_n} \leq 1$$

such that $T_n \subset T_{n+1}$ and $\bigcup T_n$ is dense in T. Let μ_n have mass $\mu((t_{n,i-1}, t_{n,i}] \cap T)$ at the point t_{ni}. If $\gamma^{(n)}$ is the process γ with the time-set cut back to T_n, then $L(\gamma^{(n)}) \uparrow L(\gamma)$ by the right-continuity of the paths. Since each $\gamma^{(n)}$ comes under Case 3, we have

$$\mathsf{P}[L(\gamma) \geq \lambda] \leq \mathsf{P}(\liminf_n [L(\gamma^{(n)} \geq \lambda - \epsilon]) \leq \frac{K}{(\lambda - \epsilon)^{4\beta}} \mu^{2\alpha}(T).$$

Let ϵ go to 0. □

A Further Inequality

The last theorem of this section will be used in Chapter 3. Suppose that $\gamma = [\gamma_t : t \in T]$ has right-continuous paths, as before, and for positive δ define $L(\gamma, \delta)$ by (10.14), but with the supremum further restricted by the inequality $t - r \le \delta$.

Theorem 10.4. *Suppose that $\alpha > \frac{1}{2}$ and $\beta \ge 0$ and that μ is a finite measure on T such that*

$$(10.20) \quad \mathsf{P}[m_{rst} \ge \lambda] \le \frac{1}{\lambda^{4\beta}} \mu^{2\alpha}(T \cap (r, t]), \quad r \le s \le t, \quad t - r < 2\delta$$

for $\lambda > 0$ and $r, s, t \in T$. Then

$$(10.21) \qquad \mathsf{P}[L(\gamma, \delta) \ge \lambda] \le \frac{2K}{\lambda^{4\beta}} \mu(T) \sup_{0 \le t \le 1 - 2\delta} \mu^{2\alpha - 1}(T \cap [t, t + 2\delta])$$

for $\lambda > 0$, where $K = K_{\alpha,\beta}$ depends only on α and β.

This theorem is somewhat analogous to Theorem 7.4 and its corollary. The K's in (10.16) and (10.21) are the same.

PROOF. Take $v = \lfloor 1/\delta \rfloor$, $t_i = i\delta$ for $0 \le i < v$, and $t_v = 1$. If $|r - t| \le \delta$, then r and t lie in the same $[t_{i-1}, t_i]$ or in adjacent ones, and hence they lie in the same $[t_{i-1}, t_{i+1}]$ for some i, $1 \le i \le v - 1$. If l_i is the supremum of m_{rst} for r, s, t ranging over $T \cap [t_{i-1}, t_{i+1}]$ $(r \le s \le t)$, then $\mathsf{P}[l_i \ge \lambda] \le K\mu^{2\alpha}(T \cap [t_{i-1}, t_{i+1}])/\lambda^{4\beta}$ by Theorem 10.3. If $p_i = \mu(T \cap [t_{i-1}, t_{i+1}])$, then

$$\sum_i p_i^{2\alpha} \le \sum_i p_i \times \max_i p_i^{2\alpha - 1} \le 2\mu(t) \max_i p_i^{2\alpha - 1},$$

and (10.20) follows from this. □

Problems

10.1. Prove a theorem standing to Theorem 10.3 as Theorem 10.2 stands to Theorem 10.1.

10.2. Weaken (10.11) by assuming only that it holds for $i = 0$ and $j = n$, but compensate by assuming that $S_1, \ldots, S_n$ is a martingale and $4\beta \ge 1$. Show that $\mathsf{P}[M_n \ge \lambda] \le \lambda^{-4\beta}(\sum_{l \le n} u_l)^{2\alpha}$, an inequality of essentially the same strength as (10.10).

10.3. Adapt the proof of Menshov's inequality (Doob [19], p. 156) to show that, if (10.12) holds with $\alpha \ge 1/2$ and $\beta \ge 1/4$, then

$$\mathsf{E}[L_n^{4\beta}] \le (\log_2 2n)^{4\beta} \left(\sum_{l \le n} u_l \right)^{2\alpha}.$$

Now deduce that, if (10.11) holds with $\alpha \geq 1$ and $\beta \geq 1/2$, then

$$\mathsf{E}[M_n^{2\beta}] \leq (\log_2 4n)^{2\beta}\left(\sum\nolimits_{l\leq n} u_l\right)^{\alpha}.$$

If $\alpha = \beta = 1$, this gives Menshov's inequality again.

10.4. Use Theorem 10.2 to show that, if $\xi_1, \xi_2, \ldots$ satisfies (10.11) for $0 \leq i \leq j < \infty$, where $\beta \geq 0$ and $\alpha > 1/2$, then $\sum_i \xi_i$ converges with probability 1.

SECTION 11. TRIGONOMETRIC SERIES*

In this section we prove functional limit theorems for lacunary series and for series defined in terms of incommensurable arguments.

Lacunary Series

A trigonometric series $\sum_{k=1}^{\infty}(a_k \cos m_k x + b_k \sin m_k x)$ is called *lacunary* if the m_k are positive integers increasing at an exponential rate: $m_{k+1}/m_k \geq q > 1$. If x is chosen at random from $[0, 2\pi]$, then the series has some of the properties of a sum of independent random variables. Here we prove a functional central limit theorem for the partial sums of a pure cosine series.

In this theorem, P and E will refer to normalized Lebesgue measure on the Borel sets in $\Omega = [0, 2\pi]$. Since the cosine integrates to 0 over $[0, 2\pi]$, it follows from the relation

$$\cos\theta \cdot \cos\theta' = \frac{1}{2}\cos(\theta + \theta') + \frac{1}{2}\cos(\theta - \theta') \tag{11.1}$$

that

$$\mathsf{E}[\cos m\omega] = 0, \quad \mathsf{E}[\cos^2 m\omega] = \frac{1}{2}, \quad \mathsf{E}[\cos m\omega \cdot \cos m'\omega] = 0, \quad 1 \leq m < m'. \tag{11.2}$$

Because of the $\frac{1}{2}$ here, it is probabilistically natural to multiply the cosine by $\sqrt{2}$; this makes the variance equal to 1. And so, define $\zeta_k(\omega) = \sqrt{2}a_k \cos m_k\omega$ and consider the sums

$$S_n(\omega) = \sum_{k=1}^{n} \zeta_k(\omega) = \sqrt{2}\sum_{k=1}^{n} a_k \cos m_k\omega. \tag{11.3}$$

By (11.2),

$$\text{(11.4)} \qquad \mathsf{E}\zeta_k = 0, \quad \mathsf{E}\zeta_k^2 = a_k^2, \quad \mathsf{E}S_n = 0, \quad s_n^2 = \mathsf{E}S_n^2 = \sum_{k=1}^n a_k^2.$$

Define a random element Y^n of C by

$$\text{(11.5)} \quad Y_t^n(\omega) = \frac{\sqrt{2}}{s_n}\sum_{i=1}^k a_i \cos m_i\omega + \frac{s_n^2 t - s_k^2}{a_{k+1}^2}\frac{\sqrt{2}a_{k+1}\cos m_{k+1}\omega}{s_n}$$

$$\text{for } \frac{s_k^2}{s_n^2} \le t < \frac{s_{k+1}^2}{s_n^2}, \quad 0 \le k < n.$$

This is a trigonometric analogue of the random function (8.5) of Donsker's theorem: Y^n is the polygonal function resulting from linear interpolation between its values $Y_t^n = S_k/s_n$ at the points $t = s_k^2/s_n^2$, $k = 0, 1, \ldots, n$.† Strengthen the lacunarity condition to

$$\text{(11.6)} \qquad m_{k+1}/m_k \ge 2.$$

Theorem 11.1. *If* (11.6) *holds and*

$$\text{(11.7)} \qquad s_n^2 \to \infty, \quad \max_{k \le n} a_k^2/s_n^2 \to 0,$$

then $Y^n \Rightarrow W$.

To show that the finite-dimensional distributions converge it is best to consider a triangular array first: Let $\zeta_{nk}(\omega) = \sqrt{2}a_{nk}\cos m_k\omega$ for $1 \le k \le n$. In the following lemma and its proof, k ranges from 1 to n in the sums, products, and maxima.

Lemma 1. *If* (11.6) *holds, if* $\sum_k a_{nk}^2 \to a^2 > 0$, *and if* $\alpha_n^2 = \max_k a_{nk}^2 \to 0$, *then* $\sum_k \zeta_{nk} \Rightarrow aN$.

PROOF. The hypothesis of convergence to a^2 we can strengthen to equality, and we can arrange that $a = 1$: If $\theta_n^2 = a^{-2}\sum_k a_{nk}^2$ and $\zeta'_{nk}(\omega) = \sqrt{2}\theta_n^{-1}a^{-1}a_{nk}\cos m_k\omega$, then $\theta_n \to 1$, $\sum_k(\theta_n^{-1}a^{-1}a_{nk})^2 = 1$, and (Example 3.2) $\sum_k \zeta'_{nk} \Rightarrow N$ implies $\sum_k \zeta_{nk} \Rightarrow aN$. Assume, therefore, that $\sum_k a_{nk}^2 = 1$.

---

† If $a_{k+1}^2 = 0$, there is a division by 0 in (11.5); but in that case, no t can satisfy the defining inequality anyway. Or simply assume all the a_k^2 to be positive.

For $|x| < 1$, the principal value of $\log(1 + ix)$ can be defined as $ix + \frac{1}{2}x^2 - r(x)$, where $r(x) = \sum_{u=3}^{\infty}(-1)^u(ix)^{u+1}/u$ and $|r(x)| \leq |x|^3$ for $|x| \leq \frac{1}{2}$. This gives

$$e^{ix} = (1 + ix)\exp[-x^2/2 + r(x)].$$

For fixed t, write

$$T_n = \prod_k (1 + it\zeta_{nk}),$$
$$Z_n = \exp\left(-\frac{1}{2}t^2 \sum_k \zeta_{nk}^2 + \sum_k r(t\zeta_{nk})\right),$$
$$\Delta_n = T_n(Z_n - e^{t^2/2}).$$

Then

$$\exp\left[it\sum_k \zeta_{nk}\right] = T_n e^{-t^2/2} + \Delta_n, \tag{11.8}$$

and so $\sum_k \zeta_{nk} \Rightarrow N$ will follow if we prove

$$\mathsf{E}\Delta_n \to 0 \tag{11.9}$$

and

$$\mathsf{E}T_n = 1. \tag{11.10}$$

We first show that

$$\sum_k \zeta_{nk}^2 \Rightarrow 1. \tag{11.11}$$

From the assumption that $\sum_k a_{nk}^2 = 1$, it follows by the double-angle formula that

$$\sum_k \zeta_{nk}^2(\omega) = \sum_k 2a_{nk}^2 \cos^2 m_k\omega = 1 + \sum_k a_{nk}^2 \cos 2m_k\omega.$$

And now (11.2)—orthogonality—implies

$$\mathsf{E}\left[\left(\sum_k \zeta_{nk}^2 - 1\right)^2\right] = \sum_k a_{nk}^4 \mathsf{E}[\cos^2 2m_k\omega]$$
$$\leq \sum_k a_{nk}^4 \leq \alpha_n^2 \sum_k a_{nk}^2 = \alpha_n^2 \to 0,$$

and (11.11) follows from this.

From $|\zeta_{nk}| \le \sqrt{2}|a_{nk}|$ it follows that, for all sufficiently large n, we have

$$\sum\nolimits_k |r(t\zeta_{nk})| \le \sum\nolimits_k |\sqrt{2}t|^3 a_{nk}^3 \le 4|t|^3 \alpha_n \sum\nolimits_k a_{nk}^2 \to 0.$$

By this and (11.11), $Z_n \Rightarrow e^{t^2/2}$. But since $1 + u \le e^u$, we also have $|T_n| \le \prod_k (1 + t^2 a_{nk}^2)^{1/2} \le \exp(\sum_k t^2 a_{nk}^2/2) \le e^{t^2}$, and so $\Delta_n \Rightarrow 0$. Finally, by (11.8), $|\Delta_n| \le 1 + |T_n| \le 1 + e^{t^2}$, and so, by Theorem 3.5, we can integrate to the limit: (11.9) follows.

There remains the proof of (11.10). Expanding the product that defines T_n gives

$$\mathsf{E}T_n = 1 + \sum (\sqrt{2}it)^v a_{j_1} \cdots a_{j_v} \mathsf{E}\left[\cos m_{j_1}\omega \cdots \cos m_{j_v}\omega\right], \tag{11.12}$$

where the sum extends over $v = 1, \ldots, n$ and $n \ge j_1 > \cdots > j_v \ge 1$. And repeated use of (11.1) leads to

$$\begin{aligned} \mathsf{E}[\cos m_{j_1}\omega \cdots \cos m_{j_v}\omega] \\ = \frac{1}{2^{v-1}} \sum \mathsf{E}[\cos(m_{j_1} \pm m_{j_2} \pm \cdots \pm m_{j_v})\omega], \end{aligned} \tag{11.13}$$

where here the sum extends over the 2^{v-1} choices of sign. The expected value in the sum is 0 unless

$$m_{j_1} \pm m_{j_2} \pm \cdots \pm m_{j_v} = 0. \tag{11.14}$$

But if $j_1 > \cdots > j_v$, then by (11.6),

$$m_{j_1} \pm m_{j_2} \pm \cdots \pm m_{j_v} \ge m_{j_1}\left(1 - \frac{1}{2} - \frac{1}{2^2} - \cdots - \frac{1}{2^{v-1}}\right) > 0, \tag{11.15}$$

and so (11.14) is impossible. The expected values on the right in (11.13) and (11.12) are therefore 0, and (11.10) follows. □

We need a second lemma for the proof of tightness. Define S_n by (11.3), as before.

Lemma 2. *If* (11.6) *holds, then*

$$\mathsf{P}\left[\max_{k \le n} |S_k| > \lambda\right] \le \frac{C}{\lambda^4}\left(\sum_{k \le n} a_k^2\right)^2, \tag{11.16}$$

where C is a universal constant.

PROOF. The proof uses (10.11) and Theorem 10.2 for $\alpha = \beta = 1$. By (11.13) for $v = 4$,

$$\text{(11.17)} \quad \mathsf{E}[S_n^4] = \sum{}' a_{j_1} \cdots a_{j_4} \sum{}'' \frac{1}{8} \mathsf{E}[\cos(m_{j_1} \pm m_{j_2} \pm m_{j_3} \pm m_{j_4})\omega],$$

where in $\sum'$ the four indices range independently from 1 to n, and $\sum''$ extends over the eight choices of sign. In place of (11.6) temporarily make the stronger assumption that $m_{k+1}/m_k \geq 4$. We must sort out the cases where the coefficient of ω in (11.17) vanishes for some choice of signs. Suppose that $j_1 \geq j_2 \geq j_3 \geq j_4$ and consider whether $L := m_{j_1} \pm m_{j_2} = \pm m_{j_3} \pm m_{j_4} =: R$ is possible. It is if $j_1 = j_2$ and $j_3 = j_4$, but this is the only case: If $j_1 > j_2$, then $m_{j_1} \geq 4m_{j_2}$ and hence $L \geq 4m_{j_2} \pm m_{j_2} > 2m_{j_2} \geq m_{j_3} + m_{j_4} \geq R$. Suppose on the other hand that $j_1 = j_2$ but $j_3 > j_4$. Then $L = m_{j_1} \pm m_{j_2}$ is either 0 or $2m_{j_2}$; but $R \neq 0$ rules out the first case and $R \leq m_{j_3} + m_{j_4} < 2m_{j_3} \leq 2m_{j_2}$ rules out the second.

Therefore (if $m_{k+1}/m_k \geq 4$), the only sets of indices $j_1, \ldots, j_4$ that contribute to the sum in (11.17) are those consisting of a single integer repeated four times and those consisting of two distinct integers each occuring twice. Since in any case the inner sum has modulus at most 1,

$$\text{(11.18)} \qquad \mathsf{E}[S_n^4] \leq \sum_{j \leq n} a_j^4 + 3 \sum_{\substack{j,k \leq n \\ j \neq k}} a_j^2 a_k^2 \leq 3\Big(\sum_{k \leq n} a_k^2\Big)^2.$$

Since (11.6) implies $m_{k+2}/m_k \geq 4$, it implies that (11.18) holds if k is restricted to even values in the sum S_n and also in the sum on the right. The same is true for odd values, and so by Minkowski's inequality, (11.18) holds if the right side is multiplied by 2^4. Finally, the same inequality holds with $S_j - S_i$ in place of S_n:

$$\mathsf{E}[|S_j - S_i|^4] \leq 16 \cdot 3 \cdot \Big(\sum_{i < l \leq j} a_l^2\Big)^2.$$

And now (11.16) follows by Theorem 10.2 if $C = 16 \cdot 3 \cdot K'$. □

PROOF OF THEOREM 11.1. We prove $Y^n \Rightarrow W$ by means of Theorem 7.5 (with Y^n and W in the roles of X^n and X), and we turn first to (7.13). Let $\alpha_n = \max_{k \leq n} |a_k|/s_n$. Taking $a_{nk} = a_k/s_n$, $k \leq n$, in Lemma 1 gives $Y_1^n = S_n/s_n \Rightarrow W_1$. Fix a t in (0,1) and let $k_n(t)$

be the largest k for which $s_k^2/s_n^2 \leq t$. Then ts_n^2 lies between $s_{k_n(t)}^2$ and $s_{k_n(t)+1}^2$, and therefore,

$$t - \alpha_n^2 \leq \frac{s_{k_n(t)}^2}{s_n^2} \leq t. \tag{11.19}$$

By (11.5), Y_t^n is within $\sqrt{2}\alpha_n$ of $S_{k_n(t)}/s_n$, and it converges in distribution to W_t by (11.19) and the lemma. And $(Y_t^n, Y_1^n) \Rightarrow (W_t, W_1)$ will follow by the Cramèr-Wold method [PM.383] if we show that $uY_t^n + vY_1^n \Rightarrow uW_t + vW_1$ for all u, v. But this also follows from the lemma if we take $a_{nk} = (u+v)a_k/s_n$ for $k \leq k_n(t)$ and $a_{nk} = va_k/s_n$ for $k_n(t) < k \leq n$. This argument extends to the case of three or more time points, and therefore $(Y_{t_1}^n, \ldots, Y_{t_k}^n) \Rightarrow_n (W_{t_1}, \ldots, W_{t_k})$ for all $t_1, \ldots, t_k$.

We turn next to (7.14), which we prove by means of (7.10). Let $\lambda = \epsilon s_n$. If $l = 0$ and $m = n$, then (11.16) can be written

$$\mathsf{P}\left[\max_{l<k\leq m} \frac{|S_k - S_l|}{s_n} \geq \epsilon\right] \leq \frac{C}{\epsilon^4}\left(\frac{s_m^2}{s_n^2} - \frac{s_l^2}{s_n^2}\right)^2. \tag{11.20}$$

But this holds for any pair of integers satisfying $0 \leq l \leq m \leq n$, because it depends only on the condition (11.6). Suppose now that t_i are points satisfying $0 = t_0 < \cdots < t_v = 1$ and (7.9), so that, by (7.10),

$$w(Y^n, \delta) \leq 3 \max_{1\leq i\leq v} \sup_{t_{i-1}\leq s\leq t_i} |Y_s^n - Y_{t_{i-1}}^n|. \tag{11.21}$$

Suppose further that we can choose the t_i in such a way that $t_i = s_{m_i}^2/s_n^2$ for integers m_i satisfying $0 = m_0 < \cdots < m_v = n$. Since the random function Y^n is a polygon with break-points over the t_i, (11.21) implies

$$w(Y^n, \delta) \leq 3 \max_{1\leq i\leq v} \max_{u_{i-1}\leq k\leq u_i} \frac{|S_k - S_{u_{i-1}}|}{s_n};$$

the inner maximum here equals the supremum in (11.21). It follows by (11.20) that

$$\mathsf{P}[w(Y^n, \delta) \geq 3\epsilon] \leq \frac{C}{\epsilon^4}\sum_{i=1}^{v}(t_i - t_{i-1})^2 \leq \frac{C}{\epsilon^4}\max_{1\leq i\leq v}(t_i - t_{i-1}). \tag{11.22}$$

Suppose we can, for all sufficiently large n, arrange that

$$t_i = s_{m_i}^2/s_n^2, \quad \min_{1<i<v}(t_i - t_{i-1}) \geq \delta, \quad \max_{1\leq i\leq v}(t_i - t_{i-1}) \leq 3\delta. \tag{11.23}$$

The first two of these conditions will ensure that (11.22) holds for large n, and it will then follow from the third one that $\mathsf{P}[w(Y^n, \delta) \geq 3\epsilon] \leq 3\delta C/\epsilon^4$. And of course this implies (7.14), which will complete the proof.

Integers m_i satisfying (11.23) can be chosen this way: Take $v = \lceil 1/(2\delta) \rceil$ and $m_v = n$, and for $0 \leq i \leq v-1$, take $m_i = k_n(2\delta i)$, where $k_n(t)$ is defined as before. For large n, (11.23) then follows from (11.19). □

Incommensurable Arguments

Replace the integers $m_1, m_2, \ldots$ in (11.3) by a sequence $\lambda_1, \lambda_2, \ldots$ of linearly independent real numbers, as at the end of Section 3. And for simplicity take $a_k \equiv 1$. That is, replace $\zeta_k(\omega) = \sqrt{2} a_k \cos m_k \omega$ by $\zeta'_k(\omega) = \sqrt{2} \cos \lambda_k \omega$ and replace (11.3) by

$$S'_n(\omega) = \sqrt{2} \sum_{k=1}^{n} \cos \lambda_k \omega. \tag{11.24}$$

For each $\omega \in R^1$, define a function $Z^n(\omega)$ in C by

$$Z^n_t(\omega) = \sqrt{\frac{2}{n}} \sum_{i=1}^{k} \cos\lambda_i \omega + (nt - k) \frac{\sqrt{2} \cos \lambda_{k+1} \omega}{\sqrt{n}} \tag{11.25}$$

$$\text{for } \frac{k}{n} \leq t \leq \frac{k+1}{n}, \ 0 \leq k < n.$$

This, a second trigonometric analogue of (8.5), is a polygonal function with values $Z^n_{k/n}(\omega) = S'_k(\omega)/\sqrt{n}$ at the corner points. The objective is to extend (3.33) to a functional central limit theorem.

Let ξ_k be the random variables in (3.31). If we put $v_n(\omega) = (\sqrt{2}\cos \lambda_1 \omega, \ldots, \sqrt{2} \cos \lambda_n \omega)$ and $V_n = (\xi_1, \ldots, \xi_n)$, then what (3.33) says is that $\mathsf{P}_T v_n^{-1} \Rightarrow_T \mathsf{P} V_n^{-1}$, where P refers to the space the independent sequence $\{\xi_k\}$ is defined on. Define $\tau_n \colon R^n \to C$ this way: If $z = (z_1, \ldots, z_n)$, take $\tau_n z = x$, where $x(0) = 0$, $x(k/n) = \sum_{i=1}^{k} z_i/\sqrt{n}$ for $1 \leq k \leq n$, and x is linear between these points. Since τ_n is continuous, it follows by the mapping theorem that $\mathsf{P}_T v_n^{-1} \tau_n^{-1} \Rightarrow_T \mathsf{P} V_n^{-1} \tau_n^{-1}$. But $\tau_n(v_n(\omega))$ is exactly $Z^n(\omega)$ as defined by (11.25), and $X^n = \tau_n V_n$ is a special case of the random function (8.5) of Donsker's theorem ($\sigma = 1$). Therefore, Donsker's theorem applies:

$$\mathsf{P}_T (Z^n)^{-1} \Rightarrow_T \mathsf{P}(X^n)^{-1} \Rightarrow_n \mathsf{P} W^{-1},$$

where the P at the right refers to some space supporting a Wiener random function W.

Now suppose that $h: C \to R^1$ is measurable, and let D_h be the set of its points of discontinuity. If

$$\mathsf{P}[X^n \in D_h] = \mathsf{P}[W \in D_h] = 0, \quad n \geq n_0, \tag{11.26}$$

for some n_0, then two further applications of the mapping theorem give

$$\mathsf{P}_T(Z^n)^{-1}h^{-1} \Rightarrow_T \mathsf{P}(X^n)^{-1}h^{-1} \Rightarrow_n \mathsf{P}W^{-1}h^{-1}.$$

Finally, if H is a linear Borel set and

$$\mathsf{P}[h(X^n) \in \partial H] = 0 = \mathsf{P}[h(W) \in \partial H], \quad n \geq n_0, \tag{11.27}$$

then

$$\mathsf{P}_T[h(Z^n(\omega)) \in H] \to_T \mathsf{P}[h(X^n) \in H] \to_n \mathsf{P}[h(W) \in H].$$

What the argument shows is this: If (11.26) and (11.27) hold, then

$$\lim_{n\to\infty} \mathsf{P}_\infty[h(Z^n(\omega)) \in H] = \mathsf{P}[h(W) \in H]. \tag{11.28}$$

Take $H = (-\infty, x]$. Then (11.27) holds for all but countably many x, and we arrive at our theorem.

Theorem 11.2. *If $\lambda_1, \lambda_1, \ldots$ are linearly independent and $Z^n(\omega)$ is defined over R^1 by* (11.25), *and if* (11.26) *holds, then*

$$\lim_{n\to\infty} \mathsf{P}_\infty[h(Z^n(\omega)) \leq x] = \mathsf{P}[h(W) \leq x] \tag{11.29}$$

for all but countably many x.

If $h(x) = \sup_t x(t)$, for example, then (11.26) holds because h is continuous everywhere on C. On the other hand, h has discontinuities if $h(x)$ is $\sup[t: x(t) = 0]$, for example, or if $h(x)$ is the Lebesgue measure of $[t: x(t) > 0]$; but $\mathsf{P}[W \in D_h]$ is 0 for these functions as well [M15]. Since the random variables ξ_k in the definition of X^n have continuous distributions in this particular case (see (3.31)), we also have $\mathsf{P}[X^n \in D_h] = 0$ as long as $n > 1$. Therefore, Theorem 11.2 gives the folded normal distribution as the limiting distribution of the maximum of the sums $\sqrt{2/n}\sum_{i\leq k} \cos \lambda_i\omega$, $1 \leq k \leq n$, as well as an arc sine law for the fraction of positive partial sums, and so on.

Problem

11.1. Prove a theorem that stands to (3.41) as Theorem 11.2 stands to (3.33).

CHAPTER 3

THE SPACE D

SECTION 12. THE GEOMETRY OF D

The space C is unsuitable for the description of processes that, like the Poisson process and unlike Brownian motion, must contain jumps. In this chapter we study weak convergence in a space that includes certain discontinuous functions.

The Definition

Let $D = D[0,1]$ be the space of real functions x on $[0,1]$ that are right-continuous and have left-hand limits:

(i) For $0 \le t < 1$, $x(t+) = \lim_{s \downarrow t} x(s)$ exists and $x(t+) = x(t)$.
(ii) For $0 < t \le 1$, $x(t-) = \lim_{s \uparrow t} x(s)$ exists.

Functions having these two properties are called *cadlag* (an acronym for "continu à droite, limites à gauche") functions. A function x is said to have a *discontinuity of the first kind* at t if $x(t-)$ and $x(t+)$ exist but differ and $x(t)$ lies between them. Any discontinuities of a cadlag function—of an element of D—are of the first kind; the requirement $x(t) = x(t+)$ is a convenient normalization. Of course, C is a subset of D.

With very little change, the theory can be extended to functions on $[0,1]$ taking values in metric spaces other than R^1. What changes are needed will be indicated along the way. Denote the space, metric, and Borel σ-field by V, v, and $\mathcal{V}$; V will always be assumed *separable and complete.* The definition of the cadlag property needs no change at all.

For $x \in D$ and $T \subset [0,1]$, put

$$w_x(T) = w(x,T) = \sup_{s,t \in T} |x(s) - x(t)|. \tag{12.1}$$

The modulus of continuity of x, defined by (7.1), can be written as

$$w_x(\delta) = w(x, \delta) = \sup_{0 \le t \le 1-\delta} w_x[t, t+\delta]. \tag{12.2}$$

A continuous function on $[0,1]$ is uniformly continuous. The following lemma gives the corresponding uniformity idea for cadlag functions.

Lemma 1. *For each x in D and each positive ϵ, there exist points $t_0, t_1, \ldots, t_v$ such that*

$$0 = t_0 < t_1 < \cdots < t_v = 1 \tag{12.3}$$

and

$$w_x[t_{i-1}, t_i) < \epsilon, \quad i = 1, 2, \ldots, v. \tag{12.4}$$

PROOF. Let t° be the supremum of those t in $[0,1]$ for which $[0,t)$ can be decomposed into finitely many subintervals $[t_{i-1}, t_i)$ satisfying (12.4). Since $x(0) = x(0+)$, we have $t^\circ > 0$; since $x(t^\circ-)$ exists, $[0, t^\circ)$ can itself be so decomposed; $t^\circ < 1$ is impossible because $x(t^\circ) = x(t^\circ+)$ in that case. □

From this lemma it follows that there can be at most finitely many points t at which the jump $|x(t) - x(t-)|$ exceeds a given positive number; therefore, x has at most countably many discontinuities. It follows also that x is bounded:

$$\|x\| = \sup_t |x(t)| < \infty. \tag{12.5}$$

Finally, it follows that x can be uniformly approximated by simple functions constant over intervals, so that it is Borel measurable.

If x takes its values in the metric space V, replace the $|x(s) - x(t)|$ in (12.1) by $v(x(s), x(t))$; the magnitude of the jump in x at t is $v(x(t), x(t-))$. Lemma 1 and its proof need no change, and in place of (12.5) we have the fact that the range of x has compact closure. Indeed, for given points $x(t_n)$ in the range, choose a sequence $\{n_i\}$ and a t so that either $t_{n_i} \downarrow t$ or $t_{n_i} \uparrow t$; in the first case, $x(t_{n_i}) \to x(t)$, and in the second, $x(t_{n_i}) \to x(t-)$. As for measurability, a function that assumes only finitely many values is measurable $\mathcal{B}/\mathcal{V}$, where $\mathcal{B}$ is the σ-field of Borel sets in $[0,1]$, and [M10] a limit of functions measurable $\mathcal{B}/\mathcal{V}$ is itself measurable $\mathcal{B}/\mathcal{V}$.

We need a modulus that plays in D the role the modulus of continuity plays in C. Call a set $\{t_i\}$ satifying (12.3) *δ-sparse* if it satisfies $\min_{1 \le i \le v}(t_i - t_{i-1}) > \delta$. And now define, for $0 < \delta < 1$,

$$w'_x(\delta) = w'(x, \delta) = \inf_{\{t_i\}} \max_{1 \le i \le v} w_x[t_{i-1}, t_i), \tag{12.6}$$

where the infimum extends over all δ-sparse sets $\{t_i\}$. Lemma 1 is equivalent to the assertion that $\lim_\delta w'_x(\delta) = 0$ holds for every x in D. Notice that $w'_x(\delta)$ is unaffected if the value $x(1)$ is changed.

Even if x does not lie in D, the definition of $w'_x(\delta)$ makes sense. Just as $\lim_\delta w_x(\delta) = 0$ is necessary and sufficient for an arbitrary function x on $[0,1]$ to lie in C, $\lim_\delta w'_x(\delta) = 0$ is necessary and sufficient for x to lie in D.

Now to compare $w'_x(\delta)$ with $w_x(\delta)$. Since $[0,1)$ can, for each $\delta < \frac{1}{2}$, be split into subintervals $[t_{i-1}, t_i)$ satisfying $\delta < t_i - t_{i-1} \le 2\delta$, we have

$$w'_x(\delta) \le w_x(2\delta), \quad \text{if } \delta < 1/2. \tag{12.7}$$

There can be no general inequality in the opposite direction because of the fact that $w_x(\delta)$ does not go to 0 with δ if x has discontinuities. But consider the maximum (absolute) jump in x:

$$j(x) = \sup_{0<t\le 1} |x(t) - x(t-)|; \tag{12.8}$$

the supremum is achieved because only finitely many jumps can exceed a given positive number. We have

$$w_x(\delta) \le 2w'_x(\delta) + j(x). \tag{12.9}$$

To see this, choose a δ-sparse $\{t_i\}$ such that $w_x[t_{i-1}, t_i) < w'_x(\delta) + \epsilon$ for each i. If $|s - t| \le \delta$, then s and t lie in the same $[t_{i-1}, t_i)$ or else in adjacent ones, and $|x(s) - x(t)|$ is at most $w'_x(\delta) + \epsilon$ in the first case and at most $2w'_x(\delta) + \epsilon + j(x)$ in the second. Letting $\epsilon \to 0$ gives (12.9). Since $j(x) = 0$ if x is continuous, we also have

$$w_x(\delta) \le 2w'_x(\delta), \quad \text{if } x \in C. \tag{12.10}$$

Because of (12.7) and (12.10), the moduli $w_x(\delta)$ and $w'_x(\delta)$ are essentially the same for continuous functions x.

The Skorohod Topology

Two funtions x and y are near one another in the uniform topology used for C if the graph of $x(t)$ can be carried onto the graph of $y(t)$ by a uniformly small perturbation of the ordinates, with the abscissas kept fixed. In D, we allow also a uniformly small deformation of the time scale. Physically, this amounts to the recognition that we cannot measure time with perfect accuracy any more than we can position. The following topology, devised by Skorohod, embodies this idea.

Let Λ denote the class of strictly increasing, continuous mappings of $[0,1]$ onto itself. If $\lambda \in \Lambda$, then $\lambda 0 = 0$ and $\lambda 1 = 1$. For x and y in D, define $d(x,y)$ to be the infimum of those positive ϵ for which there exists in Λ a λ satisfying

$$\sup_t |\lambda t - t| = \sup_t |t - \lambda^{-1} t| < \epsilon \tag{12.11}$$

and

$$\sup_t |x(t) - y(\lambda t)| = \sup_t |x(\lambda^{-1} t) - y(t)| < \epsilon. \tag{12.12}$$

To express this in more compact form, let I be the identity map on $[0,1]$ and use the notation (12.5). Then the definition becomes

$$d(x,y) = \inf_{\lambda \in \Lambda} \{\|\lambda - I\| \vee \|x - y\lambda\|\}. \tag{12.13}$$

By (12.5), $d(x,y)$ is finite (take $\lambda t \equiv t$). Of course, $d(x,y) \geq 0$; and $d(x,y) = 0$ implies that for each t either $x(t) = y(t)$ or $x(t) = y(t-)$, which in turn implies $x = y$. If λ lies in Λ, so does λ^{-1}; $d(x,y) = d(y,x)$ follows from (12.11) and (12.12). If λ_1 and λ_2 lie in Λ, so does their composition $\lambda_1\lambda_2$; the triangle inequality follows from $\|\lambda_1\lambda_2 - I\| \leq \|\lambda_1 - I\| + \|\lambda_2 - I\|$ together with $\|x - z\lambda_1\lambda_2\| \leq \|x - y\lambda_2\| + \|y - z\lambda_1\|$. Thus d is a metric.

This metric defines the Skorohod topology. The uniform distance $\|x - y\|$ between x and y may be defined as the infimum of those positive ϵ for which $\sup_t |x(t) - y(t)| < \epsilon$. The λ in (12.11) and (12.12) represents the uniformly small deformation of the time scale referred to above.

Elements x_n of D converge to a limit x in the Skorohod topology if and only if there exist functions λ_n in Λ such that $\lim_n x_n(\lambda_n t) = x(t)$ uniformly in t and $\lim_n \lambda_n t = t$ uniformly in t. If x_n goes uniformly to x, then there is convergence in the Skorohod topology (take $\lambda_n t \equiv t$). On the other hand, there is convergence $x_n = I_{[0,\alpha+1/n)} \to x = I_{[0,\alpha)}$ in the Skorohod topology ($0 \leq \alpha < 1$), whereas $x_n(t) \to x(t)$ fails in this case at $t = \alpha$. Since

$$|x_n(t) - x(t)| \leq |x_n(t) - x(\lambda_n t)| + |x(\lambda_n t) - x(t)|, \tag{12.14}$$

Skorohod convergence does imply that $x_n(t) \to x(t)$ holds for continuity points t of x and hence for all but countably many t. Moreover, it follows from (12.14) that if x is (uniformly) continuous on all of $[0,1]$, then Skorohod convergence implies uniform convergence: $\|x_n - x\| \leq \|x_n - x\lambda_n\| + w_x(\|\lambda_n - I\|)$. Therefore:

The Skorohod topology relativized to C coincides with the uniform topology there.

Example 12.1. Consider again $j(x)$, the maximum jump in x. Clearly, $|j(x) - j(y)| < \epsilon$ if $\|x - y\| < \epsilon/2$, and so $j(\cdot)$ is continuous in the uniform topology. But it is also continuous in the Skorohod topology: Since $j(y) = j(y\lambda)$ for each λ, $d(x, y) < \epsilon/2$ will imply for an appropriate λ that $|j(x) - j(y)| = |j(x) - j(y\lambda)| < \epsilon$. □

The space D is not complete under the metric d:

Example 12.2. Let x_n be the indicator of $[0, 1/2^n)$. Suppose that $\lambda_n(1/2^n) = 1/2^{n+1}$. If λ_n is linear on $[0, 1/2^n]$ and on $[1/2^n, 1]$, then $\|x_{n+1}\lambda_n - x_n\| = 0$ and $\|\lambda_n - I\| = 1/2^{n+1}$. On the other hand, if λ_n does not map $1/2^n$ to $1/2^{n+1}$, then $\|x_{n+1} - x_n\|$ is 1 instead of 0, and it follows that $d(x_n, x_{n+1}) = 1/2^{n+1}$. Therefore, $\{x_n\}$ is d-fundamental. Since $x_n(t) \to 0$ for $t > 0$ and the distance from x_n to the 0-function is 1, the sequence $\{x_n\}$ is not d-convergent.† □

We can define in D another metric d°—a metric that is equivalent to d (gives the Skorohod topology) but under which D is complete. Completeness facilitates characterizing the compact sets. The idea in defining d° is to require that the time-deformation λ in the definition of d be near the identity function in a sense which at first appears more stringent than (12.11); namely, we are going to require that the slope $(\lambda t - \lambda s)/(t - s)$ of each chord be close to 1 or, what is the same thing and analytically more convenient, that its logarithm be close to 0.

If λ is a nondecreasing function on $[0, 1]$ satisfying $\lambda 0 = 0$ and $\lambda 1 = 1$, put

$$\|\lambda\|^\circ = \sup_{s<t} \left| \log \frac{\lambda t - \lambda s}{t - s} \right|. \tag{12.15}$$

If $\|\lambda\|^\circ$ is finite, then the slopes of the chords of λ are bounded away from 0 and infinity and therefore it is both continuous and strictly increasing and hence is a member of Λ. Although $\|\lambda\|^\circ$ may be infinite even if λ does lie in Λ, these elements of Λ do not enter into the following definition.

Let $d^\circ(x, y)$ be the infimum of those positive ϵ for which Λ contains some λ such that $\|\lambda\|^\circ < \epsilon$ and (12.12) holds. In other words, let

$$d^\circ(x, y) = \inf_{\lambda \in \Lambda} \{\|\lambda\|^\circ \vee \|x - y\lambda\|\}. \tag{12.16}$$

---

† We can replace $[0, 1/2^n)$ by $[t_0, t_0 + 1/2^n)$ here, the point being that the example does not depend on special properties of 0 (like $\lambda 0 = 0$).

Since $|u-1| \le e^{|\log u|} - 1$ for $u > 0$,

(12.17) $$\sup_{0\le t\le 1} |\lambda t - t| = \sup_{0<t\le 1} t\left|\frac{\lambda t - \lambda 0}{t-0} - 1\right| \le e^{\|\lambda\|^\circ} - 1.$$

And since $v \le e^v - 1$ for all v, it follows by (12.13) that

(12.18) $$d(x,y) \le e^{d^\circ(x,y)} - 1.$$

Symmetry and the triangle inequality for d° follow from $\|\lambda^{-1}\|^\circ = \|\lambda\|^\circ$ and the inequality

(12.19) $$\|\lambda_1\lambda_2\|^\circ \le \|\lambda_1\|^\circ + \|\lambda_2\|^\circ.$$

That $d^\circ(x,y) = 0$ implies $x = y$ follows from (12.18) together with the corresponding property for d. Therefore, d° is a metric.

Because of (12.18), $d^\circ(x_n, x) \to 0$ implies $d(x_n, x) \to 0$. The reverse implication is a consequence of the following lemma.

Lemma 2. *If $d(x,y) < \delta^2$ and $\delta \le 1/2$, then $d^\circ(x,y) \le 4\delta + w'_x(\delta)$.*

In connection with Example 12.1, note that, if $\lambda_n(1/2^n) = 1/2^{n+1}$, then the chord from $(0, \lambda_n 0)$ to $(1/2^n, \lambda_n(1/2^n))$ has slope 1/2, from which it follows that $d^\circ(x_n, x_{n+1}) = \|\lambda_n\|^\circ = \log 2$. Therefore, $\{x_n\}$ is not d°-fundamental, as must be the case if D is to be d°-complete. And if δ is just greater than $\sqrt{1/2^n}$, then $d(x_n, x_{n+1}) < \delta^2$, and the lemma applies; but since $w'_{x_n}(\delta) = 1$, all that follows from this is the fact that $\log 2 \le 2^{2-2/n} + 1$.

PROOF. Take $\epsilon < \delta$ and then take $\{t_i\}$ to be a δ-sparse set satisfying $w_x[t_{i-1}, t_i) < w'_x(\delta) + \epsilon$ for each i. Now choose from Λ a μ such that

(12.20) $$\sup_t |x(t) - y(\mu t)| = \sup_t |x(\mu^{-1}t) - y(t)| < \delta^2$$

and

(12.21) $$\sup_t |\mu t - t| < \delta^2.$$

We want to define in Λ a λ that will be near μ but will not, as μ may, have chords with slopes far removed from 1. Take λ to agree with μ at the points t_i and to be linear in between. Since the composition

$\mu^{-1}\lambda$ fixes the t_i and is increasing, t and $\mu^{-1}\lambda t$ always lie in the same subinterval $[t_{i-1}, t_i)$. Therefore, by (12.20) and the choice of $\{t_i\}$,

$$|x(t) - y(\lambda t)| \le |x(t) - x(\mu^{-1}\lambda t)| + |x(\mu^{-1}\lambda t) - y(\lambda t)| \\ < w'_x(\delta) + \epsilon + \delta^2 < 4\delta + w'_x(\delta).$$

It is now enough to prove $\|\lambda\|^\circ \le 4\delta$. Since λ agrees with μ at the t_i, it follows by the condition (12.21) and the inequality $t_i - t_{i-1} > \delta$ that $|(\lambda t_i - \lambda t_{i-1}) - (t_i - t_{i-1})| < 2\delta^2 < 2\delta(t_i - t_{i-1})$. From this we can deduce that $|(\lambda t - \lambda s) - (t - s)| \le 2\delta|t - s|$ for all s and t: Because of the polygonal character of λ, this is clear for s and t in the same $[t_{i-1}, t_i]$. And it follows in the general case because, for every triple u_1, u_2, u_3,

$$|(\lambda u_3 - \lambda u_1) - (u_3 - u_1)| \le |(\lambda u_2 - \lambda u_1) - (u_2 - u_1)| \\ + |(\lambda u_3 - \lambda u_2) - (u_3 - u_2)|.$$

Therefore,

$$\log(1 - 2\delta) \le \log \frac{\lambda t - \lambda s}{t - s} \le \log(1 + 2\delta).$$

Since $|\log(1 \pm u)| \le 2|u|$ for $|u| \le 1/2$, it follows that $\|\lambda\|^\circ \le 4\delta$. □

By Lemmas 1 and 2 together, $d(x_n, x) \to 0$ implies $d^\circ(x_n, x) \to 0$; and as already observed, the reverse implication follows from (12.18):

Theorem 12.1. *The metrics d and d° are equivalent.*

Separability and Completeness of D

The following lemma will simplify several proofs. Suppose that the set $\sigma = \{s_u\}$ satisfies $0 = s_0 < s_1 < \cdots < s_k = 1$, and define a map $A_\sigma: D \to D$ in the following way. Take $A_\sigma x$ to have the constant value $x(s_{u-1})$ over the interval $[s_{u-1}, s_u)$ for $1 \le u \le k$ and to agree with x at $t = 1$. If x is sufficiently regular, then $A_\sigma x$ will be close to x in the metric d:

Lemma 3. *If* $\max(s_u - s_{u-1}) \le \delta$, *then* $d(A_\sigma x, x) \le \delta \vee w'_x(\delta)$.

PROOF. Write $A_\sigma x = \hat{x}$ to simplify the notation. Let ζt be the s_v "just to the left" of t, in the sense that $\zeta 1 = s_k = 1$ and $\zeta t = s_{u-1}$ for $t \in [s_{u-1}, s_u)$. Then $\hat{x}(t) = x(\zeta t)$. Given ϵ, find a δ-sparse set $\{t_i\}$ such that $w_x[t_{i-1}, t_i) < w'_x(\delta) + \epsilon$ for each i. Let λt_i be the s_v "just to the right" of t_i, in the sense that $\lambda t_0 = s_0 = 0$ and $\lambda t_i = s_v$ for

$t_i \in (s_{v-1}, s_v]$. Since $t_i - t_{i-1} > \delta \geq s_v - s_{v-1}$, the λt_i increase with i. Extend λ to an element of Λ by interpolating linearly between the t_i. Obviously, $\sup_t |\lambda t - t| \leq \delta$.

It is now enough to show that $|\hat{x}(t) - x(\lambda^{-1}t)| = |x(\zeta t) - x(\lambda^{-1}t)| \leq w'_x(\delta) + \epsilon$ (let $\epsilon \to 0$ at the end). This holds if t is 0 or 1, and it is enough to show that, for $0 < t < 1$, ζt and $\lambda^{-1}t$ always lie in the same $[t_{i-1}, t_i)$. To prove this it is enough to show that $t_j \leq \zeta t$ is equivalent to $t_j \leq \lambda^{-1}t$ (and hence $\zeta t < t_j$ is equivalent to $\lambda^{-1}t < t_j$). The case $t_j = 0$ being trivial, suppose that $t_j \in (s_{v-1}, s_v]$. In this case, since ζt is one of the s_i, $t_j \leq \zeta t$ is equivalent to $s_v \leq \zeta t$, which in turn is, by the definition of ζ, equivalent to $s_v \leq t$. But from $t_j \in (s_{v-1}, s_v]$ it also follows that $\lambda t_j = s_v$, and hence $s_v \leq t$ is equivalent to $\lambda t_j \leq t$, or $t_j \leq \lambda^{-1}t$. □

Theorem 12.2. *The space D is separable under d and d° and is complete under d°.*

PROOF. *Separability.* Since d and d° are equivalent and separability is a topological property, we can work with d. Let B_k be the set of functions having a constant, rational value over each $[(u-1)/k, u/k)$ and a rational value at $t = 1$. Then $B = \bigcup_k B_k$ is countable. Given x and ϵ, choose k so that $k^{-1} < \epsilon$ and $w'_x(k^{-1}) < \epsilon$. Apply Lemma 3 with $\sigma = \{u/k\}$: $d(x, A_\sigma x) < \epsilon$. Clearly, $d(A_\sigma x, y) < \epsilon$ for some y in B_k: $d(x, y) < 2\epsilon$ and $y \in B$.

Completeness. It is enough to show that each d°-fundamental sequence contains a subsequence that is d°-convergent. If $\{x_k\}$ is d°-fundamental, it contains a subsequence $\{y_n\} = \{x_{k_n}\}$ such that $d^\circ(y_n, y_{n+1}) < 1/2^n$. Then Λ contains a μ_n for which

$$\|\mu_n\|^\circ < 1/2^n \tag{12.22}$$

and

$$\sup_t |y_n(t) - y_{n+1}(\mu_n t)| = \sup_t |y_n(\mu_n^{-1}t) - y_{n+1}(t)| < 1/2^n. \tag{12.23}$$

The problem is to find a function y in D and functions λ_n in Λ for which $\|\lambda_n\|^\circ \to 0$ and $y_n(\lambda_n^{-1}t) \to y(t)$ uniformly in t.

A heuristic argument: Suppose that $y_n(\lambda_n^{-1}t)$ does go uniformly to a limit $y(t)$. By (12.23), $y_n(\mu_n^{-1}\lambda_{n+1}^{-1}t)$ is within $1/2^n$ of $y_{n+1}(\lambda_{n+1}^{-1}t)$, and therefore it, like $y(\lambda_n^{-1}t)$, must go uniformly to $y(t)$. This suggests trying to choose λ_n in such a way that $y_n(\mu_n^{-1}\lambda_{n+1}^{-1}t)$ is in fact identically equal to $y_n(\lambda_n^{-1}t)$, or $\mu_n^{-1}\lambda_{n+1}^{-1} = \lambda_n^{-1}$, or $\lambda_n = \lambda_{n+1}\mu_n = \lambda_{n+2}\mu_{n+1}\mu_n = \cdots$, and this in turn suggests trying to define λ_n as an infinitely iterated composition: $\lambda_n = \cdots \mu_{n+1}\mu_n$. If this idea works at all, λ_n should be near the identity for large n, just as the tail of a convergent infinite product is near 1.

Since $e^u - 1 \le 2u$ for $0 \le u \le 1/2$, it follows by (12.17) and (12.22) that

$$\sup_t |\mu_{n+m+1}\mu_{n+m}\cdots\mu_{n+1}\mu_n t - \mu_{n+m}\cdots\mu_{n+1}\mu_n t|$$
$$= \sup_s |\mu_{n+m+1}s - s| \le 2\|\mu_{n+m+1}\|^\circ < 1/2^{n+m}.$$

This means that, for fixed n, the functions $\mu_{n+m}\cdots\mu_{n+1}\mu_n t$ are uniformly fundamental as $m \to \infty$. Therefore, the sequence converges uniformly to a limit

$$\lambda_n t = \lim_m \mu_{n+m}\cdots\mu_{n+1}\mu_n t. \tag{12.24}$$

The function λ_n is continuous and nondecreasing and fixes 0 and 1. If we prove that $\|\lambda_n\|^\circ$ is finite, it will follow that λ_n is strictly increasing and hence is a member of Λ. By (12.19) and (12.22),

$$\left|\log\frac{\mu_{n+m}\cdots\mu_{n+1}\mu_n t - \mu_{n+m}\cdots\mu_{n+1}\mu_n s}{t-s}\right| \le \|\mu_{n+m}\cdots\mu_{n+1}\mu_n\|^\circ$$
$$\le \|\mu_{n+m}\|^\circ + \cdots + \|\mu_{n+1}\|^\circ + \|\mu_n\|^\circ \le 1/2^{n-1}.$$

Letting $m \to \infty$ in the first member of this inequality shows that $\|\lambda_n\|^\circ \le 1/2^{n-1}$; in particular, $\|\lambda_n\|^\circ$ is finite and $\lambda_n \in \Lambda$.

From (12.24) follows $\lambda_n = \lambda_{n+1}\mu_n$, which is the same thing as $\lambda_{n+1}^{-1} = \mu_n\lambda_n^{-1}$. Therefore, by (12.23),

$$\sup_t |y_n(\lambda_n^{-1}t) - y_{n+1}(\lambda_{n+1}^{-1}t)| = \sup_s |y_n(s) - y_{n+1}(\mu_n s)| < 1/2^n.$$

It follows that the functions $y_n(\lambda_n^{-1}t)$, which are elements of D, are uniformly fundamental and hence converge uniformly to a limit function $y(t)$. It is easy to show that y must be an element of D. Since $\|\lambda_n\|^\circ \to 0$, y_n is d°-convergent to y. □

It is interesting to observe that if d° is replaced by d and $\|\lambda\|^\circ$ is temporarily redefined as $\sup_t |\lambda t - t|$, then this proof of completeness goes through word for word, except at one place: $\sup_t |\lambda_n t - t| \le 1/2^{n-1}$ does not imply that λ_n is strictly increasing. In fact, if μ_n carries $1/2^n$ to $1/2^{n+1}$ and is linear over

$[0, 1/2^n]$ and $[1/2^n, 1]$ (this is the function of Example 12.1 in new notation), then $\mu_{n+m} \cdots \mu_{n+1}\mu_n(1/2^n) = 1/2^{n+m+1}$; (12.24) now implies that $\lambda_n(1/2^n) = 0$.

Compactness in D

We turn now to the problem of characterizing compact sets in D. Using the modulus $w'_x(\delta)$ defined by (12.6), we can prove an analogue of Theorem 7.2, the Arzelà-Ascoli theorem:

Theorem 12.3. *A necessary and sufficient condition for a set A to be relatively compact in the Skorohod topology is that*

$$\sup_{x \in A} \|x\| < \infty \tag{12.25}$$

and

$$\lim_{\delta \to 0} \sup_{x \in A} w'_x(\delta) = 0. \tag{12.26}$$

This theorem differs from the Arzelà-Ascoli theorem in that for no single t do $\sup_{x \in A} |x(t)| < \infty$ and (12.26) together imply (12.25) (consider $x_n = nI_{[\alpha,1)}$). The important part of the theorem is the sufficiency, which will be used to prove tightness.

PROOF OF SUFFICIENCY. Let α be the supremum in (12.25). Given ϵ, choose a finite ϵ-net H in $[-\alpha, \alpha]$ and choose δ so that $\delta < \epsilon$ and $w'_x(\delta) < \epsilon$ for all x in A. Apply Lemma 3 for any $\sigma = \{s_u\}$ satisfying $\max(s_u - s_{u-1}) < \delta$: $x \in A$ implies $d(x, A_\sigma x) < \epsilon$. Take B to be the finite set of y that assume on each $[s_{u-1}, s_u)$ a constant value from H and satisfy $y(1) \in H$. Since B obviously contains a y for which $d(A_\sigma x, y) < \epsilon$, it is a finite 2ϵ-net for A in the sense of d. Thus A is totally bounded in the sense of d.

But we must show that A is totally bounded in the sense of d°, since this is the metric under which D is complete. Given (a new) ϵ, choose (a new) δ so that $0 < \delta \leq 1/2$ and so that $4\delta + w'_x(\delta) < \epsilon$ holds for all x in A. We have already seen that A is d-totally bounded, and so there exists a finite set B' that is a δ^2-net for A in the sense of d. But then, by Lemma 2, B' is an ϵ-net for A in the sense of d°. □

The proof of necessity requires a lemma.

Lemma 4. *For fixed δ, $w'(x, \delta)$ is upper-semicontinuous in x.*

PROOF. Let x, δ, and ϵ be given. Let $\{t_i\}$ be a δ-sparse set such that $w_x[t_{i-1}, t_i) < w'_x(\delta) + \epsilon$ for each i. Now choose η small enough that $\delta+2\eta < \min(t_i - t_{i-1})$ and $\eta < \epsilon$. Suppose that $d(x,y) < \eta$. Then, for some λ in Λ, we have $\sup_t |y(t) - x(\lambda t)| < \eta$ and $\sup_t |\lambda^{-1}t - t| < \eta$. Let $s_i = \lambda^{-1}t_i$. Then $s_i - s_{i-1} > t_i - t_{i-1} - 2\eta > \delta$. Moreover, if s and t both lie in $[s_{i-1}, s_i)$, then λs and λt both lie in $[t_{i-1}, t_i)$ and hence $|y(s) - y(t)| < |x(\lambda s) - x(\lambda t)| + 2\eta \le w'_x(\delta) + \epsilon + 2\eta$. Thus $d(x,y) < \eta$ implies $w'_y(\delta) < w'_x(\delta) + 3\epsilon$. □

PROOF OF NECESSITY IN THEOREM 12.3. If A^- is compact, then it is d-bounded (has finite diameter), and since $\sup_t |x(t)|$ is the d-distance from x to the 0-function, (12.25) follows.

By Lemma 1, $w'(x,\delta)$ goes to 0 with δ for each x. But since $w'(\,\cdot\,,\delta)$ is upper semicontinuous, the convergence is uniform on compact sets [M8]. □

For the general range space V, write $\sup_t v(x(t), y(t))$ in place of $\|x - y\|$ in all the definitions and arguments (it is mostly a matter of using the triangle inequality in V rather than on the line); $\|\lambda - I\|$ and $\|\lambda\|^\circ$ need no change. In Example 12.1, take x_n to have value a on $[0, 1/2^n)$ and value b on $[1/2^n, 1]$, where a and b are distinct points of V. In the argument for separability, any countable set dense in V can replace the rationals. The completeness argument, which uses the assumed completeness of V, goes through as before. In place of (12.25) in the condition for compactness, assume that the set $[x(t): x \in A, t \in [0,1]]$ has compact closure, and in the proof of sufficiency use an ϵ-net from this set.

A Second Characterization of Compactness

Although the modulus $w'_x(\delta)$ leads to a complete characterization of compactness, a different one is in some ways easier to work with. This is

$$w''_x(\delta) = w''(x,\delta) = \sup_{\substack{t_1 \le t \le t_2 \\ t_2 - t_1 \le \delta}} \{|x(t) - x(t_1)| \wedge |x(t_2) - x(t)|\}, \tag{12.27}$$

the supremum extending over all triples t_1, t, t_2 in $[0,1]$ satisfying the constraints. Suppose that $w'_x(\delta) < w$, and let $\{\tau_i\}$ be a δ-sparse set such that $w_x[\tau_{i-1}, \tau_i) < w$ for all i. If $t_2 - t_1 \le \delta$, then t_1 and t_2 cannot lie on opposite sides of any of the subintervals $[\tau_{i-1}, \tau_i)$, and therefore, if $t_1 \le t \le t_2$, either t_1 and t lie in a common subinterval or else t and t_2 do, so that either $|x(t) - x(t_1)| < w$ or else $|x(t_1) - x(t)| < w$. Therefore (let $w \downarrow w'_x(\delta)$),

$$w''_x(\delta) \le w'_x(\delta). \tag{12.28}$$

There can be no inequality in the opposite direction: For the functions

$$x_n = I_{[0,n^{-1})}, \quad x_n = I_{[1-n^{-1},1]}, \tag{12.29}$$

we have $w''_{x_n}(\delta) \equiv 0$, whereas $w'_{x_n}(\delta) = 1$ for $n \geq \delta^{-1}$. In neither case does $\{x_n\}$ have compact closure, and so there can be no compactness condition involving (in addition to (12.25)) a restriction on $w''_x(\delta)$ alone. It is possible, however, to formulate a condition in terms of $w''_x(\delta)$ and the behavior of x near 0 and 1.

Theorem 12.4. *A necessary and sufficient condition for a set A to have compact closure in the Skorohod topology is that it satisfy* (12.25) *and*

$$\begin{cases} \lim_{\delta\to 0} \sup_{x\in A} w''(\delta) = 0, \\ \lim_{\delta\to 0} \sup_{x\in A} |x(\delta) - x(0)| = 0, \\ \lim_{\delta\to 0} \sup_{x\in A} |x(1-) - x(1-\delta)| = 0. \end{cases} \tag{12.30}$$

PROOF. It is enough to show that (12.30) is equivalent to (12.26). That (12.30) follows from (12.26) is clear from (12.28). In fact,

$$w''_x(\delta) \vee |x(\delta) - x(0)| \vee |x(1-) - x(1-\delta)| \leq w'_x(2\delta). \tag{12.31}$$

We can prove the reverse implication by showing that

$$w'_x\left(\frac{1}{2}\delta\right) \leq 24\{w''_x(\delta) \vee |x(\delta) - x(0)| \vee |x(1-) - x(1-\delta)|\}. \tag{12.32}$$

We first show that

$$|x(s) - x(t_1)| \vee |x(t_2) - x(t)| \leq 2w''_x(\delta) \quad \text{if } t_1 \leq s \leq t \leq t_2,\ t_2 - t_1 \leq \delta. \tag{12.33}$$

Indeed, if $|x(s) - x(t_1)| > w''_x(\delta)$, then, by the definition, $|x(t) - x(s)| \leq w''_x(\delta)$ and $|x(t_2) - x(s)| \leq w''_x(\delta)$, so that $|x(t_2) - x(t)| \leq 2w''_x(\delta)$.

Next,

$$w_x[t_1, t_2) \leq 2(w''_x(\delta) + |x(t_2) - x(t_1)|) \quad \text{if } t_2 - t_1 \leq \delta. \tag{12.34}$$

To see this, note that, if $t_1 \leq t < t_2$ and $|x(t) - x(t_1)| > w''_x(\delta)$, then $|x(t_2) - x(t)| \leq w''_x(\delta)$, and therefore, $|x(t) - x(t_1)| \leq |x(t) - x(t_2)| + |x(t_2) - x(t_1)| \leq w''_x(\delta) + |x(t_2) - x(t_1)|$. Hence (12.34).

From (12.34) it follows that $w_x[0, \delta) \leq 2(w''_x(\delta) + |x(\delta) - x(0)|)$ and $w_x[1-\delta, 1) \leq 2(w''_x(\delta) + |x(1-) - x(1-\delta)|)$ (for the second inequality, take $t_1 = 1 - \delta$ and let t_2 increase to 1). Because of these two inequalities, (12.32) will follow if we show that

$$w'_x\left(\frac{1}{2}\delta\right) \leq 6\{w''_x(\delta) \vee w_x[0, \delta) \vee w_x[1-\delta, 1)\}. \tag{12.35}$$

Let α exceed the maximum on the right in (12.35). It will be enough to show that $w'_x(\delta/2) \leq 6\alpha$. Suppose that x has jumps exceeding 2α at u_1 and u_2. If $u_2 - u_1 < \delta$, then there are disjoint intervals (t_1, s) and (t, t_2) containing u_1 and u_2, respectively, and satisfying $t_2 - t_1 < \delta$, and if these intervals are short enough, we have a contradiction with (12.33). Thus $[0, 1]$ cannot contain two points, within δ of one another, at each of which x jumps by more than 2α. And neither $[0, \delta)$ nor $[1 - \delta, 1)$ can contain a point at which x jumps by more than 2α.

Thus there exist points s_i, satisfying $0 = s_0 < s_1 < \cdots < s_r = 1$, such that $s_i - s_{i-1} \geq \delta$ and such that any point at which x jumps by more than 2α is one of the s_i. If $s_j - s_{j-1} > \delta$ for a pair of adjacent points, enlarge the system $\{s_i\}$ by including their midpoint. Continuing in this way leads to a new, enlarged system $s_0, \ldots, s_r$ (with a new r) that satisfies

$$\frac{1}{2}\delta < s_i - s_{i-1} \leq \delta, \quad i = 1, 2, \ldots, r.$$

Now (12.35) will follow if we show that

$$w_x[s_{i-1}, s_i) \leq 6\alpha \tag{12.36}$$

for each i. Suppose that $s_{i-1} \leq t_1 < t_2 < s_i$. Then $t_2 - t_1 < \delta$. Let σ_1 be the supremum of those points σ in $[t_1, t_2]$ for which the inequality $\sup_{t_1 \leq u \leq \sigma} |x(u) - x(t_1)| \leq 2\alpha$ holds. Let σ_2 be the infimum of those σ in $[t_1, t_2]$ for which $\sup_{\sigma \leq u \leq t_2} |x(t_2) - x(u)| \leq 2\alpha$. If $\sigma_1 < \sigma_2$, then there exist points s just to the right of σ_1 satisfying $|x(s) - x(t_1)| > 2\alpha$ and there exist points t just to the left of σ_2 satisfying $|x(t_2) - x(t)| > 2\alpha$; since we can arrange that $s < t$, this contradicts (12.33). Therefore, $\sigma_2 \leq \sigma_1$ and it follows that $|x(\sigma_1-) - x(t_1)| \leq 2\alpha$ and $|x(t_2) - x(\sigma_1)| \leq 2\alpha$. Since $t_1 < \sigma_1 \leq t_2$, σ_1 is interior to (s_{i-1}, s_i), and so the jump at σ_1 is at most 2α. Thus $|x(t_2) - x(t_1)| \leq 6\alpha$. This establishes (12.36), hence (12.35), and hence (12.32), which proves the theorem. □

Finite-Dimensional Sets

Finite-dimensional sets play in D the same role they do in C. For $0 \leq t_1 < \cdots < t_k \leq 1$, define the natural projection $\pi_{t_1 \cdots t_k}$ from D to R^k as usual:

$$\pi_{t_1 \cdots t_k}(x) = (x(t_1), \ldots x(t_k)). \tag{12.37}$$

Since each function in Λ fixes 0 and 1, π_0 and π_1 are continuous. Suppose that $0 < t < 1$. If points x_n converge to x in the Skorohod topology and x is continuous at t, then (see (12.14)) $x_n(t) \to x(t)$. Suppose, on the other hand, that x is discontinuous at t. If λ_n is the element of Λ that carries t to $t - 1/n$ and is linear on $[0, t]$ and $[t, 1]$, and if $x_n(s) \equiv x(\lambda_n s)$, then x_n converges to x in the Skorohod topology but $x_n(t)$ does not converge to $x(t)$. Therefore: *If* $0 < t < 1$, *then* π_t *is continuous at* x *if and only if* x *is continuous at* t.

We must prove that $\pi_{t_1 \cdots t_k}$ is measurable with respect to the Borel σ-field $\mathcal{D}$. We need consider only a single time point t (since a mapping into R^k is measurable if each component mapping is), and we may assume $t < 1$ (since π_1 is continuous). Let $h_\epsilon(x) = \epsilon^{-1} \int_t^{t+\epsilon} x(s)\, ds$. If $x_n \to x$ in the Skorohod topology, then $x_n(s) \to x(s)$ for continuity points s of x and hence for points s outside a set of Lebesgue measure 0; since the x_n are uniformly bounded, $\lim_n h_\epsilon(x_n) = h_\epsilon(x)$ follows. Thus $h_\epsilon(\,\cdot\,)$ is continuous in the Skorohod topology. By right-continuity, $h_{m^{-1}}(x) \to \pi_t(x)$ for each x as $m \to \infty$. Therefore: *Each* π_t *is measurable.*

Having proved the $\pi_{t_1 \cdots t_k}$ measurable, we may, as in C, define in $\mathcal{D}$ the class $\mathcal{D}_f$ of finite-dimensional sets—those of the form $\pi_{t_1 \cdots t_k}^{-1} H$ for $k \geq 1$ and $H \in \mathcal{R}^k$. If T is a subset of $[0, 1]$, let $p[\pi_t : t \in T]$ be the class of sets $\pi_{t_1 \cdots t_k}^{-1} H$, where k is arbitrary, the t_i are points of T, and $H \in \mathcal{R}^k$; these are the finite-dimensional sets based on time-points in T. Then $p[\pi_t : t \in T]$ is a π-system (see the argument in Example 1.3). Let $\sigma[\pi_t : t \in T]$ be the σ-field generated by the real functions π_t for $t \in T$; it is also generated by the class $p[\pi_t : t \in T]$.

Theorem 12.5. (i) *The projections* π_0 *and* π_1 *are continuous; for* $0 < t < 1$, π_t *is continuous at* x *if and only if* x *is continuous at* t.
(ii) *Each* π_t *is measurable* $\mathcal{D}/\mathcal{R}^1$, *and each* $\pi_{t_1 \cdots t_k}$ *is measurable* $\mathcal{D}/\mathcal{R}^k$.
(iii) *If* T *contains* 1 *and is dense in* $[0, 1]$, *then* $\sigma[\pi_t : t \in T] = \mathcal{D}$ *and* $p[\pi_t : t \in T]$ *is a separating class.*

PROOF. Only (iii) remains to be proved. By right-continuity and the assumption that T is dense, it follows that π_0 is measurable with respect to $\sigma[\pi : t \in T]$, and so we may as well assume that T contains 0. For each m, choose points $s_0, \ldots, s_k$ of T in such a way that $0 = s_0 < \cdots < s_k = 1$ and $\max(s_u - s_{u-1}) < m^{-1}$, and take $\sigma_m = \{s_u\}$; here k and the s_u are functions of m. For $\alpha = (\alpha_0, \ldots, \alpha_k)$ in R^{k+1}, let $V_m \alpha$ be the element of D taking the constant value α_{u-1} on $[s_{u-1}, s_u)$, $1 \leq u \leq k$, and taking the value α_k at $t = 1$. Clearly, $V_m : R^{k+1} \to D$ is continuous (relative to the Euclidean and Skorohod topologies) and hence is measurable with respect to $\mathcal{R}^{k+1}/\mathcal{D}$. Since $\pi_{s_0 \cdots s_k}$ is measurable $\sigma[\pi_t: t \in T]/\mathcal{R}^{k+1}$, the composition $V_m \pi_{s_0 \cdots s_k}$ is measurable $\sigma[\pi_t: t \in T]/\mathcal{D}$. But this composition is just the map A_{σ_m} of Lemma 3, which is therefore measurable $\sigma[\pi_t: t \in T]/\mathcal{D}$. Since $d(x, A_{\sigma_m}) \leq m^{-1} \vee w'_x(m^{-1})$ by the lemma, $x = \lim_m A_{\sigma_m} x$ for each x. This means [M10] that the identity function is measurable $\sigma[\pi_t: t \in T]/\mathcal{D}$, which implies $\mathcal{D} \subset \sigma[\pi_t: t \in T]$. Since $p[\pi_t: t \in T]$ is therefore a π-system generating $\mathcal{D}$, it is a separating class. □

It follows from Lemma 4 that $w'(\cdot, \delta)$ is $\mathcal{D}$-measurable. Since the functions in D are right-continuous, the supremum in (12.27) is unchanged if t_1, t, t_2 are restricted to the rationals; since the individual π_t are $\mathcal{D}$-measurable, it follows that $w''(\cdot, \delta)$ is also $\mathcal{D}$-measurable.

Random Functions in D

Just as in the case of C, the finite-dimensional distributions of a probability measure P on $(D, \mathcal{D})$ are the measures $P\pi_{t_1 \cdots t_k}^{-1}$. Since the projections are not everywhere continuous on D, there is this difference with C: $P_n \Rightarrow P$ does not always imply $P_n \pi_{t_1 \cdots t_k}^{-1} \Rightarrow P\pi_{t_1 \cdots t_k}^{-1}$; see the next section. On the other hand, D is like C in that if the finite-dimensional distributions do converge weakly, the measures on D they come from may not; in fact, Example 2.5 applies to D just as well as to C.

If $(\Omega, \mathcal{F}, \mathsf{P})$ is a probability space and X maps Ω into D, then X is a random element of D—in the sense that it is measurable $\mathcal{F}/\mathcal{D}$—if and only if each $X_t(\omega) = \pi_t(X(\omega))$ defines a random variable; the argument is as for C (see p. 84). Finally, each coordinate function $\pi_t(x) = x(t) = x_t$ can be viewed as a random variable on $(D, \mathcal{D})$ (p. 86).

The Poisson Limit*

A probability measure P_α on D describes a Poisson process with rate α if the increments are independent under P_α and have Poisson distributions:

$$P_\alpha[x_t - x_s = i] = e^{-\alpha(t-s)} \frac{(\alpha(t-s))^i}{i!}. \tag{12.38}$$

To prove the existence of such a measure, first construct the corresponding random function: Call a function in D a *count path* if it is nondecreasing, takes integers as values, and has jumps of exactly 1 at its points of discontinuity. Let D_c be the set of count paths. The idea is that the elements of D_c are the possible paths for a point process, a process that counts events. Take a Poisson process $[X_t: 0 \leq t \leq 1]$, arrange that each sample path $X(\cdot,\omega)$ lies in D_c [PM.297], and let P_α be the distribution of $X(\omega)$.

The following theorem shows that weak convergence to P_α is just a matter of the convergence of the finite-dimensional distributions. Let P_n, P be probability measures on $(D,\mathcal{D})$.

Theorem 12.6. *Suppose that $E \in \mathcal{D}$ and T_0 is a countable, dense set in $[0,1]$. Suppose further that, if $x, x_n \in E$ and $x_n(t) \to x(t)$ for $t \in T_0$, then $x_n \to x$ in the Skorohod topology. If $P_nE = PE = 1$ and $P_n\pi^{-1}_{t_1,\ldots,t_k} \Rightarrow_n P\pi^{-1}_{t_1,\ldots,t_k}$ for all k-tuples in T_0, then $P_n \Rightarrow_n P$.*

The set D_c of count paths satisfies the hypothesis on E, as will be shown.

PROOF. The idea is, in effect, to embed E in R^∞ with the product topology; see Example 2.4. Let $T_0 = \{t_1, t_2, \ldots\}$ and define $\pi\colon D \to R^\infty$ by $\pi(x) = (x(t_1), x(t_2), \ldots)$. If π_k is the natural projection from R^∞ to R^k, defined by $\pi_k(z_1, z_2, \ldots) = (z_1, \ldots, z_k)$, then $\pi_k\pi$ is the natural projection $\pi_{t_1\cdots t_k}$ from D to R^k. Therefore, $\pi^{-1}(\pi_k^{-1}H) = \pi^{-1}_{t_1\cdots t_k}H \in \mathcal{D}$ for $H \in R^k$, and since the sets $\pi_k^{-1}H$, the elements of $\mathcal{R}^\infty_f$, generate $\mathcal{R}^\infty$, it follows that π is measurable $\mathcal{D}/\mathcal{R}^\infty$.

For $A \subset D$, define $A^* = \pi^{-1}(\pi A)^-$. Then $A^* \in \mathcal{D}$ and $A \subset A^*$. If x lies in $(A\cap E)^*$, so that $\pi x \in (\pi(A\cap E))^-$, then there is a sequence $\{x_n\}$ in $A\cap E$ such that $\pi x_n \to \pi x$. If x also lies in E, then by the hypothesis, x_n converges to x in the Skorohod topology. Therefore, $(A \cap E)^* \cap E$ is contained in the closure A^- of A in the Skorohod topology. Since $P_n\pi^{-1}_{t_1\cdots t_k} \Rightarrow_n \pi^{-1}_{t_1\cdots t_k}$ by hypothesis, and since $\pi_{t_1\cdots t_k} = \pi_k\pi$, we have $P_n\pi^{-1}\pi_k^{-1} \Rightarrow_n P\pi^{-1}\pi_k^{-1}$. But since in R^∞ weak convergence is the same thing as weak convergence of the finite-dimensional distributions, it follows that $P_n\pi^{-1} \Rightarrow P\pi^{-1}$ on R^∞. Therefore, if $A \in \mathcal{D}$,

$$\begin{aligned}\limsup_n P_nA &\le \limsup_n P_nA^* \\ &= \limsup_n P_n\pi^{-1}(\pi A)^- \le P\pi^{-1}(\pi A)^- = PA^*.\end{aligned}$$

And now, since $P_nE = PE = 1$ and $(A\cap E)^* \cap E \subset A^-$,

$$\begin{aligned}\limsup_n P_nA &= \limsup_n P_n(A\cap E) \\ &\le P(A\cap E)^* = P((A\cap E)^*\cap E) \le PA^-.\end{aligned}$$

Therefore, $P_n \Rightarrow P$. □

It is not hard to see that the set D_c of count paths is closed in the Skorohod topology. Let T_0 be any countable, dense set in $[0,1]$, and suppose that $x, x_n \in D_c$ and that, for each t in T_0, $x_n(t) \to x(t)$ (which means that $x_n(t) = x(t)$ for large n). A function in D_c has only finitely many discontinuities; order those for x: $0 < t_1 < \cdots < t_k \le 1$. For a given ϵ, choose points u_i and v_i in T_0 in such a way that $u_i < t_i \le v_i$, $v_i - u_i < \epsilon$, and the intervals $[v_{i-1}, u_i]$ are disjoint. Then, for n exceeding some n_0, x_n agrees with x over each $[v_{i-1}, u_i]$ and has a single jump in each $[u_i, v_i]$. If λ_n (in Λ) carries t_i to the point in $[u_i, v_i]$ where x_n has a jump (and is defined elsewhere by linearity, say), then $\sup_t |\lambda_n t - t| \le \epsilon$ and $x_n(\lambda_n t) \equiv x(t)$. Therefore, x_n converges to x in the Skorohod topology, and so D_c satisfies the hypothesis on E in Theorem 12.6.

Example 12.3. Suppose that, for each n, $\xi_{n1}, \ldots, \xi_{nn}$ are independent and take the values 1 and 0 with probabilities α/n and $1 - \alpha/n$. Define a random function X^n by $X_t^n(\omega) = \sum_{i \le nt} \xi_{ni}(\omega)$. Then X^n has independent increments; since $X_t^n - X_s^n$ has the binomial distribution for $\lfloor nt \rfloor - \lfloor ns \rfloor$ trials, with probability of success α/n at each, it has in the limit the Poisson distribution with mean $\alpha(t-s)$. It follows easily from the theorem that $X^n \Rightarrow P_\alpha$. This can be extended in various ways, for example to the $\langle \lambda_i \omega \rangle$ of (3.34) and (3.35). □

Problems

12.1. Let D^+ be the class of functions on $[0,1]$ that have only discontinuities of the first kind in (0,1) and have right-hand limits at 0 and left-hand limits at 1. Convert D^+ into a pseudo-metric space in such a way that (i) x and y are at distance 0 if and only if they agree at their common continuity points and at 0 and 1, and (ii) D is isometric to the standard quotient space (see Kelley [41], p. 123).

12.2. Under the Skorohod topology and pointwise addition of functions, D is not a topological group.

12.3. The set C is nowhere dense in D.

12.4. Put

$$w_x'''(\delta) = \sup_{t_2 - t_1 < \delta} \sup_{t_1 < t < t_2} \{w_x(t_1, t) \wedge w_x(t, t_2)\}.$$

Show that

$$w_x'''(\delta) = \sup_{t_2 - t_1 < \delta} \inf_{t_1 < t < t_2} \{w_x(t_1, t) \vee w_x(t, t_2)\}$$

and

$$w_x''(\delta) \le w_x'''(\delta) \le 2 w_x''(\delta).$$

12.5. Suppose that ξ is uniformly distributed over $[\frac{1}{3}, \frac{2}{3}]$, and consider the random functions

$$X = 2I_{[\xi,1]}, \quad X^n = I_{[\xi - n^{-1}, 1]} + I_{[\xi + n^{-1}, 1]}.$$

Show that $X^n \not\Rightarrow X$, even though $X^n_{t_1 \cdots t_k} \Rightarrow X_{t_1 \cdots t_k}$ for all $t_1, \ldots, t_k$. Why does Theorem 12.6 not apply?

SECTION 13. WEAK CONVERGENCE AND TIGHTNESS IN D

Prove weak convergence in function space by proving weak convergence of the finite-dimensional distributions and then proving tightness—this was the technique so useful in C, and we want to adapt it to D. Since D is separable and complete under the metric d°, a family of probability measures on $(D, \mathcal{D})$ is relatively compact if and only if it is tight, and so there is no difficulty on that point. On the other hand, the fact that the natural projections are not continuous complicates matters somewhat.

Finite-Dimensional Distributions

For probability measures P on $(D, \mathcal{D})$, denote by T_P the set of t in $[0,1]$ for which the projection π_t is continuous except at points forming a set of P-measure 0. Since π_0 and π_1 are everywhere continuous, the points 0 and 1 always lie in T_P. If $0 < t < 1$, then π_t is continuous at x if and only if x is continuous at t, and it follows that $t \in T_P$ if and only if $PJ_t = 0$, where

$$J_t = [x \colon x(t) \neq x(t-)]. \tag{13.1}$$

(This equivalence holds only in the interior of the unit interval: 0 lies in T_P, and each function x is continuous at 0; 1 lies in T_P, but an x may or may not be continuous at 1.)

An element of D has at most countably many jumps. Let us prove the corresponding fact that $PJ_t > 0$ is possible for at most countably many t. Let $J_t(\epsilon) = [x \colon |x(t) - x(t-)| > \epsilon]$. For fixed, positive ϵ and δ, there can be at most finitely many t for which $P(J_t(\epsilon)) \geq \delta$, since if this held for infinitely many distinct t_n, then $P(\limsup_n J_{t_n}(\epsilon)) \geq \delta$ would follow, contradicting the fact that for a single x the saltus can exceed ϵ at only finitely many places. Since $P(J_t(\epsilon)) \uparrow PJ_t$ as $\epsilon \downarrow 0$, the result follows:

The set T_P contains 0 and 1, and its complement in $[0,1]$ is at most countable.

If $t_1, \ldots, t_k$ all lie in T_P, then $\pi_{t_1 \cdots t_k}$ is continuous on a set of P-measure 1, and it follows by the mapping theorem that

$$P_n \Rightarrow P \tag{13.2}$$

implies

$$P_n\pi_{t_1\cdots t_k}^{-1} \Rightarrow P\pi_{t_1\cdots t_k}^{-1}. \tag{13.3}$$

If some t_i lies outside T_P, then (13.3) may not follow from (13.2): Take P to be a unit mass at $I_{[0,t)}$ and P_n to be a unit mass at $I_{[0,t+n^{-1})}$.

Here is the analogue of Theorem 7.1:

Theorem 13.1. *If $\{P_n\}$ is tight, and if $P_n\pi_{t_1\cdots t_k}^{-1} \Rightarrow P\pi_{t_1\cdots t_k}^{-1}$ holds whenever $t_1, \ldots t_k$ all lie in T_P, then $P_n \Rightarrow P$.*

PROOF. By the corollary to Theorem 5.1, it is enough to show that, if a subsequence $\{P_{n_i}\}$ converges weakly to some Q, then Q must coincide with P. Assume that $P_{n_i} \Rightarrow_i Q$. If $t_1, \ldots, t_k$ lie in T_P, then $P_{n_i}\pi_{t_1\cdots t_k}^{-1} \Rightarrow_i P\pi_{t_1\cdots t_k}^{-1}$ by the hypothesis of the theorem. If $t_1, \ldots, t_k$ lie in T_Q, then $P_{n_i}\pi_{t_1\cdots t_k}^{-1} \Rightarrow_i Q\pi_{t_1\cdots t_k}^{-1}$ by the assumption. Therefore, if $t_1, \ldots, t_k$ lie in $T_P \cap T_Q$, then $P\pi_{t_1\cdots t_k}^{-1} = Q\pi_{t_1\cdots t_k}^{-1}$. But by Theorem 12.5, this implies that $P = Q$, because $T_P \cap T_Q$ contains 0 and 1 and has countable complement—since T_P and T_Q both do. □

Tightness

The analysis of tightness in C began with a result (Theorem 7.3) which substituted for compactness its Arzelá-Ascoli characterization. Theorem 12.3, which characterizes compactness in D, gives the following result. Let $\{P_n\}$ be a sequence of probability measures on $(D, \mathcal{D})$.

Theorem 13.2. *The sequence $\{P_n\}$ is tight if and only if these two conditions hold:*

(i) *We have*

$$\lim_{a\to\infty} \limsup_n P_n[x\colon \|x\| \ge a] = 0. \tag{13.4}$$

(ii) *For each ϵ,*

$$\lim_\delta \limsup_n P_n[x\colon w'_x(\delta) \ge \epsilon] = 0. \tag{13.5}$$

PROOF. Conditions (i) and (ii) here are exactly conditions (i) and (ii) of Theorem 7.3 with $\|x\|$ in place of $|x(0)|$ and w' in place of w. Since D is separable and complete, a single probability measure on D is tight, and so the previous proof goes through. □

There are two convenient alternative forms for (13.4). The first controls $\|x\|$ by controlling $|x(t)|$ for the individual values of t, the second by controlling $|x(0)|$ and $j(x)$ (defined by (12.8)).

Corollary. *Either of the following two conditions can be substituted for* (i) *in Theorem* 13.2:

(i′) *For each t in a set T that is dense in* $[0,1]$ *and contains* 1,

$$\lim_{a\to\infty}\limsup_n P_n[x: |x(t)| \ge a] = 0. \tag{13.6}$$

(i″) *The relation* (13.6) *holds for $t = 0$, and*

$$\lim_{a\to\infty}\limsup_n P_n[x: j(x) \ge a] = 0. \tag{13.7}$$

The proof will show that (i), (i′), and (i″) are all equivalent in the presence of (ii).

PROOF. It is clear that (i) implies (i′) and (i″). It is therefore enough to show that, in the presence of (ii), (i) follows from (i′) and also from (i″).

Assume (ii) and (i′). Let $\{t_0,\dots t_v\}$ be a δ-sparse set for which $w_x[t_{i-1},t_i) < w'_x(\delta)+1$, $i \le v$. Choose from T points s_j such that $0 = s_0 < s_1 < \cdots < s_k = 1$ and $s_j - s_{j-1} < \delta$. (The t_i depend on x, but the s_j do not.) If $m(x) = \max_{0\le j\le k}|x(s_j)|$, then, since each $[t_{i-1},t_i)$ contains an s_j, $\|x\| \le m(x) + w'_x(\delta) + 1$. If (13.5) and (13.6) hold, then there exist a δ and an a such that $P_n[x: w'_x(\delta) \ge 1] < \eta$ and $P_n[x: m(x) \ge a] < \eta$ for large n. But then $\mathsf{P}_n[x: \|x\| \ge a+2] < 2\eta$ for large n, which proves (13.4). Thus (ii) and (i′) imply (i).

Assume (ii) and (i″). Let $\{t_i\}$ be the δ-sparse set (depending on x) described above. Now $|x(t_i) - x(t_{i-1})| \le w'_x(\delta) + 1 + j(x)$, and so $\max_{i\le v}|x(t_i)| \le |x(0)| + v(w'_x(\delta)+1+j(x))$ and (since $\delta v \le 1$) $\|x\| \le |x(0)| + \delta^{-1}(w'_x(\delta)+1+j(x)) + w'_x(\delta)$. If (13.5) holds, there is a δ such that $P_n[x: w'_x(\delta) \ge 1] < \eta$ for large n. Further, if (13.6) holds for $t = 0$, and if (13.7) holds, then there is an a such that $P_n[x: |x(0)| \ge a] < \eta$ and $P_n[x: \delta^{-1}(2+j(x)) + 1 \ge a] < \eta$ for large n. And then $P_n[x: \|x\| \ge 2a] < 3\eta$ for large n. Thus (ii) and (i″) imply (i). □

We can use Theorem 12.4 in place of Theorem 12.3 here. In fact, the inequalities (12.31) and (12.32) make it clear that we can replace

(13.5) by the condition that, for each positive ϵ and η, there exist a δ, $0 < \delta < 1$, and an integer n_0 such that

$$(13.8) \qquad \begin{cases} P_n[x: w''_x(\delta) \geq \epsilon] \leq \eta, \\ P_n[x: |x(\delta) - x(0)| \geq \epsilon] \leq \eta, \\ P_n[x: |x(1-) - x(1-\delta)| \geq \epsilon] \leq \eta, \end{cases}$$

for $n \geq n_0$. As the next theorem shows, the second and third conditions in (13.8) are automatically satisfied in the most important cases.

Theorem 13.3. *Suppose that $P_n \pi_{t_1 \cdots t_k}^{-1} \Rightarrow P\pi_{t_1 \cdots t_k}^{-1}$ whenever the t_i all lie in T_P. Suppose further that, for every ϵ,*

$$(13.9) \qquad \lim_{\delta \to 0} P[x: |x(1) - x(1-\delta)| \geq \epsilon] = 0,$$

and that, for positive ϵ and η, there exist a δ, $0 < \delta < 1$, and an n_0 such that

$$(13.10) \qquad P_n[x: w''_n(\delta) \geq \epsilon] \leq \eta, \quad n \geq n_0.$$

Then $P_n \Rightarrow P$.

PROOF. We need only prove that $\{P_n\}$ is tight, and for this it is enough to check (13.8) and (13.6) with T_P in the role of T. For each t in T_P, the weakly convergent sequence $\{P_n \pi_t^{-1}\}$ is tight, which implies (13.6).

As for (13.8), we need consider only the second and third conditions, since of course (13.10) takes care of the first one. By right-continuity we have $P[x: |x(\delta) - x(0)| \geq \epsilon] < \eta$ for small enough δ; by hypothesis, $P_n \pi_{0,\delta}^{-1} \Rightarrow P\pi_{0,\delta}^{-1}$ if $\delta \in T_P$, and it then follows that the middle condition in (13.8) holds for all sufficiently large n. With one change, the symmetric argument works for the third condition. From the fact that cadlag functions have left-hand limits at 1, it follows that $P[x: |x(1-) - x(1-\delta)| \geq \epsilon] \to 0$ as $\delta \to 0$, and from (13.9) it follows further that $P[x: |x(1) - x(1-)| \geq \epsilon] = 0$—that is, $PJ_1 = 0$. Therefore, there is continuity from the left at 1 outside a set of P-measure 0, and the symmetric argument does go through. □

Let $\psi_t: D \to R^2$ carry x to $(x(t), x(1-))$. One can replace (13.9) in this theorem by the condition that $P_n \psi_t^{-1} \Rightarrow P\psi_t^{-1}$ for $t \in T_P$. But some restriction on the oscillations near 1 is needed: Consider the case where P_n is a unit mass at $I_{[1-n^{-1},1]}$.

Theorem 13.3 can be stated in terms of random elements of D; compare Theorem 7.5. The second proof of Theorem 7.5 makes very little use of the general theory of weak convergence, and the convergence-in-distribution version of Theorem 13.3 can be given a similar proof.[†]

Let X_n and X be random elements of D.

Theorem 13.4. *Suppose that $X_n \Rightarrow X$. Then $\mathsf{P}[X \in C] = 1$ if and only if $j(X_n) \Rightarrow 0$.*

PROOF. By Example 12.1 and the mapping theorem, $j(X_n) \Rightarrow j(X)$. And $j(X) = 0$ if and only if $X \in C$. □

A corollary shows what happens if we use the conditions for weak convergence in C.

Corollary. *Suppose the conditions (7.6) and (7.7) hold for probability measures on $(D, \mathcal{D})$. Suppose further that $P_n\pi_{t_1\cdots t_k}^{-1} \Rightarrow P\pi_{t_1\cdots t_k}^{-1}$ for all $t_1, \ldots, t_k$. Then $P_n \Rightarrow P$, and PC=1.*

PROOF. The proof of Theorem 7.3 shows that condition (i) of Theorem 13.2 holds, and by (12.7), condition (ii) holds as well. Therefore, $P_n \Rightarrow P$, or $X_n \Rightarrow X$ for the corresponding random functions. And since $j(X_n) \le w(X_n, \delta)$ for each δ, (7.7) implies that $j(X_n) \Rightarrow 0$. □

A Criterion for Convergence

Write T_X for T_P, where P is the distribution of X.

Theorem 13.5. *Suppose that*

$$(X_{t_1}^n, \ldots, X_{t_k}^n) \Rightarrow_n (X_{t_1}, \ldots, X_{t_k}) \tag{13.11}$$

for points t_i of T_X, that

$$X_1 - X_{1-\delta} \Rightarrow_{\delta \to 0} 0, \tag{13.12}$$

and that, for $r \le s \le t$, $n \ge 1$, and $\lambda > 0$,

$$\mathsf{P}[|X_s^n - X_r^n| \wedge |X_t^n - X_s^n| \ge \lambda] \le \frac{1}{\lambda^{4\beta}}[F(t) - F(r)]^{2\alpha}, \tag{13.13}$$

where $\beta \ge 0$ and $\alpha > \frac{1}{2}$, and F is a nondecreasing, continuous function on $[0, 1]$. Then $X^n \Rightarrow_n X$.

---

† See Problem 13.1.

There is a more restrictive version of (13.13) involving moments, namely

$$\mathsf{E}[|X_s^n - X_r^n|^{2\beta}|X_t^n - X_s^n|^{2\beta}] \le [F(t) - F(r)]^{2\alpha}. \tag{13.14}$$

PROOF. By Theorem 13.3, it is enough to show that for each positive ϵ and η there is a δ such that $\mathsf{P}[w''(X^n, \delta) \ge \epsilon] \le \eta$ for all n. We can prove this by applying Theorem 10.4 to each X^n, since the paths are right-continuous and $w''(X^n, \delta) = L(X^n, \delta)$. Let $T = [0, 1]$ and define μ by $\mu(s, t] = F(t) - F(s)$. With X^n in the role of γ, (10.20) is the same thing as (13.13). It follows by (10.21) that

$$\begin{aligned}\mathsf{P}[w''(X^n, \delta) \ge \epsilon] &\le \frac{2K}{\epsilon^{4\beta}}\mu[0, 1] \sup_{0 \le t \le 1-2\delta} \mu^{2\alpha-1}[t, t + 2\delta] \\ &= \frac{2K}{\epsilon^{4\beta}}\mu[0, 1](w_F(2\delta))^{2\alpha-1},\end{aligned}$$

where w_F is the modulus of continuity. Since F is uniformly continuous and $\alpha > \frac{1}{2}$, it is possible, for given ϵ and η, to choose δ so that the right side here is less than η. □

For applications of Theorem 13.5, see the next section.

A Criterion for Existence*

These ideas lead to a condition for the existence in D of a random element with specified finite-dimensional distributions. For each k-tuple $t_1, \ldots, t_k$, let $\mu_{t_1 \cdots t_k}$ let be a probability measure on $(R^k, \mathcal{R}^k)$, and assume that these measures satisfy the consistency conditions of Kolmogorov's existence theorem.

Theorem 13.6. *There exists in D a random element with finite-dimensional distributions $\mu_{t_1 \cdots t_k}$, provided these distributions are consistent; provided that, for $t_1 \le t \le t_2$,*

$$\mu_{t_1 t t_2}[(u_1, u, u_2)\colon |u - u_1| \wedge |u_2 - u| \ge \lambda] \le \frac{1}{\lambda^{4\beta}}[F(t_2) - F(t_1)]^{2\alpha}, \tag{13.15}$$

where $\beta \ge 0$, $\alpha > \frac{1}{2}$, and F is a nondecreasing, continuous function on $[0, 1]$; and provided that

$$\lim_{h \downarrow 0} \mu_{t, t+h}[(u_1, u_2)\colon |u_2 - u_1| \ge \epsilon] = 0, \quad 0 \le t < 1. \tag{13.16}$$

By right-continuity, (13.16) is in fact necessary for the existence of such a random element of D.

PROOF. For each n, consider the points $t_i^n = i/2^n$ for $i = 0, \ldots, 2^n$, and let X^n be a random function that is constant over each $[t_{i-1}^n, t_i^n)$ and for which $(X_{t_0}^n, \ldots, X_{t_{2^n}}^n)$ has distribution $\mu_{t_0 \cdots t_{2^n}}$. Let Y^n be the process X^n with the time-set cut back to $T_n = \{t_i^n\}$. Let μ_n have mass $F(t_i^n) - F(t_{i-1}^n)$ at t_i^n for $i = 1, \ldots, 2^n$, and apply Theorem 10.4. By (13.15),

$$\mathsf{P}[|Y_s^n - Y_r^n| \wedge |Y_t^n - Y_s^n| \geq \lambda] \leq \frac{1}{\lambda^{4\beta}} \mu_n^{2\alpha}(T_n \cap (r, t]),$$

and it follows by (10.21) that

$$\mathsf{P}[L(Y^n, \delta) \geq \epsilon] \leq \frac{2K}{\epsilon^{4\beta}} \mu_n(T_n) \sup_{0 \leq t \leq 1-2\delta} \mu_n^{2\alpha-1}(T_n \cap (t, t+2\delta]).$$

Suppose that $r \leq s \leq t$ and $t - s \leq \delta$. Move r, s, t to the left endpoints r', s', t' of the intervals $[t_{i-1}^n, t_i^n)$ containing them ($t' = 1$ if $t = 1$). Then $|X_s^n - X_r^n| \wedge |X_t^n - X_s^n| = |Y_{s'}^n - Y_{r'}^n| \wedge |Y_{t'}^n - Y_{s'}^n|$ and $t' - s' \leq \delta + 2^{-n}$. It follows that, if $2^{-n} \leq \delta$, then $w''(X^n, \delta) \leq L(Y^n, 2\delta)$. And, again if $2^{-n} \leq \delta$, $\mu_n(T_n \cap [t, t+4\delta]) \leq F(t + 4\delta + 2^{-n}) - F(t - 2^{-n}) \leq w_F(4\delta + 2 \cdot 2^{-n}) \leq w_F(6\delta)$. Since $\mu_n(T_n) = F(1) - F(0)$, we arrive at

$$\mathsf{P}[w''(X^n, \delta) \geq \epsilon] \leq \frac{2K}{\epsilon^{4\beta}} [F(1) - F(0)](w_F(6\delta))^{2\alpha-1}.$$

From the uniform continuity of F, it follows that, for given ϵ and η, there exists a δ such that

$$\mathsf{P}[w''(X^n, \delta) \geq \epsilon] \leq \eta \tag{13.17}$$

for n large enough that $2^{-n} \leq \delta$.

The distributions of the X^n will be tight if they satisfy (13.4) and (13.8). If $2^{-k} \leq \delta$, then

$$\|X^n\| \leq \max_{i \leq 2^{-k}} |X_{i2^{-k}}^n| + w''(X^n, \delta).$$

Since the distributions of the first term on the right all coincide for $n \geq k$, it follows by (13.17) that (13.4) is satisfied.

The first condition in (13.8) of course holds because of (13.17). To take care of the second and third conditions, we temporarily assume that, for some positive δ_0, $h \le \delta_0$ implies

$$\mu_{0,h}[(u_1, u_2): u_1 = u_2] = 1, \quad \mu_{1-h,1}[(u_1, u_2): u_1 = u_2] = 1. \tag{13.18}$$

Under this assumption, the second and third conditions in (13.8) hold, so that $\{X^n\}$ is tight. Suppose that X is the limit in distribution of some subsequence. Because of the consistency hypothesis, the distribution of $(X_{t_1}, \ldots, X_{t_k})$ is $\mu_{t_1 \cdots t_k}$ for dyadic rational t_i, and the general case follows via (13.16) by approximation from above.

It remains to remove the restriction (13.18). For a $\delta_0 < \frac{1}{2}$, take $f(t)$ to be 0 to the left of δ_0, 1 to the right of $1-\delta_0$, and $(t-\delta_0)/(1-2\delta_0)$ in between. Now define $\nu_{t_1 \cdots t_k}$ as $\mu_{s_1 \cdots s_k}$ for $s_i = f(t_i)$. Then the $\nu_{t_1 \cdots t_k}$ satisfy the conditions of the theorem with a new F, as well as the special condition (13.18), so that there is a random element Z of D with these finite-dimensional distributions. We now need only define X by $X(t) = Z(\delta_0 + t(1 - 2\delta_0))$ for $0 \le t \le 1$. □

Example 13.1. We can use this theorem to construct a random function representing an additive process. Suppose that F is nondecreasing and continuous over $[0, 1]$, and for $0 \le t \le 1$, let ν_t be a measure on the line for which $\nu_t(R^1) = F(t)$. Suppose now that, for $s \le t$, $\nu_s(A) \le \nu_t(A)$ for all A, so that $\nu_t - \nu_s$ is a measure with total mass $F(t) - F(s)$. By the general theory [PM.372], there is an infinitely divisible distribution having characteristic function

$$\phi_{s,t}(u) = \exp \int_{R^1} (e^{iux} - 1 - iux) \frac{1}{x^2} (\nu_t - \nu_s)(dx); \tag{13.19}$$

the mean and variance are 0 and $F(t) - F(s)$.

We are to construct a random element of D for which the increments are independent and $X_t - X_s$ has characteristic function (13.19). Since $\phi_{r,t} = \phi_{r,s}\phi_{s,t}$, the distributions specified for $X_s - X_r$ and $X_t - X_s$ convolve to that specified for $X_t - X_r$, and so the implied finite-dimensional distributions are consistent. Further, by Chebyshev's inequality and the assumed independence, the left side of (13.15) is at most

$$\frac{1}{\lambda^2}[F(t) - F(t_1)] \cdot \frac{1}{\lambda^2}[F(t_2) - F(t)] \le \frac{1}{\lambda^4}[F(t_2) - F(t_1)]^2.$$

And (13.16) follows by another application of Chebyshev's inequality.

In the case of Brownian motion, ν_t consists of a mass of t at the origin. If ν_t consists of a mass of $F(t)$ at 1, then (13.19) is the characteristic function of a Poisson variable with its mean $F(t) - F(s)$ subtracted away. We can add the means back in, which gives a Poisson process with rate function $F(t)$: The increments are independent, and $X_t - X_s$ has the Poisson distribution with mean $F(t) - F(s)$. □

Problem

13.1. Prove Theorem 13.3 by the method of the second proof of Theorem 7.5. In place of the M_u of the previous proof, use the A_σ of Lemma 3 in Section 12.

SECTION 14. APPLICATIONS

This is a sampling of the applications of weak-convergence theory in D. Of the results here, only Theorems 14.2 and 14.4 are used in later proofs.

The identity map i from C to D is continuous and is therefore measurable $\mathcal{C}/\mathcal{D}$. If W is Wiener measure on $(C, \mathcal{C})$, then Wi^{-1} is a probability measure on $(D, \mathcal{D})$, and it is easy to see that W and Wi^{-1} have the same finite-dimensional distributions. From now on, we denote this new measure by W rather than Wi^{-1}; W will also denote a random element of D having this distribution.

Donsker's Theorem Again

Given random variables $\xi_1, \xi_2, \ldots$ on an $(\Omega, \mathcal{F}, \mathsf{P})$, with partial sums $S_n = \xi_1 + \cdots + \xi_n$, let $X^n(\omega)$ be the function in D with value

$$X_t^n(\omega) = \frac{1}{\sigma\sqrt{n}} S_{\lfloor nt \rfloor}(\omega) \tag{14.1}$$

at t. This random element of D (each X_t^n is measurable $\mathcal{D}/\mathcal{R}^1$) is the analogue of (8.5); it is slightly easier to analyze.

Theorem 14.1. *Suppose the ξ_n are independent and identically distributed with mean 0 and variance σ^2. Then the random functions defined by (14.1) satisfy $X^n \Rightarrow W$.*

PROOF. The proof in Section 8 that the finite-dimensional distributions converge carries over with essentially no change. Since W is continuous at 1, we can apply Theorem 13.5. By independence,

$$\begin{aligned}\mathsf{E}[|X_t^n - X_{t_1}^n|^2|X_{t_2}^n - X_t^n|^2] &= \frac{1}{n^2}(\lfloor nt\rfloor - \lfloor nt_1\rfloor)(\lfloor nt_2\rfloor - \lfloor nt\rfloor) \\ &\le \left(\frac{\lfloor nt_2\rfloor - \lfloor nt_1\rfloor}{n}\right)^2\end{aligned}$$

for $t_1 \le t \le t_2$. If $t_2 - t_1 \ge 1/n$, the right side here is at most $4(t_2 - t_1)^2$. If $t_2 - t_1 < 1/n$, then either t_1 and t lie in the same subinterval $[(i-1)/n, i/n)$ or else t and t_2 do; in either case the left side vanishes. Therefore, (13.14) holds for $\alpha = \beta = 1$ and $F(t) = 2t$. □

The equations (9.2) and (9.10) through (9.14) can be derived as before. Some functions of the partial sums S_i can be more simply expressed in terms of the random element of D defined by (14.1) than in terms of the random element of C defined by (8.5). For example, for $x \in D$, let $h(x)$ be the Lebesgue measure of the set of t for which $x(t) > 0$. Then [M15] h is measurable $\mathcal{D}/\mathcal{R}^1$ and is continuous on a set of Wiener measure 1. If X^n is defined by (14.1), then $h(X^n)$ is exactly n^{-1} times the number of positive partial sums among $S_1, \ldots, S_{n-1}$, which leads to a simple derivation of the arc sine law.

An Extension

Let X be the random function constructed in Example 13.1. Suppose that, for each n, $\xi_{n1}, \ldots, \xi_{nr_n}$ are independent random variables with mean 0 and variances σ_{nk}^2. Let $s_{nk}^2 = \sum_{i=1}^k \sigma_{ni}^2$ and $M_n = \max \sigma_{nk}^2$, and assume that $s_{nr_n}^2 = 1$ and $M_n \to 0$. Let $k_n(t)$ be the maximum k for which $s_{nk}^2 \le t$, and define measures $\nu_{n,t}$ and random functions X^n by

$$\nu_{n,t}(-\infty, x] = \sum_{k \le k_n(t)} \int_{\xi_{nk} \le x} \xi_{nk}^2 d\mathsf{P}, \qquad X_t^n = \sum_{k \le k_n(t)} \xi_{nk}.$$

If $\nu_{n,t}$ converges vaguely to ν_t (in the sense that $\nu_{n,t}(a,b] \to_n \nu_t(a,b]$ if $\nu_t\{a\} = \nu_t\{b\} = 0$), then it follows by the theory of infinitely divisible distributions [PM.375] that $X_t^n \Rightarrow X_t$. Since $\nu_{n,t} - \nu_{n,s}$ then converges vaguely to $\nu_t - \nu_s$ for $s < t$, it follows also that $X_t^n - X_s^n \Rightarrow X_t - X_s$; and by the independence of the increments, the finite-dimensional distributions of X^n converge weakly to those of X.

We can easily prove that

$$X^n \Rightarrow X \tag{14.2}$$

if we assume that $\nu_t(R^1) \equiv t$ and that, if $m_n = \min_k \sigma_{nk}^2$, then $M_n \leq Km_n$ for some K. The variance of $X_t^n - X_s^n$ is $\sum_{k_n(s)<k\leq k_n(t)} \sigma_{nk}^2 \leq (t - s + M_n)$, and so

$$\mathsf{E}[|X_t^n - X_{t_1}^n|^2 |X_{t_2}^n - X_t^n|^2] \leq (t-t_1+M_n)(t_2-t+M_n) \leq (t_2-t_1+2M_n)^2.$$

Now $t_2 - t_1 \geq m_n$ implies that the right side here does not exceed $(2K+1)^2(t_2-t_1)^2$, while $t_2 - t_1 < m_n$ implies that the left side is 0. This proves (13.14) for $\alpha = \beta = 1$ and $F(t) = (2K+1)t$, and (13.12) holds because $X_1 - X_{1-\delta}$ has variance δ. Hence (14.2). This extends Theorem 14.1 to triangular arrays satisfying the Lindeberg condition. It also covers Example 12.3.

Dominated Measures

The random variables ξ_n in Theorem 14.1 are defined on a space $(\Omega, \mathcal{F}, \mathsf{P})$, and we can replace P by any probability measure P_0 absolutely continuous with respect to it. Suppose, for example, that $\Omega = [0,1]$, $\mathcal{F}$ consists of the Borel subsets of $[0,1]$, and ξ_n is the nth Rademacher function: $\xi_n(\omega) = 2\omega_n - 1$, where ω_n is the nth digit in the dyadic representation of ω. Theorem 14.1 applies to $\{\xi_n\}$ if ω is drawn from $[0,1]$ according to Lebesgue measure, but this is also true if ω is drawn according to a distribution having a density with respect to Lebesgue measure:

Theorem 14.2. *Theorem* 14.1 *remains true if* P *is replaced by a* P_0 *absolutely continuous with respect to it.*

PROOF. Define $\bar{X}^n$ by

$$\bar{X}_t^n(\omega) = \frac{1}{\sigma\sqrt{n}} \sum_{p_n \leq i \leq nt} \xi_i(\omega), \tag{14.3}$$

where the p_n are integers going to infinity slowly enough that $p_n = o(n)$. Since $\|X^n - \bar{X}^n\| \leq \sum_{i=1}^{p_n} |\xi_n|/\sigma\sqrt{n}$, it follows by Chebyshev's inequality that (d is Skorohod distance)

$$d(X^n, \bar{X}^n) \leq \|X^n - \bar{X}^n\| \Rightarrow 0, \tag{14.4}$$

where this is interpreted in the sense of P (that is, P governs the distribution of the ξ_n). By Theorem 14.1,

$$X^n \Rightarrow W \tag{14.5}$$

in the sense of P, and it follows by (14.4) and Theorem 3.1 that

$$\bar{X}^n \Rightarrow W \tag{14.6}$$

in the sense of P.

Let A be a W-continuity set (in D), temporarily fixed; (14.6) implies that $\mathsf{P}[\bar{X}^n \in A] \to W(A)$. Let $\mathcal{F}_0$ be the field of cylinders—sets of the form $[(\xi_1, \ldots, \xi_k) \in H]$. If $E \in \mathcal{F}_0$, then, since $p_n \to \infty$, $\mathsf{P}([\bar{X}^n \in A] \cap E) = \mathsf{P}[\bar{X}^n \in A]\mathsf{P}(E)$ for large n, and therefore

$$\mathsf{P}([\bar{X}^n \in A] \cap E) \to W(A)\mathsf{P}(E). \tag{14.7}$$

Since the events $[\bar{X}^n \in A]$ all lie in $\sigma(\mathcal{F}_0)$, it follows by Rényi's theorem on mixing sequences [M21] that $\mathsf{P}_0[\bar{X}^n \in A] \to W(A)$. This holds for each W-continuity set A, and so (14.6) holds in the sense of P_0 (as well as P). Since P dominates P_0, it follows that (14.4) holds in the sense of P_0 (think of the ϵ-δ version of absolute continuity), and by another application of Theorem 3.1, so does (14.5). □

Empirical Distribution Functions

The empirical distribution for observations $\xi_1(\omega), \ldots, \xi_n(\omega)$ in $[0,1]$ is defined as $F_n(t,\omega) = n^{-1}\sum_{i=1}^n I_{[0,t]}(\xi_i(\omega))$. Assume the ξ_n are independent and have a common distribution function F over $[0,1]$, and consider the random function defined by

$$Y_t^n(\omega) = \sqrt{n}(F_n(t,\omega) - F(t)). \tag{14.8}$$

This describes the *empirical process*.

Theorem 14.3. *If $\xi_1, \xi_2, \ldots$ are independent and have a common distribution function F over $[0,1]$, and if (14.8) defines Y^n, then $Y^n \Rightarrow Y$, where Y is the Gaussian random element of D specified by $\mathsf{E}Y_t = 0$ and $\mathsf{E}[Y_sY_t] = F(s)(1 - F(t))$ for $s \le t$.*

That such a random function Y exists will be part of the proof.

PROOF. We first prove the theorem under the assumption that $F(t) \equiv t$, in which case Y is the Brownian bridge W° (extended from C to D). Let U_t^n be the number among $\xi_1, \ldots, \xi_n$ that lie in $[0,t]$. For $0 = t_0 < t_1 < \cdots < t_k = 1$, the $U_{t_i}^n - U_{t_{i-1}}^n$ have the multinomial distribution with parameters $t_i - t_{i-1}$, and it follows by the central limit theorem for multinomial trials that the finite-dimensional distributions of the Y^n converge weakly to those of W°. By Theorem 13.5, it suffices to prove that

$$\mathsf{E}[|Y_t^n - Y_{t_1}^n|^2 |Y_{t_2}^n - Y_t^n|^2] \le 6(t-t_1)(t_2-t) \le 6(t_2-t_1)^2 \tag{14.9}$$

for $t_1 \le t \le t_2$.

Let $p_1 = t - t_1$, $p_2 = t_2 - t$ and $p_3 = 1 - p_1 - p_2$; for $1 \le i \le n$, let α_i be $1 - p_1$ or $-p_1$ as ξ_i lies in $(t_1, t]$ or not, and let β_i be $1 - p_2$ or $-p_2$ as ξ_i lies in $(t, t_2]$ or not. Then the first inequality in (14.9) is equivalent to

$$\mathsf{E}\Big[\Big(\sum_{i=1}^n \alpha_i\Big)^2 \Big(\sum_{i=1}^n \beta_i\Big)^2\Big] \le 6n^2 p_1 p_2. \tag{14.10}$$

Since $\mathsf{E}\alpha_i = \mathsf{E}\beta_i = 0$, considerations of symmetry show that the left side of (14.10) is

$$n\mathsf{E}[\alpha_1^2\beta_1^2] + n(n-1)\mathsf{E}[\alpha_1^2]\mathsf{E}[\beta_2^2] + 2n(n-1)\mathsf{E}[\alpha_1\beta_1]\mathsf{E}[\alpha_2\beta_2].$$

And now (14.10) follows from

$$\begin{cases} \mathsf{E}[\alpha_1^2\beta_1^2] = p_1(1-p_1)^2p_2^2 + p_2p_1^2(1-p_2)^2 + p_3p_1^2p_2^2 \le 3p_1p_2, \\ \mathsf{E}[\alpha_1^2]\mathsf{E}[\beta_2^2] = p_1(1-p_1)p_2(1-p_2) \le p_1p_2, \\ \mathsf{E}[\alpha_1\beta_1]\mathsf{E}[\alpha_2\beta_2] = p_1^2p_2^2 \le p_1p_2. \end{cases}$$

That proves the theorem for the uniform case. The quantile function $\phi(s) = \inf[t: s \le F(t)]$ satisfies $\phi(s) \le t$ if and only if $s \le F(t)$. If η_n is uniformly distributed over $[0,1]$, then $\phi(\eta_n)$ has distribution function F, and so we can use the representation $\xi_n = \phi(\eta_n)$, where the η_n are independent and uniformly distributed.

If $G_n(\cdot, \omega)$ is the empirical distribution for $\eta_1(\omega), \dots, \eta_n(\omega)$ and $Z_t^n(\omega) = \sqrt{n}[G_n(t,\omega) - t]$, then $Z^n \Rightarrow W^\circ$ by the case already treated. But $F_n(t,\omega) = G_n(F(t),\omega)$, so that Y^n as defined by (14.8) satisfies $Y_t^n(\omega) = Z_{F(t)}^n(\omega)$. Define $\psi: D \to D$ by $(\psi x)(t) = x(F(t))$. If x_n converges to x in the Skorohod topology and $x \in C$, then the convergence is uniform, so that ψx_n converges to ψx uniformly and hence in the Skorohod topology. From the mapping theorem and the fact that $Z^n \Rightarrow W^\circ$, it follows that $Y^n = \psi(Z^n) \Rightarrow \psi(W^\circ)$; since $\psi(W^\circ)$ is Gaussian and has the means and covariances specified for Y, the proof is complete. □

For an example, apply the mapping theorem:

$$\sqrt{n}\sup_t |F_n(t,\omega) - F(t)| = \sup_t |Y_t^n(\omega)| \Rightarrow \sup_t |W^\circ_{F(t)}|.$$

If F is continuous, the limiting varable here has the same distribution as $\sup_t |W_t^\circ|$—namely that given by (9.40). If $F(t) \equiv t$, we can also apply (9.39) through (9.43).

Random Change of Time

If $S_n/\sigma\sqrt{n}$ is approximately normal for large n, and if ν is a random index that is large with high probability, then perhaps $S_\nu/\sigma\sqrt{n}$ or $S_\nu/\sigma\sqrt{\nu}$ will be approximately normal. Since $S_\nu/\sigma\sqrt{n} = X^n_{\nu/n}$ if X^n is defined by (14.1), this leads to the question of what happens to a random function if the time scale is subjected to a random change, a question best considered first in a general context.

Let D_0 consist of those elements ϕ of D that are nondecreasing and satisfy $0 \le \phi(t) \le 1$ for all t. Such a ϕ represents a transformation of the time interval $[0,1]$. We topologize D_0 by relativizing the Skorohod topology of D. Since $D_0 \in \mathcal{D}$, as is easily shown, the σ-field $\mathcal{D}_0$ of Borel sets in D_0 consists of the subsets of D_0 that lie in $\mathcal{D}$ [M10]. For $x \in D$ and $\phi \in D_0$, the composition $x \circ \phi$, with value $x(\phi(t))$ at t, is clearly an element of D. Define $\psi: D \times D_0 \to D$ by

$$\psi(x,\phi) = x \circ \phi; \tag{14.11}$$

then [M16] ψ is measurable $\mathcal{D} \times \mathcal{D}_0/\mathcal{D}$.

Let X be a random element of D and let Φ be a random element of D_0. We assume X and Φ have the same domain, so that (X,Φ) is a random element of $D \times D_0$ with the product topology [M6]. If $X \circ \Phi$ has value $X(\omega)\circ\Phi(\omega)$ at ω—that is, if $X\circ\Phi = \psi(X,\Phi)$—then $X\circ\Phi$ is the random element of D that results from subjecting X to the time-change represented by Φ. Suppose that, in addition to X and Φ, we have, for each n, random elements X^n and Φ^n—of D and D_0—having a common domain (which may vary with n).

Lemma. *If* $(X^n,\Phi^n) \Rightarrow (X,\Phi)$ *and* $\mathsf{P}[X \in C] = 1$, *then* $X^n \circ \Phi^n \Rightarrow X \circ \Phi$.

PROOF. This will follow by the mapping theorem if we show that ψ (defined by (14.11)) is continuous at (x,ϕ) for $x \in C$. Suppose that $x_n \to x$ and $\phi_n \to \phi$ in the Skorohod topology, and choose $\lambda_n \in \Lambda$ in such a way that $\|\lambda_n - I\| \to 0$ and $\|\phi_n - \phi\lambda_n\| \to 0$. Then, since x lies in C,

$$\begin{aligned}|x_n\phi_n t - x\phi\lambda_n t| &\le |x_n\phi_n t - x\phi_n t| + |x\phi_n t - x\phi\lambda_n t| \\ &\le \|x_n - x\| + w_x(\|\phi_n - \phi\lambda_n\|) \to 0. \qquad \square\end{aligned}$$

Return now to a consideration of sums $S_n = \xi_1 + \cdots + \xi_n$. Let ν_n be positive-integer-valued random variables defined on the same space as the ξ_n. Define X^n by (14.1), and define Y^n by

$$Y_t^n(\omega) = \frac{1}{\sigma\sqrt{\nu_n(\omega)}} S_{\lfloor \nu_n(\omega)t \rfloor}(\omega) = X_t^{\nu_n(\omega)}(\omega). \tag{14.12}$$

Theorem 14.4. *If*

$$\frac{\nu_n}{a_n} \Rightarrow \theta, \tag{14.13}$$

where θ is a positive constant and the a_n are constants going to infinity, then $X^n \Rightarrow W$ implies $Y^n \Rightarrow W$.

PROOF. There is no loss in generality in assuming that $0 < \theta < 1$ (this can be arranged by passing to new constants a_n) and that the a_n are integers. Define

$$\Phi_t^n(\omega) = \begin{cases} t\nu_n(\omega)/a_n & \text{if } \nu_n(\omega)/a_n \le 1, \\ t\theta & \text{otherwise.} \end{cases} \tag{14.14}$$

Since

$$\sup_t |\Phi_t^n - t\theta| \le \left| \frac{\nu_n}{a_n} - \theta \right| \Rightarrow 0, \tag{14.15}$$

the Skorohod distance from Φ^n to the element $\phi(t) = \theta t$ of D_0 goes to 0 in probability. Because of the assumptions $X^n \Rightarrow W$ and $a_n \to \infty$ it follows by Theorem 3.9 that $(X^{a_n}, \Phi^n) \Rightarrow (W, \phi)$ and hence by the lemma above that $X^{a_n} \circ \Phi^n \Rightarrow W \circ \phi$. If

$$R_t^n(\omega) = \frac{1}{\sigma\sqrt{a_n}} S_{\lfloor \nu_n(\omega)t \rfloor}(\omega),$$

then $X^{a_n} \circ \Phi^n$ and R^n have the same value at ω if $\nu_n(\omega)/a_n < 1$, the probability of which goes to 1 by (14.13) and the fact that $\theta < 1$. Therefore, $R^n \Rightarrow W \circ \phi$. Now

$$\sup_t \left| \frac{1}{\sqrt{\theta}} R_t^n(\omega) - Y_t^n(\omega) \right| = \left| \frac{1}{\sqrt{\theta}} - \sqrt{\frac{a_n}{\nu_n(\omega)}} \right| \sup_t |R_t^n(\omega)| \Rightarrow 0,$$

and hence $Y^n \Rightarrow \theta^{-1/2}(W \circ \phi)$. Since this limit has the same distribution as W, $Y^n \Rightarrow W$. $\square$

Of course, $Y^n \Rightarrow W$ implies $S_{\nu_n}/\sigma\sqrt{\nu_n} \Rightarrow N$, as well as an arc sine law and limit theorems for maxima and so on. Since the hypothesis requires only (14.13) and $X^n \Rightarrow W$, it applies to various dependent sequences $\{\xi_n\}$.

If the limit in (14.13) is not constant, we need more specific assumptions about the ξ_n. Again define Y^n by (14.12).

Theorem 14.5. *Suppose $\xi_1, \xi_2, \ldots$ are independent and identically distributed with mean 0 and variance σ. If*

$$\left| \frac{\nu_n}{a_n} - \theta \right| \Rightarrow 0, \tag{14.16}$$

where θ is a positive random variable and the a_n are constants going to infinity, then $Y^n \Rightarrow W$.

PROOF. Assume first that there is a constant K such that $0 < \theta < K$ with probability 1. We may adjust the a_n so that they are integers and $K < 1$.

If we define Φ^n by (14.14), and if Φ is the random element of D_0 defined by $\Phi_t = \theta t$, then (d is the Skorohod distance) $d(\Phi^n, \Phi) \Rightarrow 0$. From this and (14.16), it follows that the distance in $D_0 \times R^1$ between $(\Phi^n, \nu_n/n)$ and (Φ, θ) goes to 0 in probability, and it follows by the corollary to Theorem 3.1 that

$$(\Phi^n, \nu_n/a_n) \Rightarrow (\Phi, \theta). \tag{14.17}$$

Define $\bar{X}^n$ by (14.3), where $p_n \to \infty$ and $p_n = o(n)$. As before, we have (14.4) and(14.6)—and hence $\bar{X}^{a_n} \Rightarrow W$—and (14.7) holds for E in the field $\mathcal{F}_0$ of cylinders. And now, by (14.17) and Theorem 3.10,

$$(\bar{X}^{a_n}, \Phi^n, \nu_n/a_n) \Rightarrow (W, \Phi^\circ, \theta^\circ)$$

in the sense of the product topology on $D \times (D_0 \times R^1)$, where θ° is independent of W and $\Phi_t^\circ = \theta^\circ t$. By (14.4),

$$(X^{a_n}, \Phi^n, \nu_n/a_n) \Rightarrow (W, \Phi^\circ, \theta^\circ).$$

The mapping that carries the point (x, ϕ, α) to $\alpha^{-1/2}(x \circ \phi)$ is continuous at that point if $x \in C$, $\phi \in D_0$, and $\alpha > 0$. By the mapping theorem, therefore,

$$(\nu_n/a_n)^{-1/2}(X^{a_n} \circ \Phi^n) \Rightarrow (\theta^\circ)^{-1/2}(W \circ \Phi^\circ).$$

Since θ° and W are independent, $(\theta^\circ)^{-1/2}(W \circ \Phi^\circ)$ has the same distriubtion as W. Moreover, $(\nu_n/a_n)^{-1/2}(X^{a_n} \circ \Phi^\circ)$ coincides with Y^n if $\nu_n/a_n < 1$, the probability of which goes to 1 because $K < 1$. Thus $Y^n \Rightarrow W$ if θ is bounded.

Suppose θ is not bounded. For positive u, define $\theta_u = \theta$ and $\nu_{u,n} = \nu_n$ if $\theta \leq u$, and define $\theta_u = u$ and $\nu_{u,n} = a_n u$ if $\theta > u$. Then, for each u, $|\nu_{u,n}/a_n - \theta_u| \Rightarrow 0$ as $n \to \infty$, and by the case already treated, if

$$Y_t^{u,n}(\omega) = \frac{1}{\sigma\sqrt{\nu_{u,n}(\omega)}} S_{\lfloor \nu_{u,n}(\omega)t \rfloor}(\omega),$$

then $Y^{u,n} \Rightarrow_n W$. Since $\mathsf{P}[Y^{u,n} \neq Y^n] \leq \mathsf{P}[\theta > u]$, it follows by Theorem 3.2 that $Y^n \Rightarrow W$. □

Renewal Theory

These ideas can be used to derive a functional central limit theorem connected with renewal theory. Let $\eta_1, \eta_2, \ldots$ be positive random variables and define

$$\nu_t = \max\,[k\colon \sum_{i=1}^{k} \eta_i \leq t], \quad t \geq 0; \tag{14.18}$$

take $\nu_t = 0$ if $t < \eta_1$. If η_k is the length of time between the occurrences of the $(k-1)$st and kth events in a series, then ν_t is the number of occurrences up to time t.

We assume the existence of positive constants μ and σ such that, if

$$X_t^n(\omega) = \frac{1}{\sigma\sqrt{n}} \sum_{i=1}^{\lfloor nt \rfloor} (\eta_i(\omega) - \mu),$$

then $X^n \Rightarrow W$. This will be true if, as in the usual renewal setting, the η_n are independent and identically distributed with mean μ and variance σ^2. Define Z^n by

$$Z_t^n(\omega) = \frac{\nu_{nt}(\omega) - nt/\mu}{\sigma\mu^{-3/2}\sqrt{n}}.$$

Theorem 14.6. *If $X^n \Rightarrow W$, then $Z^n \Rightarrow W$.*

PROOF. We assume in the proof that $\mu > 1$, since this is only a mater of scale. We first show that

$$\sup_{0 \leq v \leq u} \left| \frac{\nu_v}{u} - \frac{1}{\mu}\frac{v}{u} \right| \Rightarrow_u 0. \tag{14.19}$$

The hypothesis $X^n \Rightarrow W$ implies

$$\sup_{0 \le t \le s} \frac{1}{s} \left| \sum_{i=1}^{\lfloor t \rfloor} \eta_i - \mu t \right| \Rightarrow_s 0, \tag{14.20}$$

since replacing s^{-1} by $s^{-1/2}$ would give convergence in distribution. By the definition (14.18), $\nu_v > t$ implies $\sum_{i=1}^{\lfloor t \rfloor} \eta_i \le v$. Therefore,

$$\sup_{0 \le v \le u} \left(\frac{\nu_v}{u} - \frac{1}{\mu} \frac{v}{u} \right) > \epsilon$$

implies

$$\sup_{0 \le t \le u(\mu^{-1}+\epsilon)} \left| \sum_{i=1}^{\lfloor t \rfloor} \eta_i - \mu t \right| \ge \mu u \epsilon. \tag{14.21}$$

Similary, if $\epsilon < \mu^{-1}$, then

$$\inf_{0 \le v \le u} \left(\frac{\nu_v}{u} - \frac{1}{\mu} \frac{v}{u} \right) < -\epsilon$$

implies

$$\sup_{0 \le t \le u(\mu^{-1}-\epsilon)} \left| \sum_{i=1}^{\lfloor t \rfloor} \eta_i - \mu t \right| \ge \mu u \epsilon. \tag{14.22}$$

By (14.20), the probabilities of (14.21) and (14.22) go to 0 as $u \to \infty$, which proves (14.19).

Put

$$\Phi_t^n(\omega) = \begin{cases} \nu_{tn}(\omega)/n & \text{if } \nu_n(\omega)/n \le 1, \\ t/\mu & \text{otherwise.} \end{cases}$$

By (14.19), $\Phi^n \Rightarrow \phi$, where $\phi(t) = t/\mu$, so that, by the lemma (p. 151) again, $X^n \circ \Phi^n \Rightarrow W \circ \phi$. Let

$$Y_t^n(\omega) = \frac{1}{\sigma\sqrt{n}} \sum_{i=1}^{\nu_{tn}(\omega)} (\eta_i(\omega) - \mu);$$

$Y^n = X^n \circ \Phi^n$ if $\nu_n/n \le 1$, the probability of which goes to 1 by (14.19) and the assumption $\mu > 1$. Therefore, $Y^n \Rightarrow W \circ \phi$.

By the definition (14.18),

$$Y_t^n \le \frac{nt - \mu\nu_{nt}}{\sigma\sqrt{n}} \le Y_t^n + \frac{\eta_{\nu_{nt}+1}}{\sigma\sqrt{n}}.$$

From $\max_{i\le n}|\eta_i|/\sqrt{n} \Rightarrow 0$ it follows that $\sup_{t\le 1}|\eta_{\nu_{nt}+1}|/\sigma\sqrt{n} \Rightarrow 0$, which in turn implies that, if

$$R_t^n = \frac{nt - \mu\nu_{nt}}{\sigma\sqrt{n}},$$

then $R^n \Rightarrow W \circ \phi$. Therefore, $\mu^{1/2}R^n \Rightarrow W$ (recall that $\phi(t) = t/\mu$), from which $Z^n \Rightarrow W$ follows because of the symmetry of W. □

Problems

14.1. Let

$$x = I_{[1/4,1]} + I_{[1/2,1]}, \quad \phi = \frac{1}{2}I_{[1/2,1]}, \quad \phi_n(t) = tI_{[(1/2)-n^{-1},1/2)}(t) + \frac{1}{2}I_{[1/2,1]}.$$

Show that $\phi_n \to \phi$ (in the Skorohod topology) but that $x\circ\phi_n \not\to x\circ\phi$. Relate this to the lemma in this section.

14.2. Under the hypotheses of the Lindeberg-Lévy theorem, there are limiting distributions for

$$(14.23) \qquad \begin{cases} \text{(a)} \quad n^{-3/2}\sum_{k=1}^n |S_k|, \\ \text{(b)} \quad n^{-2}\sum_{k=1}^n S_k^2, \\ \text{(c)} \quad n^{-1/2}\min_{\beta n\le k\le n} S_k, \quad 0 < \beta < 1. \end{cases}$$

Construct the relevant mappings from D to R^1 and prove that they are measurable and continuous on a set of W-measure 1. For the forms of the limiting distributions, see Donsker [18] for (a) and (b) and Mark [47] for (c).

14.3. Let ξ_n be the Rademacher functions on the unit interval, and let P be Lebesgue measure, so that Theorem 14.1 applies. Show that the requirement in Theorem 14.2 that P_0 be dominated by P is essential: The therorem fails if P_0 is a point mass or (more interesting) if P_0 is Cantor measure.

SECTION 15. UNIFORM TOPOLOGIES*

In this section we use the weak-convergence theory for the ball σ-field (Section 6) to prove two results in empirical-process theory.

The Uniform Metric on $D[0,1]$

Let d be the Skorohod metric on D and let u be the uniform metric there:

$$u(x,y) = \|x-y\| = \sup_t |x(t) - y(t)|. \tag{15.1}$$

Clearly u is finer than d; and in fact, since $u(I_{[s,1]}, I_{[t,1]}) = 1$ for $s \neq t$, the u-topology is not separable. If $\mathcal{D}$ is the σ-field of d-Borel sets, and if $\mathcal{D}_f$ is the class of finite-dimensional sets as defined above Theorem 12.5, then, by that theorem, $\mathcal{D} = \sigma(\mathcal{D}_f)$; and since D is d-separable, $\mathcal{D}$ coincides with the d-ball σ-field $\mathcal{D}_0$.

Let $\mathcal{U}_0$ and $\mathcal{U}$ be the ball and Borel σ-fields for the uniform metric u. The five classes of sets are related by

$$\mathcal{D}_0 = \mathcal{D} = \sigma(\mathcal{D}_f) = \mathcal{U}_0 \subset \mathcal{U}, \tag{15.2}$$

where the inclusion is strict. We have already noted the first two equalities. For the third equality, observe first that the closed u-ball $B_u(x,\epsilon)^-$ is by right-continuity the intersection over rational r of the $\mathcal{D}_f$-sets $[y: x(r) - \epsilon \leq y(r) \leq x(r) + \epsilon]$, so that $\mathcal{U}_0 \subset \sigma(\mathcal{D}_f)$. The reverse inclusion will follow if we prove for each t and α that

$$[x: \pi_t(x) = x(t) > \alpha] = \bigcup_n B_u(x_n, n)^- \tag{15.3}$$

for an appropriate $\{x_n\}$ depending on t and α. If $t < 1$, take $x_n = \alpha + (n + n^{-1}) I_{[t, t+n^{-1})}$. It is easy to show that the right side is contained in the left side. Suppose on the other hand that $x(t) > \alpha$ and choose n so that $x(s) > \alpha + n^{-1}$ for $s \in [t, t+n^{-1})$ and $|x(s)| < n - |\alpha|$ for all s; that $|x(s) - x_n(s)| \leq n$ for all s then follows by separate consideration of the cases where s does and does not lie in $[t, t+n^{-1})$. If $t = 1$, take $x_n = \alpha + (n + n^{-1}) I_{\{t\}}$ and use a similar argument. Hence (15.3), which proves the right-hand equality in (15.2).

Let $\mathcal{B}$ be the class of ordinary Borel sets in $[0,1]$ and define $\phi : [0,1] \to D$ by $\phi(t) = I_{[t,1]}$. For each s and α, $\phi^{-1}[x: x(s) < \alpha]$ lies in $\mathcal{B}$, and since the finite-dimensional sets $[x: x(s) < \alpha]$ generate $\mathcal{D}$, ϕ is measurable $\mathcal{B}/\mathcal{D}$. But ϕ is not measurable $\mathcal{B}/\mathcal{U}$, because the set $A = \bigcup_{t \in H} B_u(\phi(t), \frac{1}{2})$ is u-open for each H, while $\phi^{-1} A = H$ need not lie in $\mathcal{B}$. This proves that the inclusion in (15.2) is strict.

Consider next the empirical process, the random function Y^n defined by (14.8) for random variables ξ_i on a probability space $(\Omega, \mathcal{F}, \mathsf{P})$.

Now Y^n is a random element of D with the Skorohod metric d, because it is measurable $\mathcal{F}/\mathcal{D}$. But Y^n is not a random element of D with the uniform metric u (is not measurable $\mathcal{F}/\mathcal{U}$), as follows by essentially the same argument that proves the inclusion relation in (15.2) to be strict. By Theorem 14.3, $Y^n \Rightarrow Y$, and so, if P_n is the distribution on D of Y^n and P is the distribution on D of Y, then $P_n \Rightarrow P$ holds in the sense of d:

$$P_n \Rightarrow P \quad \text{re}\, d. \tag{15.4}$$

Since these measures are defined on $\mathcal{U}_0$, one can ask whether there is weak° convergence in the sense of Section 6:

$$P_n \Rightarrow^{\circ} P \quad \text{re}\, u. \tag{15.5}$$

But here Theorem 6.6 applies. Since (15.2) holds, and since d-convergence to a point of the separable set C of continuous functions implies u-convergence, it is enough to ask whether C supports P, which is true if the distribution function F common to the ξ_i is continuous. For in that case, $j(Y^n) = 1/\sqrt{n}$ (see (12.8)), and Theorem 13.4 implies that $PC = \mathsf{P}[Y \in C] = 1$. Therefore, (15.4) and (15.5) both hold. As explained in Section 6, (15.5) contains more information than (15.4) does.

A Theorem of Dudley's

Let (T, h) be a compact metric space and let $\mathcal{T}$ be the Borel σ-field for T. Take $M = M(T)$ to be the space of all bounded, real, $\mathcal{T}$-measurable functions on T, with the metric u defined (on M now) by (15.1). Note that, for $T = [0, 1]$, M is much larger than $D[0, 1]$. In Theorem 15.2 below, h will be the Hausdorff metric [M17] on the space T of closed, convex sets in the unit square. But for now, (T, h) can be any compact space. Let $C = C(T)$ be the space of continuous functions on T. Since an element of C is measurable $\mathcal{T}$ and is bounded because of compactness, C is a closed subset of M. And C is separable: For each integer q, decompose T into finitely many $\mathcal{T}$-sets A_{qi} of diameter less than $1/q$. Consider the class of functions that, for some q, have a constant, rational value over each A_{qi}; it is countable, and it is dense in C because the continuous functions are uniformly continuous.

Let $\mathcal{M}_0$ and $\mathcal{M}$ be the ball σ-field and Borel σ-field for (M, u). We can define the modulus of continuity for elements of M in the usual way:

$$w_x(\delta) = w(x, \delta) = \sup_{h(s,t) \le \delta} |x(s) - x(t)|. \tag{15.6}$$

It is easy enough to see that $w(\cdot, \delta)$ is u-continuous and hence $\mathcal{M}$-measurable, but conceivably it is not always $\mathcal{M}_0$-measurable.† On the other hand, $\|\cdot\|$ is obviously $\mathcal{M}_0$-measurable.

Let P_n be probability measures on $(M, \mathcal{M}_0)$, and let P_n^* be the corresponding outer measures (on the class of all subsets of M).

Theorem 15.1. *Suppose that for every η there exist an a and an n_0 such that*

$$P_n[x: \|x\| \geq a] < \eta, \qquad n \geq n_0. \tag{15.7}$$

Suppose further that for every ϵ and η there exist a δ and an n_0 such that

$$P_n^*[x: w_x(\delta) \geq \epsilon] < \eta, \qquad n \geq n_0. \tag{15.8}$$

Then $\{P_n\}$ is tight°, and C supports the limit of every weak°ly convergent subsequence.

This, of course, looks like Theorem 7.3. The outer measure in (15.8) covers the possibility that the set in question does not belong to the σ-field $\mathcal{M}_0$ on which the P_n are defined. In order to prove Theorem 15.1, we need a generalization of the Arzelà-Ascoli theorem.

Lemma. *A set A in M is relatively compact if*

$$\sup_{x \in A} \|x\| < \infty \tag{15.9}$$

and

$$\lim_{\delta \to 0} \sup_{x \in A} w_x(\delta) = 0. \tag{15.10}$$

In Theorem 7.2, it was sufficient to control $|x(0)|$ (see (7.3)), but (15.9) is needed here; consider $T = \{a, b\}$ and $A = [x: x(a) = 0]$. It is a consequence of (15.10) that A^- is in fact contained in C.

PROOF. Since T is compact by assumption, it contains a countable, dense subset T_0. Given a sequence $\{x_n\}$ in A, first use (15.9) to replace it by a subsequence along which $x_n(t)$ converges to a limit for each t in T_0. Given ϵ, choose δ so that $w_x(\delta) < \epsilon$ for all x in A, and choose in T_0 a finite δ-net $\{t_1, \ldots, t_k\}$ for T. Finally, choose n_0 so that $m, n \geq n_0$ implies that $|x_m(t_i) - x_n(t_i)| \leq \epsilon$ for each i. For each s in T, there is an i such that $h(s, t_i) \leq \delta$ and hence $|x_n(s) - x_n(t_i)| \leq \epsilon$ for all n. But then, $m, n \geq n_0$ implies $|x_m(s) - x_n(s)| \leq 3\epsilon$. Thus $\{x_n\}$ is uniformly fundamental and hence converges uniformly to a continuous function on T. □

† Whether this can in fact happen seems to be unknown.

PROOF OF THEOREM 15.1. We must show that, for each ϵ, M contains a compact set K with the property that, for each positive γ, $P_n K^\gamma \geq 1-\epsilon$ for all large enough n. Given the ϵ, first choose an a and an n_0 such that $a > 1$ and

$$P_n[x\colon \|x\| \geq a] \leq \frac{\epsilon}{2} \quad \text{for } n \geq n_0. \tag{15.11}$$

Then choose a sequence $\{\alpha_i\}$ such that $0 < \alpha_{i+1} < \alpha_i$ and a sequence $\{n_0(i)\}$ of positive integers such that $n_0 < n_0(i)$ and

$$P_n^*\left[x\colon w_x(\alpha_i) \geq \frac{1}{2^i}\right] \leq \frac{\epsilon}{2^i} \quad \text{for } n \geq n_0(i). \tag{15.12}$$

And now define

$$\delta_i = \frac{\alpha_i}{a2^{i+1}} < \frac{\alpha_i}{3} \tag{15.13}$$

$(a > 1)$ and

$$K = \left[z\colon \|z\| \leq a;\ w_z(\delta_i) \leq \frac{3}{2^i},\ i \geq 1\right]^-. \tag{15.14}$$

This, a compact subset of C because of the lemma, is our K.

Let B and B_i be the sets in (15.11) and (15.12). Given γ, choose m so that $1/2^{m-1} < \gamma$ and take $G_m = B^c \cap \bigcap_{i \leq m} B_i^c$ and $n_m = \max_{i \leq m} n_0(i)$. By (15.11) and (15.12), the inner measure $(P_n)_*$ satisfies $(P_n)_*(G_m) \geq 1-\epsilon$ for $n \geq n_m$. If we prove that

$$G_m \subset K^\gamma, \tag{15.15}$$

then $P_n K^\gamma \geq (P_n)_*(G_m) \geq 1-\epsilon$ will hold for $n \geq n_m$, which will complete the proof. (In the application below, Theorem 15.2, we have $P_n K \leq P_n C = 0$, and so we definitely need the K^γ, not just K.)

Fix an x that lies in G_m. Then x satisfies

$$\|x\| \leq a, \qquad w_x(\alpha_i) \leq 1/2^i \quad \text{for } i \leq m.$$

To prove (15.15), we must find a z such that

$$z \in K, \qquad \|x - z\| < \gamma. \tag{15.16}$$

If $h(s,t) \geq \alpha_i$, then $|x(s) - x(t)| \leq 2a \leq 2ah(s,t)/\alpha_i = h(s,t)/2^i\delta_i$. And if $h(s,t) \leq \alpha_i$ and $i \leq m$, then $|x(s) - x(t)| \leq 1/2^i$. Therefore, for all s and t,

$$|x(s) - x(t)| \leq \frac{1}{2^i}\left[1 \vee \frac{h(s,t)}{\delta_i}\right], \quad i \leq m. \tag{15.17}$$

Choose a finite set T_m in T with the properties that (i) $h(s,t) \geq \delta_m$ for distinct s and t in T_m and (ii) for each s in T we have $h(s,t) < \delta_m$ for some t in T_m. (If we choose points one by one to satisfy (i), we must, by compactness, end with a finite set satisfying (ii).) If s and t are distinct points of T_m, then, since $h(s,t) \geq \delta_m$, (15.17) for $i = m$ gives $|x(s) - x(t)| \leq h(s,t)/2^m\delta_m$, which of course also holds if $s = t$. This is a Lipschitz condition for the function x restricted to T_m, and [M9] the restricted function can be extended to a function z that is defined on all of T, has the same bound, and satisfies the same Lipschitz condition for all s and t in T:

$$\|z\| \leq a, \qquad |z(s) - z(t)| \leq \frac{h(s,t)}{2^m\delta_m}. \tag{15.18}$$

This is our z.

Since $\|z\| \leq a$, $z \in K$ will follow if we show that $w_z(\delta_i) \leq 3/2^i$ for all i. If $i \geq m$ and $h(s,t) \leq \delta_i$, then by (15.18) and the fact that $2^i\delta_i$ is decreasing, $|z(s) - z(t)| \leq h(s,t)/2^m\delta_m \leq h(s,t)/2^i\delta_i \leq 1/2^i$. Now suppose that $i < m$ and $h(s,t) \leq \delta_i$. Choose points s_m and t_m in T_m such that $h(s, s_m) < \delta_m$ and $h(t, t_m) < \delta_m$. Since $h(s, s_m) < \delta_m$, (15.18) gives $|z(s) - z(s_m)| \leq h(s, s_m)/2^m\delta_m \leq 1/2^m \leq 1/2^i$. Similarly, $|z(t) - z(t_m)| \leq 1/2^i$. And since $h(s_m, t_m) \leq \delta_i + 2\delta_m \leq 3\delta_i \leq \alpha_i$ and z agrees with x on T_m, $|z(s_m) - z(t_m)| = |x(s_m) - x(t_m)| \leq w_x(\alpha_i) \leq 1/2^i$. And now the triangle inequality gives $|z(s) - z(t)| \leq 3/2^i$, and so z does lie in K.

For the other condition in (15.16), it will be enough to show that $|x(s) - z(s)| \leq 2/2^m$ for all s. Choose from T_m an s_m such that $h(s, s_m) < \delta_m$. From $x(s_m) = z(s_m)$ it follows that $|x(s) - z(s)| \leq |x(s) - x(s_m)| + |z(s_m) - z(s)|$. Since $x \in G_m$ and $h(s, s_m) < \delta_m \leq \alpha_m$, we have $|x(s) - x(s_m)| \leq w_x(\alpha_m) \leq 1/2^m$. And from (15.18) it follows that $|z(s_m) - z(s)| \leq h(s_m, s)/2^m\delta_m \leq 1/2^m$.

It remains to show that, if $P_{n_i} \Rightarrow^\circ P$, then $PC = 1$. Since C is closed and separable, it lies in $\mathcal{M}_0$ (Lemma 1, Section 6). Each of the compact sets K constructed above is contained in C, and since $P_n K^\gamma \geq 1 - \gamma$ for large n, $P((C^\gamma)^-) \geq P((K^\gamma)^-) \geq \limsup_n P_n K^\gamma \geq 1 - \epsilon$. Let $\gamma \to 0$ and then $\epsilon \to 0$: $PC = 1$. □

Empirical Processes Indexed by Convex Sets

Let T be the space of closed, convex subsets of $Q = [0,1]^2$, and let h be the Hausdorff metric [M17] on T: (T, h) is a compact metric space. Let T, M, C, $\mathcal{M}_0$, and $\mathcal{M}$ be the specializations to this case of the spaces and classes of sets defined above. Define the projection $\pi_{t_1\cdots t_k}: M \to R^k$ in the usual way: $\pi_{t_1\cdots t_k}(x) = (x(t_1), \ldots, x(t_k))$. And let $\mathcal{M}_f$ be the class of finite-dimensional sets, those of the form $\pi_{t_1\ldots t_k}^{-1} H$ for $H \in \mathcal{R}^k$. Note that (M, u) is not separable.

First, π_t is measurable $\mathcal{M}_0/\mathcal{R}^1$. To prove this, it is enough to verify (15.3) (same symbols in a new context), where now $x_n = \alpha + (n+n^{-1})I_{\{t\}}$. Since $\{t\}$ is h-closed and hence lies in $\mathcal{T}$, $x_n \in M$. That the right side of (15.3) is contained in the left is again easy. If $x(t) > \alpha$, choose n so that $x(t) \geq \alpha + n^{-1}$ and $|x(s)| \leq n - |\alpha|$ for all s; to show that $|x(s) - x_n(s)| \leq n$ for all s, consider separately the cases $s = t$ and $s \neq t$. Since each π_t is $\mathcal{M}_0$-measurable,

$$\sigma(\mathcal{M}_f) \subset \mathcal{M}_0 \subset \mathcal{M}. \tag{15.19}$$

Let M' be the set of y in M that are continuous from above in the sense that, if $t_n \downarrow t$ in the set-theoretic sense, then $y(t_n) \to y(t)$. There is [M18] a countable set T_0 in T such that, for each $t \in T$, there exists a sequence $\{t_n\}$ in T_0 for which $t_n \downarrow t$. From this it follows that

$$M' \cap B_u(x, r)^- = \bigcap_{t \in T_0} (M' \cap [y: |y(t) - x(t)| \leq r]). \tag{15.20}$$

Suppose that $(\Omega, \mathcal{F}, \mathsf{P})$ is a probability space and $Z: \Omega \to M$; let $Z(t, \omega) = \pi_t(Z(\omega))$. Call Z a random element of $(M, \mathcal{M}_0)$ if it is measurable $\mathcal{F}/\mathcal{M}_0$—because M is not separable, $\mathcal{M}_0$, not $\mathcal{M}$, is the relevant σ-field here. If Z is a random element of $(M, \mathcal{M}_0)$, then, by (15.19), each $Z(t, \cdot)$ is a random variable (measurable $\mathcal{F}/\mathcal{R}^1$). Suppose, on the other hand, that $Z(t, \cdot)$ is a random variable for each t and that $Z(\cdot, \omega)$ lies in M' for each ω. Then, by (15.20),

$$\begin{aligned}[\omega: Z(\cdot, \omega) \in B_u(x, r)^-] &= [\omega: Z(\cdot, \omega) \in M' \cap B_u(x, r)^-] \\ &= \bigcap\nolimits_{t \in T_0} [\omega: x(t) - r \leq Z(t, \omega) \leq x(t) + r],\end{aligned}$$

and therefore Z is a random element of $(M, \mathcal{M}_0)$.

Suppose that $\xi_1, \xi_2, \ldots$ are independent two-dimensional random vectors on $(\Omega, \mathcal{F}, \mathsf{P})$, each uniformly distributed over Q. Suppose that $\xi_i(\omega) \in Q$ for each i and ω, and define $\mu_n(\omega) = \mu_n(\cdot, \omega)$ by

$$\mu_n(t, \omega) = \frac{1}{n} \sum_{i=1}^{n} I_t(\xi_i(\omega)), \quad t \in T. \tag{15.21}$$

Of course, $\mu_n(t,\omega)$ is well defined for arbitrary subsets t of Q, and if it is restricted to the σ-field of ordinary Borel subsets of Q, it is a probability measure. But we restrict it further to the class T. Since $\mu_n(\cdot,\omega)$ is the restriction of a probability measure, it lies in M', and since each $\mu_n(t,\cdot)$ is a random variable, μ_n is a random element of $(M,\mathcal{M}_0)$. And if λ is Lebesgue measure on the Borel subsets of Q, then

$$X^n(t,\omega) = \sqrt{n}(\mu_n(t,\omega) - \lambda(t)), \quad t \in T \tag{15.22}$$

defines another random element of $(M,\mathcal{M}_0)$.

Theorem 15.2. *There is a random element X of $(M,\mathcal{M}_0)$ such that $X^n \Rightarrow^\circ X$; and $X \in C$ with probability* 1.

Of course, $X^n \Rightarrow^\circ X$ means that the corresponding distributions on $\mathcal{M}_0$ satisfy $P_n \Rightarrow^\circ P$.

PROOF. By the multinomial central limit theorem, the random vector $((X^n(t_1),\ldots,X^n(t_k))$ has asymptotically the centered normal distribution with covariances $\lambda(t_i \cap t_j) - \lambda(t_i)\lambda(t_j)$. Suppose we can verify the hypotheses of Theorem 15.1. It will follow by Theorems 6.5 and 15.1 that every subsequence of $\{P_n\}$ contains a further subsequence that converges weak°ly to a limit supported by C. We showed above that π_t is measurable $\mathcal{M}_0/\mathcal{R}^1$, and it is obviously u-continuous. Therefore, each $\pi_{t_1\ldots t_k}$ is measurable $\mathcal{M}_0/\mathcal{R}^k$ and u-continuous. If $P_{n_i} \Rightarrow_i^\circ P$, then it follows by Theorem 6.4 that $P_{n_i}\pi_{t_1\ldots t_k}^{-1} \Rightarrow_i P\pi_{t_1\ldots t_k}^{-1}$ and hence that $P\pi_{t_1\ldots t_k}^{-1}$ is the normal distribution specified above.

Since $C \subset M'$ ($t_n \downarrow t$ implies $h(t_n,t) \to 0$ [M17]), (15.20) holds if M' is replaced by C. Therefore, $C \cap B_u(x,r)^- \in \sigma(\mathcal{M}_f \cap C)$ and hence $\mathcal{M}_0 \cap C \subset \sigma(\mathcal{M}_f \cap C)$: $\mathcal{M}_f \cap C$ is a π-system which generates $\mathcal{M}_0 \cap C$. It follows that, if P and Q are two probability measures on $(M,\mathcal{M}_0)$, and if $PC = QC = 1$ and $P\pi_{t_1\cdots t_k}^{-1} = Q\pi_{t_1\cdots t_k}^{-1}$ for all $t_1,\ldots,t_k$, then $P = Q$. This means that there is only one limit for the weak°ly convergent subsequences—the one having the normal distributions specified above as its finite-dimensional distributions. Since this will complete the proof, it is enough to verify the hypotheses of Theorem 15.1.

Assume for the moment that (15.8) holds, and consider the δ and n_0 corresponding to an ϵ of 1, say. Let $t_1,\ldots,t_k$ be a δ-net. Since the finite-dimensional distributions converge, there is an a such that $\mathsf{P}[\max_{i\le k}|X_{t_i}^n| \ge a] \le \eta$ for all n. But then $\mathsf{P}[\|X^n\| \ge a+1] \le 2\eta$ for $n \ge n_0$, and (15.7) follows. It is therefore enough to verify (15.8).

Preliminaries: Let $\Delta^n(s,t) = |X^n(s) - X^n(t)|$. Since $I_s(\xi_i) - I_t(\xi_i)$ has variance at most $\lambda(s\Delta t)$, Bernstein's inequality [M20] gives ($x > 0$)

$$\mathsf{P}[\Delta^n(s,t) \geq x] \leq 2\exp\Big[-\frac{x^2}{2(\lambda(s\Delta t) + x/\sqrt{n})}\Big]. \tag{15.23}$$

Let T_m be the 2^{-m}-net $T(2^{-m})$ constructed in [M18]; the number N_m of points (sets) in T_m satisfies

$$\log N_m \leq A\sqrt{2^m}\log 2^m \leq Bm2^{m/2} \tag{15.24}$$

for constants A and B. For each t in T, let t'_m and t''_m be the sets in [M18(32)]:

$$t'_m \subset t \subset t''_m, \quad \lambda(t''_m - t'_m) < 2^{-m}, \quad h(t, t''_m) < 2^{-m}; \tag{15.25}$$

for each m, the t'_m and t''_m are functions of t. There are N_m pairs t'_m, t''_m. Fix a θ in the range $\frac{1}{2} < \theta < 1$, and define a function $m = m(n)$ by $m = \lfloor (2\theta \log 2)^{-1} \log n \rfloor$. Then

$$2^{\theta m} \leq \sqrt{n} < 2^{\theta(m+1)}, \quad \frac{1}{2} < \theta < 1. \tag{15.26}$$

The next step is to show that (t''_m is a function of t)

$$P_n^*\Big[x\colon \sup_{t \in T} |x(t''_m) - x(t)| > \epsilon\Big] \to_n 0 \tag{15.27}$$

for each ϵ. We have

$$\begin{aligned}\Delta^n(t''_m, t) &= \sqrt{n}|\mu_n(t''_m - t) - \lambda(t''_m - t)| \leq \sqrt{n}\mu_n(t''_m - t) + \frac{\sqrt{n}}{2^m} \\ &\leq \sqrt{n}\mu_n(t''_m - t'_m) + \frac{\sqrt{n}}{2^m} \leq \Delta^n(t''_m, t'_m) + \frac{2\sqrt{n}}{2^m},\end{aligned}$$

and the last term goes to 0 by (15.26). Let C_n be the maximum of $\Delta^n(t''_m, t'_m)$ for t ranging over T. Since there are only N_m pairs (t'_m, t''_m), it follows by (15.23) through (15.26) that

$$\mathsf{P}[C_n \geq \epsilon] \leq 2\exp\Big[Bm2^{m/2} - \frac{\epsilon^2}{2(2^{-m} + \epsilon/2^{\theta m})}\Big] \to_n 0.$$

This implies (15.27).

Because of (15.27), (15.8) will follow if we prove that

$$\lim_{\delta} \limsup_{n} \mathsf{P}\Big[\max_{h(s,t)\le\delta} \Delta^n(s''_m, t''_m) \ge \epsilon\Big] = 0, \tag{15.28}$$

where the supremum has become a maximum (and there are no measurability problems) because T_m is finite. We prove (15.28) by the method of *chaining*. We take $\delta = 2^{-k}$, where k will be specified later. Suppose that $m > k$, and consider the chain of transitions

$$s''_m \to s''_{m-1} \to \cdots \to s''_{k+1} \to s''_k \to t''_k \to t''_{k+1} \to \cdots \to t''_{m-1} \to t''_m.$$

To bound the increment in (15.28), we add the increments across the links of the chain:

$$\begin{aligned}\Delta^n(s''_m, t''_m) \le & \sum_{k<i\le m} \Delta^n(s''_i, s''_{i-1}) \\ & + \Delta^n(s''_k, t''_k) + \sum_{k\le i<m} \Delta^n(t''_i, t''_{i+1}).\end{aligned} \tag{15.29}$$

Define

$$F_n(\delta) = \max_{h(s,t)\le\delta} \Delta^n(s''_m, t''_m), \quad D_{ni} = \max_{t\in T} \Delta^n(t''_i, t''_{i-1}),$$

$$E_{nk} = \max_{h(s,t,)\le 2^{-k}} \Delta^n(s''_k, t''_k).$$

Note that $F_n(\delta) = F_n(2^{-k}) = E_{n,m(n)}$. By (15.29),

$$F_n(\delta) \le 2\sum_{i=k+1}^{n} D_{ni} + E_{nk}, \tag{15.30}$$

and we can treat the terms on the right separately.

The number of possible pairs (t''_i, t''_{i-1}) in D_{ni} exceeds N_i, because for each τ in T_i there are several t in T for which $\tau = t''_i$, and these different t's can give different values for t''_{i-1}. But the number of these pairs is certainly bounded by N_i^2. By (15.25), $\lambda(t''_i \Delta t''_{i-1}) \le \lambda(t''_i \Delta t) + \lambda(t \Delta t''_{i-1}) < 2^{-(i-2)}$. If $i \le m$, then $\sqrt{n} \ge 2^{\theta i}$, and (15.23) and (15.24) imply

$$\begin{aligned}\mathsf{P}[D_{ni} \ge i^{-2}] \le & N_i^2 \cdot 2\exp\Big[-\frac{i^{-4}}{2(2^{-(i-2)} + i^{-2}/2^{\theta i})}\Big] \\ & \le 2\exp[-b_1 2^{\theta i}] \quad \text{for } i \le m,\end{aligned} \tag{15.31}$$

where b_1 is a positive constant independent of i.

If $h(s,t) \le 2^{-k}$, then [M17(30)], $\lambda(s\Delta t) < 45 \cdot 2^{-k}$ and hence, by (15.25), $\lambda(s_k''\Delta t_k'') < 47 \cdot 2^{-k}$. Suppose that $k \le m$ and use (15.23) and (15.24) again:

$$\begin{aligned}(15.32)\qquad \mathsf{P}[E_{nk} \ge \epsilon] \le & N_k^2 \cdot 2\exp\left[-\frac{\epsilon^2}{2(47 \cdot 2^{-k} + \epsilon/2^{\theta k})}\right] \\ \le & 2\exp[-b_2 \epsilon 2^{\theta k}] \quad \text{for } k \le m,\end{aligned}$$

where b_2 is independent of k and ϵ. Given ϵ, choose k large enough that $\sum_{i \ge k} i^{-2} < \epsilon$. Since (15.31) and (15.32) both hold if $k \le i \le m$, it follows by (15.30) that

$$\mathsf{P}[F_n(2^{-k}) \ge 3\epsilon] \le \sum_{i=k+1}^{m} 2\exp[-b_1 2^{\theta i}] + 2\exp[-b_2 \epsilon 2^{\theta k}].$$

Increase the sum here by requiring only $i > k$. By further increasing k, we can ensure that the right side is less than η. Therefore, for each ϵ and η, there is a k, such that, if $\delta = 2^{-k}$, then $\mathsf{P}[F_n(\delta) \ge 3\epsilon] < \eta$ for n large enough that $m > k$. This proves (15.28). □

If the index set T in Theorem 15.2 is replaced by the smaller set consisting of the rectangles $[0,u] \times [0,v]$ for $0 \le u, v \le 1$, the theorem then has to do with the distribution of two-dimensional empirical distribution functions: The methods touched on in this section have implications for empirical-process theory.

SECTION 16. THE SPACE $D[0,\infty)$

Here we extend the Skorohod theory to the space $D_\infty = D[0,\infty)$ of cadlag functions on $[0,\infty)$, a space more natural than $D = D[0,1]$ for certain problems. This theory is needed in Chapter 4 but not in Chapter 5.

Definitions

In addition to D_∞, consider for each $t > 0$ the space $D_t = D[0,t]$ of cadlag functions on $[0,t]$. All the definitions for D_1 have obvious analogues for D_t: $\|x\|_t = \sup_{s \le t} |x(s)|$, Λ_t, $\|\lambda\|_t^\circ$, d_t°, d_t. And all the theorems carry over from D_1 to D_t in an obvious way. If x is an element of D_∞, or if x is an element of D_u and $t < u$, then x can also be regarded as an element of D_t by resticting its domain of definition. This new cadlag function will be denoted by the same symbol; it will always be clear what domain is intended.

One might try to define Skorohod convergence $x_n \to x$ in D_∞ by requiring that $d_t^\circ(x_n, x) \to_n 0$ for each finite, positive t (x_n and x restricted to $[0,t]$, of course). But in a natural theory, $x_n = I_{[0,1-1/n]}$ will converge to $x = I_{[0,1]}$ in D_∞, while $d_1^\circ(x_n, x) = 1$. The problem here is that x is discontinuous at 1, and the definition must accommodate discontinuities.

Lemma 1. *Let x_n and x be elements of D_u. If $d_u^\circ(x_n, x) \to_n 0$ and $m < u$, and if x is continuous at m, then $d_m^\circ(x_n, x) \to_n 0$.*

PROOF. We can (Theorem 12.1) work with the metrics d_u and d_m. By hypothesis, there are elements λ_n of Λ_u such that $\|\lambda_n - I\|_u \to_n 0$ and $\|x_n - x\lambda_n\|_u \to_n 0$. Given ϵ, choose δ so that $|t-m| \le 2\delta$ implies $|x(t) - x(m)| < \epsilon/2$. Now choose n_0 so that, if $n \ge n_0$ and $t \le u$, then $|\lambda_n t - t| < \delta$ and $|x_n(t) - x(\lambda_n t)| < \epsilon/2$. Then, if $n \ge n_0$ and $|t-m| \le \delta$, we have $|\lambda_n t - m| \le |\lambda_n t - t| + |t-m| < 2\delta$ and hence $|x_n(t) - x(m)| \le |x_n(t) - x(\lambda_n t)| + |x(\lambda_n t) - x(m)| < \epsilon$. Thus

$$\sup_{|t-m|\le\delta} |x(t)-x(m)| < \epsilon, \quad \sup_{|t-m|\le\delta} |x_n(t)-x(m)| < \epsilon, \text{ for } n \ge n_0. \tag{16.1}$$

If (i) $\lambda_n m < m$, let $p_n = m - n^{-1}$; if (ii) $\lambda_n m > m$, let $p_n = \lambda_n^{-1}(m - n^{-1})$; and if (iii) $\lambda_n m = m$, let $p_n = m$. In case (i), $|p_n - m| = n^{-1}$; in case (ii), $|p_n - m| \le |\lambda_n^{-1}(m-n^{-1}) - (m-n^{-1})| + n^{-1}$; and in case (iii), $p_n = m$. Therefore, $p_n \to m$. Since $|\lambda_n p_n - m| \le |\lambda_n p_n - p_n| + |p_n - m|$, we also have $\lambda_n p_n \to m$. Define $\mu_n \in \Lambda_m$ so that $\mu_n t = \lambda_n t$ on $[0, p_n]$ and $\mu_n m = m$; and interpolate linearly on $[p_n, m]$. Since $\mu_n m = m$ and μ_n is linear over $[p_n, m]$, we have $|\mu_n t - t| \le |\lambda_n p_n - p_m|$ there, and therefore, $\mu_n t \to t$ uniformly on $[0,m]$. Increase the n_0 of (16.1) so that $p_n > m - \delta$ and $\lambda_n p_n > m - \delta$ for $n \ge n_0$. If $t \le p_n$, then $|x_n(t) - x(\mu_n t)| = |x_n(t) - x(\lambda_n t)| \le \|x_n - x\lambda_n\|_u$. On the other hand, if $p_n \le t \le m$ and $n \ge n_0$, then $m \ge t \ge p_n > m - \delta$ and $m \ge \mu_n t \ge \mu_n p_n = \lambda_n p_n > m - \delta$, and therefore, by (16.1), $|x_n(t) - x(\mu_n t)| \le |x_n(t) - x(m)| + |x(m) - x(\mu_n t)| < 2\epsilon$. Thus $|x_n(t) - x(\mu_n t)| \to 0$ uniformly on $[0,m]$. □

The metric on D_∞ will be defined in terms of the metrics $d_m^\circ(x,y)$ for integral m,[†] but before restricting x and y to $[0,m]$, we transform

[†] In what follows, m will be an integer, although the m in Lemma 1 can be arbitrary.

them in such a way that they are continuous at m. Define

$$g_m(t) = \begin{cases} 1 & \text{if } t \le m-1, \\ m-t & \text{if } m-1 \le t \le m, \\ 0 & \text{if } t \ge m. \end{cases} \tag{16.2}$$

For $x \in D_\infty$, let x^m be the element of D_∞ defined by

$$x^m(t) = g_m(t)x(t), \quad t \ge 0. \tag{16.3}$$

And now take

$$d_\infty^\circ(x,y) = \sum_{m=1}^\infty 2^{-m}(1 \wedge d_m^\circ(x^m, y^m)). \tag{16.4}$$

If $d_\infty^\circ(x,y) = 0$, then $d_m^\circ(x,y) = 0$ and $x^m = y^m$ for all m, and this implies $x = y$. The other properties being easy to establish, d_∞° is a metric on D_∞; it defines the Skorohod topology there. If we replace d_m° by d_m in (16.4), we have a metric d_∞ equivalent to d_∞°.

Properties of the Metric

Let Λ_∞ be the set of continuous, increasing maps of $[0,\infty)$ onto itself.

Theorem 16.1. *There is convergence $d_\infty^\circ(x_n, x) \to 0$ in D_∞ if and only if there exist elements λ_n of Λ_∞ such that*

$$\sup_{t<\infty} |\lambda_n t - t| \to 0 \tag{16.5}$$

and, for each m,

$$\sup_{t \le m} |x_n(\lambda_n t) - x(t)| \to 0. \tag{16.6}$$

PROOF. Suppose that $d_\infty^\circ(x_n, x)$ and $d_\infty(x_n, x)$ go to 0. Then there exist elements λ_n^m of Λ_m such that

$$\epsilon_n^m = \|I - \lambda_n^m\|_m \vee \|x_n^m \lambda_n^m - x^m\|_m \to_n 0$$

for each m. Choose l_m so that $n \ge l_m$ implies $\epsilon_n^m < 1/m$. Arrange that $l_m < l_{m+1}$, and for $l_m \le n < l_{m+1}$, let $m_n = m$. Since $l_{m_n} \le n < l_{m_n+1}$, we have $m_n \to_n \infty$ and $\epsilon_n^{m_n} < 1/m_n$. Define

$$\lambda_n t = \begin{cases} \lambda_n^{m_n} t & \text{if } t \le m_n, \\ t + \lambda_n^{m_n}(m_n) - m_n & \text{if } t \ge m_n. \end{cases}$$

Then $|\lambda_n t - t| < 1/m_n$ for $t \geq m_n$ as well as for $t \leq m_n$, and therefore, $\sup_t |\lambda_n t - t| \leq 1/m_n \to_n 0$. Hence (16.5). Fix c. If n is large enough, then $c < m_n - 1$, and so $\|x_n\lambda_n - x\|_c = \|x_n^{m_n}\lambda_n^{m_n} - x^{m_n}\|_c \leq 1/m_n \to_n 0$, which is equivalent to (16.6).

Now suppose that (16.5) and (16.6) hold. Fix m. First,

$$x_n^m(\lambda_n t) = g_m(\lambda_n t)x_n(\lambda_n t) \to_n g_m(t)x(t) = x^m(t) \tag{16.7}$$

holds uniformly on $[0, m]$. Define p_n and μ_n as in the proof of Lemma 1. As before, $\mu_n t \to t$ uniformly on $[0, m]$. For $t \leq p_n$, $|x^m(t) - x_n^m(\mu_n t)| = |x^m(t) - x_n^m(\lambda_n t)|$, and this goes to 0 uniformly by (16.7). For the case $p_n \leq t \leq m$, first note that $|x^m(u)| \leq g_m(u)\|x\|_m$ for all $u \geq 0$ and hence

$$|x^m(t) - x_n^m(\mu_n t)| \leq g_m(t)\|x\|_m + g_m(\mu_n t)\|x_n\|_m. \tag{16.8}$$

By (16.5), for large n we have $\lambda_n(2m) > m$ and hence $\|x_n\|_m \leq \|x_n\lambda_n\|_{2m}$; and $\|x_n\lambda_n\|_{2m} \to_n \|x\|_{2m}$ by (16.6). This means that $\|x_n\|_m$ is bounded (m is fixed). Given ϵ, choose n_0 so that $n \geq n_0$ implies that p_n and $\mu_n p_n$ both lie in $(m-\epsilon, m]$, an interval on which g_m is bounded by ϵ. If $n \geq n_0$ and $p_n \leq t \leq m$, then t and $\mu_n t$ both lie in $(m - \epsilon, m]$, and it follows by (16.8) that $|x^m(t) - x_n^m(\mu_n t)| \leq \epsilon(\|x\|_m + \|x_n\|_m)$. Since $\|x_n\|_m$ is bounded, this implies that $|x^m(t) - x_n^m(\mu_n t)| \to_n 0$ holds uniformly on $[p_n, m]$ as well as on $[0, p_n]$. Therefore, $d_m(x_n^m, x^m) \to_n 0$ for each m, and hence $d_\infty(x_n, x)$ and $d_\infty^\circ(x_n, x)$ go to 0. □

A second characterization of convergence in D_∞:

Theorem 16.2. *There is convergence $d_\infty^\circ(x_n, x) \to 0$ in D_∞ if and only if $d_t^\circ(x_n, x) \to 0$ for each continuity point t of x.*

This theorem almost brings us back to that first, unworkable definition of convergence in D_∞.

PROOF. If $d_\infty^\circ(x_n, x) \to 0$, then $d_m^\circ(x_n^m, x^m) \to_n 0$ for each m. Given a continuity point t of x, fix an integer m for which $t < m - 1$. By Lemma 1 (with t and m in the roles of m and u) and the fact that y and y^m agree on $[0, t]$, $d_t^\circ(x_n, x) = d_t^\circ(x_n^m, x^m) \to_n 0$.

To prove the reverse implication, choose continuity points t_m of x in such a way that $t_m \uparrow \infty$. The argument now follows the first part of the proof of Theorem 16.1. Choose elements λ_n^m of Λ_{t_m} in such a way that

$$\epsilon_n^m = \|\lambda_n^m - I\|_{t_m} \vee \|x_n\lambda_n^m - x\|_{t_m} \to_n 0$$

for each m. As before, define integers m_n in such a way that $m_n \to \infty$ and $\epsilon_n^{m_n} < 1/m_n$, and this time define $\lambda_n \in \Lambda_\infty$ by

$$\lambda_n t = \begin{cases} \lambda_n^{m_n} t & \text{if } t \le t_{m_n}, \\ t & \text{if } t \ge t_{m_n}. \end{cases}$$

Then $|\lambda_n t - t| \le 1/m_n$ for all t, and if $c < t_{m_n}$, then $\|x_n \lambda_n - x\|_c = \|x_n \lambda_n^{m_n} - x\|_c \le 1/m_n \to_n 0$. This implies that (16.5) and (16.6) hold, which in turn implies that $d_\infty^\circ(x_n, x) \to 0$. □

Separability and Completeness

For $x \in D_\infty$, define $\psi_m x$ as x^m restricted to $[0, m]$. Then, since $d_m^\circ(\psi_m x_n, \psi_m x) = d_m^\circ(x_n^m, x^m)$, ψ_m is a continuous map of D_∞ into D_m. In the product space $\Pi = D_1 \times D_2 \times \cdots$, the metric $\rho(\alpha, \beta) = \sum_{m=1}^\infty 2^{-m}(1 \wedge d_m^\circ(\alpha_m, \beta_m))$ defines the product topology, that of coordinatewise convergence [M6]. Now define $\psi\colon D_\infty \to \Pi$ by $\psi x = (\psi_1 x, \psi_2 x, \ldots)$:

$$\psi_m\colon D_\infty \to D_m, \quad \psi\colon D_\infty \to \Pi.$$

Then $d_\infty^\circ(x, y) = \rho(\psi x, \psi y)$: ψ is an isometry of D_∞ into Π.

Lemma 2. *The image ψD_∞ is closed in Π.*

PROOF. Suppose that $x_n \in D_\infty$ and $\alpha \in \Pi$, and $\rho(\psi x_n, \alpha) \to_n 0$; then $d_m^\circ(x_n^m, \alpha_m) \to_n 0$ for each m. We must find an x in D_∞ such that $\alpha = \psi x$—that is, $\alpha_m = \psi_m x$ for each m.

Let T be the dense set of t such that, for every $m \ge t$, α_m is continuous at t. Since $d_m^\circ(x_n^m, \alpha_m) \to_n 0$, $t \in T \cap [0, m]$ implies $x_n^m(t) = g_m(t) x_n(t) \to_n \alpha_m(t)$. This means that, for every t in T, the limit $x(t) = \lim_n x_n(t)$ exists (consider an $m > t + 1$, so that $g_n(t) = 1$). Now $g_m(t) x(t) = \alpha_m(t)$ on $T \cap [0, m]$. It follows that $x(t) = \alpha_m(t)$ on $T \cap [0, m-1]$, so that x can be extended to a cadlag function on each $[0, m-1]$ and then to a cadlag function on $[0, \infty)$. And now, by right continuity, $g_m(t) x(t) = \alpha_m(t)$ on $[0, m]$, or $\psi_m x = x^m = \alpha_m$. □

Theorem 16.3. *The space D_∞ is separable and complete.*

PROOF. Since Π is separable and complete [M6], so are the closed subspace ψD_∞ and its isometric copy D_∞. □

Compactness

Theorem 16.4. *A set A is relatively compact in D_∞ if and only if, for each m, $\psi_m A$ is relatively compact in D_m.*

PROOF. If A is relatively compact, then A^- is compact and hence [M5] the continuous image $\psi_m A^-$ is also compact. But then, $\psi_m A$, as a subset of $\psi_m A^-$, is relatively compact.

Conversely, if each $\psi_m A$ is relatively compact, then each $(\psi_m A)^-$ is compact, and therefore [M6] $B = (\psi_1 A)^- \times (\psi_2 A)^- \times \cdots$ and (Lemma 2) $E = \psi D_\infty \cap B$ are both compact in Π. But $x \in A$ implies $\psi_m x \in (\psi_m A)^-$ for each m, so that $\psi x \in B$. Hence $\psi A \subset E$, which implies that ψA is totally bounded and so is its isometric image A. □

For an explicit analytical characterization of relative compactness, analogous to the Arzelà-Ascoli theorem, we need to adapt the $w'(x,\delta)$ of (12.6) to D_∞. For an x in D_m (or an x in D_∞ restricted to $[0,m]$), define

$$w'_m(x,\delta) = \inf \max_{1\le i\le v} w(x,[t_{i-1},t_i)), \tag{16.9}$$

where the infimum extends over all decompositions $[t_{i-1},t_i), 1 \le i \le v$, of $[0,m)$ such that $t_i - t_{i-1} > \delta$ for $1 \le i < v$. Note that the definition does not require $t_v - t_{v-1} > \delta$: Although 1 plays a special role in the theory of D_1, the integers m should play no special role in the theory of D_∞.

The exact analogue of $w'(x,\delta)$ is (16.9), but with the infimum extending only over the decompositions satisfying $t_i - t_{i-1} > \delta$ for $i = v$ as well as for $i < v$. Call this $\bar{w}_m(x,\delta)$. By an obvious extension of Theorem 12.3, a set B in D_m is relatively compact if and only if $\sup_{x\in B}\|x\|_m < \infty$ and $\lim_\delta \sup_{x\in B} \bar{w}(x,\delta) = 0$. Suppose that $A \subset D_\infty$, and transform the two conditions by giving $\psi_m A$ the role of B. By Theorem 16.4, A is relatively compact if and only if, for every m (recall that $\psi_m x = x^m$),

$$\sup_{x\in A}\|x^m\|_m < \infty \tag{16.10}$$

and

$$\lim_{\delta\to 0}\sup_{x\in A}\bar{w}_m(x^m,\delta) = 0. \tag{16.11}$$

The next step is to show that (16.10) and (16.11) are together equivalent to the condition that, for every m,

$$\sup_{x\in A}\|x\|_m < \infty \tag{16.12}$$

and

$$\lim_{\delta \to 0} \sup_{x \in A} w'_m(x, \delta) = 0. \tag{16.13}$$

The equivalence of (16.10) and (16.12) (for all m) follows easily because $\|x^m\|_m \le \|x\|_m \le \|x^{m+1}\|_{m+1}$. Suppose (16.12) and (16.13) both hold, and let K_m be the supremum in (16.12). If $x \in A$ and $\delta < 1$, then we have $|x^m(t)| \le K_m \delta$ for $m - \delta \le t < m$. Given ϵ, choose δ so that $K_m \delta < \epsilon/4$ and the supremum in (16.13) is less than $\epsilon/2$. If $x \in A$ and $m - \delta$ lies in the interval $[t_{j-1}, t_j)$ of the corresponding partition, replace the intervals $[t_{i-1}, t_i)$ for $i \ge j$ by the single interval $[t_{j-1}, m)$. This new partition shows that $\bar{w}_m(x, \delta) < \epsilon$. Hence (16.11).

That (16.11) implies (16.13) is clear because $w'_m(x, \delta) \le \bar{w}_m(x, \delta)$: An infimum increases if its range is reduced.

This gives us the following criterion.

Theorem 16.5. *A set A in D_∞ is relatively compact if and only if* (16.12) *and* (16.13) *hold for all m.*

Finite-Dimensional Sets

Let $\mathcal{D}_m$ and $\mathcal{D}_\infty$ be the Borel σ-fields in D_m and D_∞ for the metrics d_m° and d_∞°. And let $\pi_{t_1 \cdots t_k}: D_\infty \to R^k$ ($t_i \ge 0$) and $\pi^m_{t_1 \cdots t_k}: D_m \to R^k$ ($0 \le t_i \le m$) be the natural projections. We know from Theorem 12.5 that each π_t^m is measurable $\mathcal{D}_m/\mathcal{R}^1$; and ψ_m is measurable $\mathcal{D}_\infty/\mathcal{D}_m$ because it is continuous. But $\pi_t = \pi_t^m \psi_m$ for $t \le m - 1$, and therefore each π_t is measurable $\mathcal{D}_\infty/\mathcal{R}^1$ and each $\pi_{t_1 \ldots t_k}$ is measurable $\mathcal{D}_\infty/\mathcal{R}^k$. Finally, the argument following (12.37) shows that π_0 is everywhere continuous and that, for $t > 0$, π_t is continuous at x if and only if x is continuous at t.

Suppose that $T \subset [0, \infty)$. Define $\sigma[\pi_t : t \in T]$ in the usual way, and let $p[\pi_t : t \in T]$ be the π-system of sets of the form $\pi^{-1}_{t_1 \ldots t_k} H$ for $k \ge 1$, $t_i \in T$, and $H \in \mathcal{R}^k$—the finite-dimensional D_∞-sets based on time-points lying in T.

Theorem 16.6. (i) *The projection π_0 is continuous, and for $t > 0$, π_t is continuous at x if and only if x is continuous at t.*

(ii) *Each π_t is measurable $\mathcal{D}_\infty/\mathcal{R}^1$; each $\pi_{t_1 \cdots t_k}$ is measurable $\mathcal{D}_\infty/\mathcal{R}^k$.*

(iii) *If T is dense in $[0, \infty)$, then $\sigma[\pi_t : t \in T] = \mathcal{D}_\infty$ and $p[\pi_t : t \in T]$ is a separating class.*

PROOF. Only (iii) needs proof. If we show that

$$\mathcal{D}_\infty \subset \sigma[\pi_t : t \in T], \tag{16.14}$$

then there is actually equality, since each π_t is measurable $\mathcal{D}_\infty/\mathcal{R}^1$. And (16.14) will imply that the π-system $p[\pi_t: t \in T]$ generates $\mathcal{D}_\infty$ and hence is a separating class. The finite case of (16.14) is Theorem 12.5(iii), which applies to $T_m = (T \cap [0,m]) \cup \{m\}$:

$$\mathcal{D}_m = \sigma[\pi_t^m : t \in T_m]. \tag{16.15}$$

The problem is to derive (16.14) from (16.15).

If $t \le m$, then

$$\psi_m^{-1}((\pi_t^m)^{-1}H) = [x: \pi_t^m(\psi_m x) \in H] = [x: g_m(t)\pi_t x \in H]. \tag{16.16}$$

If $t = m$, then $g_m(t) = 0$, and this set is D_∞ or $\emptyset$ according as $0 \in H$ or $0 \notin H$—an element of $\sigma[\pi_t: t \in T]$ in either case. If $t < m$, then (16.16) is $[x: \pi_t x \in (g_m(t))^{-1}H]$, and this, too, lies in $\sigma[\pi_t: t \in T]$ (if $H \in \mathcal{R}^1$). Therefore, if $A = (\pi_t^m)^{-1}H$ and $t \in T_m$, then $\psi_m^{-1}A \in \sigma[\pi_t: t \in T]$. By (16.15), the sets A of this form generate $\mathcal{D}_m$, and it follows that ψ_m is measurable $\sigma[\pi_t: t \in T]/\mathcal{D}_m$.

Now $d_m^\circ(\,\cdot\,,\alpha)$ is measurable $\mathcal{D}_m/\mathcal{R}^1$ for each α in D_m, and so (composition) $d_m^\circ(\psi_m(\,\cdot\,),\alpha)$ is measurable $\sigma[\pi_t: t \in T]/\mathcal{R}^1$, as is $d^\circ(\,\cdot\,,y) = \sum 2^{-m}(1 \wedge d_m^\circ(\psi_m(\,\cdot\,),\psi_m(y)))$. This means that $\sigma[\pi_t: t \in T]$ contains the balls and hence (separability) contains $\mathcal{D}_\infty$. □

Weak Convergence

Let P_n and P be probability measures on $(D_\infty, \mathcal{D}_\infty)$.

Lemma 3. *A necessary and sufficient condition for $P_n \Rightarrow P$ (on D_∞) is that $P_n\psi_m^{-1} \Rightarrow_n P\psi_m^{-1}$ (on D_m) for every m.*

PROOF. Since ψ_m is continuous, the necessity follows by the mapping theorem.

For the sufficiency, we also need the isometry ψ of D_∞ into Π; since ψ maps D_∞ onto ψD_∞ there is the inverse isometry ξ:

$$D_\infty \xrightarrow{\psi_k} D_k, \quad D_\infty \xrightarrow{\psi} \Pi, \quad D_\infty \xleftarrow{\xi} \psi D_\infty.$$

Consider now the Borel σ-field $\mathcal{P}$ for Π with the product topology [M6]. Define the (continuous) projection $\zeta_k: \Pi \to D_1 \times \cdots \times D_k$ by $\zeta_k(\alpha) = (\alpha_1, \ldots, \alpha_k)$, and let $\mathcal{P}_f$ be the class of sets $\zeta_k^{-1}H$ for $k \ge 1$ and $H \in \mathcal{D}_1 \times \cdots \times \mathcal{D}_k$. The argument in Example 2.4 shows that $\mathcal{P}_f$ is a convergence-determining class: Given a ball $B(\alpha,\epsilon)$ in Π, take k so that $2^{-k} < \epsilon/2$ and argue as before (use Theorem 2.4) with the sets

$A_\eta = [\beta \in \Pi: d_i^\circ(\alpha_i, \beta_i) < \eta, i \le k]$ for $0 < \eta < \epsilon/2$. And now argue as in Example 2.6: $\partial \zeta_k^{-1} H = \zeta_k^{-1} \partial H$ for $H \in \mathcal{D}_1 \times \cdots \times \mathcal{D}_k$ and hence, for probability measures Q_n and Q on Π, $Q_n \zeta_k^{-1} \Rightarrow Q \zeta_k^{-1}$ for every k implies $Q_n \Rightarrow Q$.

We need one further map. For $\gamma \in D_k$, let $\rho_k(\gamma) = (\alpha_1, \ldots, \alpha_k)$ be the point of $D_1 \times \cdots \times D_k$ for which $\alpha_i(t) = g_i(t)\gamma(t)$ on $[0, i]$, $i = 1, \ldots, k$:

$$\Pi \xrightarrow{\zeta_k} D_1 \times \cdots \times D_k, \quad D_k \xrightarrow{\rho_k} D_1 \times \cdots \times D_k.$$

The map ρ_k is continuous: Suppose that $\gamma_n \to \gamma$ in D_k; then $g_i \cdot \gamma_n \to g_i \cdot \gamma$ in D_k, and this is also true if the functions are restricted to $[0, i]$ because $g_i \cdot \gamma$ is continuous at i (use Lemma 1).

The hypothesis now is that $P_n \psi_k^{-1} \Rightarrow_n P \psi_k^{-1}$ (on D_k) for each k, and by the mapping theorem this implies that $P_n \psi_k^{-1} \rho_k^{-1} \Rightarrow P \psi_k^{-1} \rho_k^{-1}$ (on $D_1 \times \cdots \times D_k$). Since $\rho_k \psi_k = \zeta_k \psi$, we have $P_n \psi^{-1} \zeta_k^{-1} \Rightarrow P \psi^{-1} \rho_k^{-1}$ (on D_k) for each k. It follows by the argument given above that $P_n \psi^{-1} \Rightarrow P \psi^{-1}$ (on Π).

Extend the isometry ξ defined above to a map η on Π by giving it some fixed value (in D_∞) outside the closed set ψD_∞. Then η is continuous when restricted to ψD_∞, and since ψD_∞ supports $P\psi^{-1}$ and the $P_n \psi^{-1}$, it follows by Example 2.10 that ($\eta\psi$ is the identity on D_∞) $P_n = P_n \psi^{-1} \eta^{-1} \Rightarrow P \psi^{-1} \eta^{-1} = P$. □

As in the case of $D[0,1]$, define T_P as the set of t for which π_t is continuous outside a set of P-measure 0. As in Section 13, $t \in T_P$ if and only if $PJ_t = 0$ where J_t is the set of x that are disconinuous at t. And as before, *T_P contains 0, and its complement is at most countable.* For x in D_∞, let $r_t x$ be the restriction of x to $[0, t]$.

We must prove that r_t is measurable $\mathcal{D}_\infty / \mathcal{D}_t$. Define $r_t^k x$ as the element of D_t having the value $x((i-1)t/k)$ on $[(i-1)t/k, it/k)$, $1 \le i \le k$, and the value $x(t)$ at t. Since the $\pi_{it/k}$ are measurable $\mathcal{D}_\infty / \mathcal{R}^1$, it follows as in the proof of Theorem 12.5(iii) that r_t^k is measurable $\mathcal{D}_\infty / \mathcal{D}_t$. By Lemma 3 of Section 12, $d_t(r_t^k x, r_t x) \le tk^{-1} \vee w_t'(x, tk^{-1}) \to_k 0$ for each x in D_∞. Therefore [M10], r_t is measurable.

Theorem 16.7 *A necessary and sufficient condition for $P_n \Rightarrow P$ is that $P_n r_t^{-1} \Rightarrow P r_t^{-1}$ for every t in T_P.*

PROOF. Given a t in T_P, fix an integer m exceeding $t + 1$; if $d_\infty^\circ(x_n, x) \to 0$, then $d_m^\circ(x_n, x) \to 0$, and if x is continuous at t, then it follows by Lemma 1 that $d_t^\circ(r_t x_n, r_t x) \to 0$. In other words, the set of points at which r_t is continuous contains J_t^c, and if $P_n \Rightarrow P$ and $t \in T_P$, then the mapping theorem gives $P_n r_t^{-1} \Rightarrow P r_t^{-1}$.

For the reverse implication, it is enough, by Lemma 3, to show that $P_n \psi_m^{-1} \Rightarrow P\psi_m^{-1}$. Choose t so that $t \in T_P$ and $t \geq m$. Let τ_m be the continuous mapping from D_m to D_∞ defined by

$$(\tau_m x)(s) = \begin{cases} g_m(s)x(s) & \text{if } s \leq t \\ 0 & \text{if } s \geq t. \end{cases}$$

Since $\psi_m = \tau_m r_t$, the mapping theorem gives $P_n \psi_m^{-1} = (P_n r_t^{-1})\tau_m^{-1} \Rightarrow (P r_t^{-1})\tau_m^{-1} = P\psi_m^{-1}$. □

Tightness

Here is the analogue of Theorem 13.2; it follows from Theorem 16.5 by the analogous argument.

Theorem 16.8. *The sequence $\{P_n\}$ is tight if and only if these two conditions hold:*

(i) *For each m,*

$$\lim_{a\to\infty} \limsup_n P_n[x\colon \|x\|_m \geq a] = 0. \tag{16.17}$$

(ii) *For each m and ϵ,*

$$\lim_\delta \limsup_n P_n[x\colon w'_m(x,\delta) \geq \epsilon] = 0. \tag{16.18}$$

And there is the corresponding corollary. Let

$$j_m(x) = \sup_{t\leq m} |x(t) - x(t-)|. \tag{16.19}$$

Corollary. *Either of the following two conditions can be substituted for* (i) *in Theorem* 16.8*:*

(i′) *For each t in a set T that is dense in $[0,\infty)$,*

$$\lim_{a\to\infty} \limsup_n P_n[x\colon |x(t)| \geq a] = 0. \tag{16.20}$$

(i″) *The relation* (16.20) *holds for $t = 0$, and for each m,*

$$\lim_{a\to\infty} \limsup_n P_n[x\colon j_m(x) \geq a] = 0. \tag{16.21}$$

PROOF. The proof is almost the same as that for the corollary to Theorem 13.2. Assume (ii) and (i′). Choose points t_i such that $0 = t_0 < t_1 < \cdots < t_v = m$, $t_i - t_{i-1} > \delta$ for $1 \le i \le v-1$ (perhaps not for $i = v$), and $w_x[t_{i-1}, t_i) < w'_m(x, \delta) + 1$ for $1 \le i \le v$. Choose from T points s_j such that $0 = s_0 < s_1 < \cdots < s_k = m$ and $s_j - s_{j-1} < \delta$ for $1 \le j \le k$. Let $m(x) = \max_{0 \le j \le k} |x(s_j)|$. If $t_v - t_{v-1} > \delta$, then $\|x\|_m \le m(x) + w'_m(x, \delta) + 1$, just as before. If $t_v - t_{v-1} \le \delta$ (and $\delta < 1$, so that $t_{v-1} > m-1$), then $\|x\|_{m-1} \le m(x) + w'_m(x, \delta) + 1$. The old argument now gives (16.17), but with $\|x\|_m$ replaced by $\|x\|_{m-1}$, which is just as good.

In the proof that (ii) and (i″) imply (i), we have $(v-1)\delta \le m$ instead of $v\delta \le 1$. But then, $v \le m\delta^{-1} + 1$, and the old argument goes through. □

Aldous's Tightness Criterion

Let X^n be random elements of D_∞. We need probabilistic conditions under which the (distributions of the) X^n are tight. First, (16.17) translates into

$$\lim_{a \to \infty} \limsup_n \mathsf{P}[\|X^n\|_m \ge a] = 0, \tag{16.22}$$

which must hold for each m. There is a useful stopping-time condition that implies (16.18).

A *stopping time* for the process X^n is a nonnegative random variable τ with the property that, for each $t \ge 0$, the event $[\tau \le t]$ lies in the σ-field $\sigma[X^n_s : s \le t]$. A stopping time also satisfies $[\tau = t] \in \sigma[X^n_s : s \le t]$. All our stopping times will be discrete in the sense that they have finite range. Consider two conditions.

Condition 1°. For each ϵ, η, m, there exist a δ_0 and an n_0 such that, if $\delta \le \delta_0$ and $n \ge n_0$, and if τ is a discrete X^n-stopping time satisfying $\tau \le m$, then

$$\mathsf{P}[|X^n_{\tau+\delta} - X^n_\tau| \ge \epsilon] \le \eta. \tag{16.23}$$

Condition 2°. For each ϵ, η, m, there exist a δ and an n_0 such that, if $n \ge n_0$ and τ_1 and τ_2 are X^n-stopping times satisfying $0 \le \tau_1 \le \tau_2 \le m$, then

$$\mathsf{P}[|X^n_{\tau_2} - X^n_{\tau_1}| \ge \epsilon,\ \tau_2 - \tau_1 \le \delta] \le \eta. \tag{16.24}$$

Note that, if (16.24) holds for a value of δ, then it holds for all smaller ones.

Theorem 16.9. *Conditions* 1° *and* 2° *are equivalent.*

PROOF. Since $\tau + \delta$ is a stopping time, 2° implies 1°. For the converse, suppose that $\tau \le m$ and choose δ_0 so that $\delta \le 2\delta_0$ (note the factor 2) and $n \ge n_0$ together imply (16.23). Fix an $n \ge n_0$ and a $\delta \le \delta_0$, and let (enlarge the probability space for X^n) θ be a random variable independent of $\mathcal{F}^n = \sigma[X_s^n : s \ge 0]$ and uniformly distributed over $J = [0, 2\delta]$. For the moment, fix an x in D_∞ and points t_1 and t_2 satisfying $0 \le t_1 \le t_2$. Let μ be the uniform distribution over J, and let $I = [0, \delta]$, $M_i = [s \in J : |x(t_i + s) - x(t_i)| < \epsilon]$, and $d = t_2 - t_1$. Suppose that

$$t_2 - t_1 \le \delta \tag{16.25}$$

and

$$\mu(M_i) = \mathsf{P}[\theta \in M_i] > \frac{3}{4}, \quad \text{for } i = 1 \text{ and } i = 2. \tag{16.26}$$

If $\mu(M_2 \cap I) \le \frac{1}{4}$, then $\mu(M_2) \le \frac{3}{4}$, a contradiction. Hence $\mu(M_2 \cap I) > \frac{1}{4}$, and $(0 \le d \le \delta)$ $\mu((M_2 + d) \cap J) \ge \mu((M_2 \cap I) + d) = \mu(M_2 \cap I) > \frac{1}{4}$. Thus $\mu(M_1) + \mu((M_2 + d) \cap J) > 1$, which implies $\mu(M_1 \cap (M_2 + d)) > 0$. There is therefore an s such that $s \in M_1$ and $s - d \in M_2$, from which follows

$$|x(t_1) - x(t_2)| < 2\epsilon. \tag{16.27}$$

Thus (16.25) and (16.26) together imply (16.27). To put it another way, if (16.25) holds but (16.27) does not, then either $\mathsf{P}[\theta \in M_1^c] \ge \frac{1}{4}$ or $\mathsf{P}[\theta \in M_2^c] \ge \frac{1}{4}$. Therefore,

$$\begin{aligned} &\mathsf{P}[|X_{\tau_2}^n - X_{\tau_1}^n| \ge 2\epsilon,\ \tau_2 - \tau_1 \le \delta] \\ &\le \sum_{i=1}^{2} \mathsf{P}\Big[\mathsf{P}[|X_{\tau_i+\theta}^n - X_{\tau_i}^n| \ge \epsilon \| \mathcal{F}^n] \ge \frac{1}{4}\Big] \le 4 \sum_{i=1}^{2} \mathsf{P}[|X_{\tau_i+\theta}^n - X_{\tau_i}^n| \ge \epsilon]. \end{aligned}$$

Since $0 \le \theta \le 2\delta \le 2\delta_0$, and since θ and $\mathcal{F}^n$ are independent, it follows by (16.23) that the final term here is at most 8η. Therefore, 1° implies 2°. □

This is Aldous's theorem:

Theorem 16.10. *If* (16.22) *and Condition* 1° *hold, then* $\{X^n\}$ *is tight.*

PROOF. By Theorem 16.8, it is enough to prove that

$$\lim_{\delta} \limsup_{n} \mathsf{P}[w'_m(X^n, \delta) \geq \epsilon] = 0. \tag{16.28}$$

Let Δ_k be the set of nonnegative dyadic rationals $j/2^k$ of order k. Define random variables $\tau_0^n, \tau_1^n, \ldots$ by $\tau_0^n = 0$ and

$$\tau_i^n = \min[t \in \Delta_k : \tau_{i-1}^n < t \leq m, |X_t^n - X_{\tau_{i-1}^n}^n| \geq \epsilon],$$

with $\tau_i^n = m$ if there is no such t. The τ_i^n depend on ϵ, m, and k as well as on i and n, although the notation does not show this. It is easy to prove by induction that the τ_i^n are all stopping times.†

Because of Theorem 16.9, we can assume that Condition 2° holds. For given ϵ, η, m, choose δ' and n_0 so that

$$\mathsf{P}[|X_{\tau_i^n}^n - X_{\tau_{i-1}^n}^n| \geq \epsilon, \tau_i^n - \tau_{i-1}^n \leq \delta'] \leq \eta$$

for $i \geq 1$ and $n \geq n_0$. Since $\tau_i^n < m$ implies that $|X_{\tau_i^n}^n - X_{\tau_{i-1}^n}^n| \geq \epsilon$, we have

$$\mathsf{P}[\tau_i^n < m, \tau_i^n - \tau_{i-1}^n \leq \delta'] \leq \eta, \quad i \geq 1, n \geq n_0. \tag{16.29}$$

Now choose an integer q such that $q\delta' \geq 2m$. There is also a δ such that (increase n_0 if necessary) $\mathsf{P}[\tau_i^n < m, \tau_i^n - \tau_{i-1}^n \leq \delta] \leq \eta/q$ for $i \geq 1$ and $n \geq n_0$. But then

$$\mathsf{P}\Big(\bigcup_{i=1}^{q} [\tau_i^n < m, \tau_i^n - \tau_{i-1}^n \leq \delta]\Big) \leq \eta, \quad n \geq n_0. \tag{16.30}$$

Although the τ_i^n depend on k, (16.29) and (16.30) hold for all k simultaneously.

By (16.29),

$$\begin{aligned} \mathsf{E}[\tau_i^n - \tau_{i-1}^n | \tau_q^n < m] &\geq \delta' \mathsf{P}[\tau_i^n - \tau_{i-1}^n \geq \delta' | \tau_q^n < m] \\ &\geq \delta'(1 - \eta/\mathsf{P}[\tau_q^n < m]), \end{aligned}$$

† This is easy because the τ_i^n are discrete; proofs of this kind are more complicated in the continuous case.

and therefore,

$$m \geq \mathsf{E}[\tau_q^n | \tau_q^n < m] = \sum_{i=1}^{q} \mathsf{E}[\tau_i^n - \tau_{i-1}^n | \tau_q^n < m] \geq q\delta'(1 - \eta/\mathsf{P}[\tau_q^n < m]).$$

Since $q\delta' \geq 2m$ by the choice of q, this leads to $\mathsf{P}[\tau_q^n < m] \leq 2\eta$. By this and (16.30),

$$\mathsf{P}\Big([\tau_q^n < m] \cup \bigcup_{i=1}^{q} [\tau_i^n < m,\ \tau_i^n - \tau_{i-1}^n \leq \delta]\Big) \leq 3\eta, \qquad \text{for} \quad k \geq 1,\ n \geq n_0. \tag{16.31}$$

Let A_{nk} be the complement of the set in (16.31). On this set, let v be the first index for which $\tau_v^n = m$. Fix an n beyond n_0. There are points t_i^k (the τ_i^n) such that $0 = t_0^k < \cdots < t_v^k = m$ and $t_i^k - t_{i-1}^k > \delta$ for $1 \leq i < v$ (but not necessarily for $i = v$). And $|X_t^n - X_s^n| < \epsilon$ if s and t lie in the same $[t_{i-1}^k, t_i^k)$ as well as in Δ_k. If $A_n = \limsup_k A_{nk}$, then $\mathsf{P}A_n \geq 1 - 3\eta$, and on A_n there is a sequence of values of k along which v is constant ($v \leq q$) and, for each $i \leq v$, t_i^k converges to some t_i. But then, $0 = t_0 < \cdots < t_v = m$, $t_i - t_{i-1} \geq \delta$ for $i < v$, and by right continuity, $|X_t^n - X_s^n| \leq \epsilon$ if s and t lie in the same $[t_{i-1}, t_i)$. It follows that $w'_m(X^n, \delta) \leq \epsilon$ on a set of probability at least $1 - 3\eta$. Hence (16.28). □

From the corollary to Theorem 16.8 follows this one:

Corollary. *If, for each m, the sequences $\{X_0^n\}$ and $\{j_m(X^n)\}$ are tight on the line, and if Condition 1° holds, then $\{X^n\}$ is tight.*

Finally, Condition 1° can be restated in a sequential form: If τ_n are discrete X^n-stopping times with a common upper bound, and if δ_n are constants converging to 0, then

$$X_{\tau_n+\delta_n}^n - X_{\tau_n}^n \Rightarrow_n 0. \tag{16.32}$$

CHAPTER 4

DEPENDENT VARIABLES

SECTION 17. MORE ON PRIME DIVISORS *

Introduction

Section 4 treats, among other things, the asymptotic distribution of the largest prime divisors of a random integer.[†] Here we study the total number of prime divisors of a random integer and the number of them in a given range, and we show that the limit is normal in some cases and Poisson in others. And we prove the associated functional limit theorems.

On the σ-field of all subsets of $\Omega = \{1, 2, \ldots\}$, let P_n be the probability measure corresponding to a mass of $1/n$ at m for $1 \le m \le n$, so that $\mathsf{P}_n A$ is the proportion among the first n integers that lie in A. And let E_n denote the corresponding expected value. For each integer a,

$$\frac{1}{a} - \frac{1}{n} < \mathsf{P}_n[m\colon a|m] = \frac{1}{n}\left\lfloor \frac{n}{a} \right\rfloor \le \frac{1}{a}: \tag{17.1}$$

the probability that a divides m is approximately $1/a$ for large n. By the fundamental theorem of arithmetic, distinct primes p and q individually divide m if and only if their product does, and this, together with (17.1), leads to the approximate equation

$$\mathsf{P}_n[m\colon p|m,\ q|m] \approx \mathsf{P}_n[m\colon p|m] \cdot \mathsf{P}_n[m\colon q|m], \tag{17.2}$$

the three probabilities having the approximate values $1/pq$, $1/p$, and $1/q$. The events "divisible by p" and "divisible by q" are therefore approximately independent, and similarly for three or more distinct

[†] The theory of Section 4 is not required here; it is used only in passing, for comparison.

primes, and this means that probability theory can be brought to bear on problems of multiplicative arithmetic.

Let $\delta_p(m)$ be 1 or 0 according as p divides m or not. By (17.2), the δ_p behave approximately like independent random variables. Now the sum

$$\sum_{u_n<p\leq v_n} \delta_p(m) \tag{17.3}$$

is the number of prime divisors p of m lying in the range $u_n < p \leq v_n$. There are many theorems about sums of stictly independent random variables, and many of these can be carried over to the arithmetic case.

A General Limit Theorem

Replace (17.3) by the more general sum

$$f_n(m) = \sum_{p\leq n} f_n(p)\delta_p(m), \quad 1 \leq m \leq n. \tag{17.4}$$

Consider independent random variables ξ_p (on some probability space), one for each prime p, that take the values 1 and 0 with probabilities p^{-1} and $1-p^{-1}$. The properties of the sum $S_n = \sum_{p\leq n} f_n(p)\xi_p$ can serve as a guide to those of f_n. Define

$$\mu_n = \sum_{p\leq n} \frac{f_n(p)}{p} = \mathsf{E}[S_n], \quad \sigma_n^2 = \sum_{p\leq n} f_n^2(p)\frac{1}{p}\Big(1-\frac{1}{p}\Big) = \mathrm{Var}[S_n]. \tag{17.5}$$

Assume that $\sigma_n^2 > 0$, at least for large n.

If m is drawn at random from $\{1, 2, \ldots, n\}$, then f_n becomes a random variable governed by P_n. Write $f_n \Rightarrow \xi$ to indicate that this random variable converges in distribution to ξ.

Theorem 17.1. *Suppose there exist positive constants α_n such that*

$$\sum_{\alpha_n<p\leq n} \frac{|f_n(p)|}{p} = o(\sigma_n), \quad \frac{1}{\sigma_n}\sum_{p\leq\alpha_n} |f_n(p)| = o(n^\epsilon), \tag{17.6}$$
$$\max_{p\leq\alpha_n} |f_n(p)| = O(\sigma_n),$$

where the second relation holds for each positve ϵ. If

$$\frac{S_n-\mu_n}{\sigma_n} = \frac{1}{\sigma_n}\sum_{p\leq n}\Big(f_n(p)\Big(\xi_p-\frac{1}{p}\Big)\Big) \Rightarrow \xi \tag{17.7}$$

and the distribution of ξ is determined by its moments, then

$$\frac{f_n - \mu_n}{\sigma_n} = \frac{1}{\sigma_n} \sum_{p \le n} \left(f_n(p) \left(\delta_p - \frac{1}{p} \right) \right) \Rightarrow \xi. \tag{17.8}$$

The proof depends on the method of moments. Consider first an application.

Example 17.1. Suppose that $f_n(p) \equiv 1$, so that $f_n(m)$ is the total number of prime divisors of m. (Even though the function f_n is the same for all n in this case, the probability mechanism changes with n, and so it is clearer to preserve it in the notation.) All that is needed from number theory is the fact that there exists a constant c such that[†]

$$\sum_{p \le x} \frac{1}{p} = \log\log x + c + \mathrm{o}(1) = \log\log x + \mathrm{O}(1). \tag{17.9}$$

From this it follows that the μ_n and σ_n of (17.5) are each $\log\log n + \mathrm{O}(1)$. If $\alpha_n = n^{1/\log\log n}$, then the first sum in (17.6) is $\log\log\log n + \mathrm{O}(1)$, and the second is $\lfloor \alpha_n \rfloor = \mathrm{o}(n^{\epsilon})$: (17.6) is satisfied. Since $\sigma_n \to \infty$ and the random variables $\xi_p - 1/p$ are uniformly bounded, it follows by the Lindeberg (or the Liapounov) central limit theorem that (17.7) holds with N in the role of ξ. Therefore, Theorem 17.1 gives the striking result of Erdös and Kac:

$$\frac{1}{\sqrt{\log\log n}} \sum_{p \le n} \left(\delta_p - \frac{1}{p} \right) \Rightarrow N, \qquad \frac{\sum_{p \le n} \delta_p - \log\log n}{\sqrt{\log\log n}} \Rightarrow N. \tag{17.10}$$

Dividing through by $\sqrt{\log\log n}$ leads to the weak law of large numbers of Hardy and Ramanujan:

$$\frac{1}{\log\log n} \sum_{p \le n} \delta_p \Rightarrow 1. \tag{17.11}$$

□

PROOF OF THEOREM 17.1. We can assume that $\sigma_n \equiv 1$ (pass from $f_n(p)$ to $f_n(p)/\sigma_n$). Let

$$g_n = \sum_{p \le \alpha_n} f_n(p)\delta_p, \quad T_n = \sum_{p \le \alpha_n} f_n(p)\xi_p, \quad \nu_n = \sum_{p \le \alpha_n} \frac{f_n(p)}{p} = \mathsf{E}[T_n].$$

[†] Hardy & Wright [37], Theorem 427, or [PM.240].

By the first condition in (17.6),

$$\mathsf{E}_n[|f_n - g_n|] \le \sum_{\alpha_n < p \le n} |f_n(p)| \frac{1}{n} \left\lfloor \frac{n}{p} \right\rfloor \le \sum_{\alpha_n < p \le n} \frac{|f_n(p)|}{p} \to 0,$$

$$|\mu_n - \nu_n| \le \sum_{\alpha_n < p \le n} \frac{|f_n(p)|}{p} \to 0.$$

Therefore, (17.8) is equivalent to

$$g_n - \nu_n \Rightarrow \xi, \tag{17.12}$$

and similarly, (17.7) is equivalent to

$$T_n - \nu_n \Rightarrow \xi. \tag{17.13}$$

It is therefore enough to show that (17.13) implies (17.12).

For each positive integer r,

$$\mathsf{E}[T_n^r] = \sum_{u=1}^{r} \sum{}' \frac{r!}{r_1! \cdots r_u!} \sum{}'' f_n^{r_1}(p_1) \cdots f_n^{r_u}(p_u) \mathsf{E}[\xi_{p_1}^{r_1} \cdots \xi_{p_u}^{r_u}] \tag{17.14}$$

by the multinomial theorem, where $\sum'$ extends over those u-tuples $r_1, \ldots, r_u$ of positive integers adding to r and $\sum''$ extends over the u-tuples $p_1, \ldots, p_u$ of primes satisfying $p_1 < \cdots < p_u \le \alpha_n$. And $\mathsf{E}_n[g_n^r]$ is the same thing with

$$\mathsf{E}[\xi_{p_1}^{r_1} \cdots \xi_{p_u}^{r_u}] = \mathsf{E}[\xi_{p_1} \cdots \xi_{p_u}] = \frac{1}{p_1 \cdots p_u}$$

replaced by

$$\mathsf{E}_n[\delta_{p_1}^{r_1} \cdots \delta_{p_u}^{r_u}] = \mathsf{E}_n[\delta_{p_1} \cdots \delta_{p_u}] = \frac{1}{n} \left\lfloor \frac{n}{p_1 \cdots p_u} \right\rfloor.$$

Since these two expected values differ by at most $1/n$, $|\mathsf{E}[T_n^r] - \mathsf{E}_n[g_n^r]| \le n^{-1}(\sum_{p \le \alpha_n} |f_n(p)|)^r$. Two binomial expansions, together with the second condition in (17.6), now give

$$\begin{aligned} &|\mathsf{E}[(T_n - \nu_n)^r] - \mathsf{E}_n[(g_n - \nu_n)^r]| \\ &\quad \le \sum_{k=0}^{r} \binom{n}{k} \frac{1}{n} \Big(\sum_{p \le \alpha_n} |f_n(p)| \Big)^k |\nu_n|^{r-k} \\ &\quad = \frac{1}{n} \Big(\sum_{p \le \alpha_n} |f_n(p)| + |\nu_n| \Big)^r \le \Big(\frac{2}{n^{1/r}} \sum_{p \le \alpha_n} |f_n(p)| \Big)^r \to_n 0. \end{aligned} \tag{17.15}$$

(Notice that the first condition in (17.6) requires large values for the α_n, while the second requires small ones.)

By (17.13), if $\mathsf{E}[(T_n - \nu_n)^r]$ is bounded in n for even r, then (see (3.18)) $\mathsf{E}[(T_n - \nu_n)^r] \to \mathsf{E}[\xi^r]$ holds for all r. It will follow by (17.15) that $\mathsf{E}_n[(g_n - \nu_n)^r] \to \mathsf{E}[\xi^r]$ holds for all r, which, by the method of moments [PM.388], will in turn imply (17.12). If $\eta_p = \xi_p - p^{-1}$, then

$$(17.16)\ \mathsf{E}[(T_n - \nu_n)^r] = \mathsf{E}\Big[\Big(\sum_{p \le \alpha_n} f_n(p)\eta_p\Big)^r\Big]$$

$$= \sum_{u=1}^{r} \sum{}' \frac{r!}{r_1! \cdots r_u!} \sum{}'' f_n^{r_1}(p_1) \cdots f_n^{r_u}(p_u) \mathsf{E}[\eta_{p_1}^{r_1} \cdots \eta_{p_u}^{r_u}].$$

Since the η_p are independent and have mean 0, we can require in $\sum'$ that each r_i exceed 1. If β_n is the maximum on the right in (17.6), then $\beta_n \le \beta$ for some β, and if we take $\beta > 1$, then $|f_n(p)\eta_p|^{r_i} \le \beta^{r_i} f_n^2(p)\eta_p^2$ for $r_i \ge 2$. Therefore, the inner sum on the right in (17.16) has modulus at most

$$\sum{}'' \mathsf{E}[\beta^{r_1} f_n^2(p_1)\eta_{p_1}^2] \cdots \mathsf{E}[\beta^{r_u} f_n^2(p_u)\eta_{p_u}^2] \le \beta^r \Big(\sum_{p \le \alpha_n} f_n^2(p)\mathsf{E}[\eta_p^2]\Big)^u.$$

Now $1 = \sigma_n^2 \ge \sum_{p \le \alpha_n} f_n^2(p)\mathsf{E}[\eta_p^2]$, and it follows by (17.16) that

$$|\mathsf{E}[(T_n - \nu_n)^r]| \le \sum_{u=1}^{r} \sum{}' \frac{r!}{r_1! \cdots r_u!} \beta^r < \infty.$$

Thus $|\mathsf{E}[(T_n - \nu_n)^r]|$ is indeed bounded for each r. □

This argument uses from number theory only the fundamental theorem of arithmetic. The applications of Theorem 17.1 require further number theory—(17.9), for example.

Example 17.2. To prove the functional version of (17.10), we need a slight generalization of the argument leading to it. Suppose that $\sup_{n,p} |f_n(p)| < \infty$ and $\sigma_n \to \infty$. Then the central limit theorem applies to S_n just as before. Take $\alpha_n = n^{1/\sigma_n^2}$. Then the first sum in (17.6) is $\log \sigma_n^2 + O(1) = o(\sigma_n)$; the second sum is $O(n^{1/\sigma_n^2})$, which is $o(n^\epsilon)$ because $\sigma_n \to \infty$. Therefore, (17.6) holds, and we can conclude that

$$(17.17)\quad \frac{f_n - \mu_n}{\sigma_n} = \frac{1}{\sigma_n} \sum_{p \le n} \Big(f_n(p)\Big(\delta_p - \frac{1}{p}\Big)\Big) \Rightarrow N. \quad □$$

An example where the limit is Poisson:

Example 17.3. Suppose that f_n counts the prime divisors between u_n and v_n:

$$f_n(p) = I_{(u_n, v_n]}(p), \quad f_n(m) = \sum_{u_n < p \le v_n} \delta_p(m). \tag{17.18}$$

Let $\zeta_\lambda = \zeta(\lambda)$ have the Poisson distribution with mean λ. If

$$\sum_{u_n < p \le v_n} \frac{1}{p} \to \lambda, \quad v_n = o(n^\epsilon), \quad u_n \to \infty, \tag{17.19}$$

where the second condition holds for each positive ϵ, then the f_n in (17.18) satisfies $f_n \Rightarrow \zeta_\lambda$. To prove this, first note that, since $\max_{u_n < p \le v_n} 1/p \to 0$ by the third condition in (17.19), it follows by a standard limit theorem for the Poisson case [PM.302] that $S_n = \sum_{u_n < p \le v_n} \xi_p \Rightarrow \zeta_\lambda$. And since μ_n and σ_n^2 go to λ, it follows further that $(S_n - \mu_n)/\sigma_n \Rightarrow \eta = (\zeta_\lambda - \lambda)/\sqrt{\lambda}$. Apply Theorem 17.1 with $\alpha_n = v_n$; the first and third conditions in (17.6) are obviously satisfied, and the second condition in (17.19) implies the second one in (17.6). Therefore, $(f_n - \mu_n)/\sigma_n \Rightarrow \eta$, which in turn implies $f_n \Rightarrow \zeta_\lambda$.

Let $u_n = e^{sc_n}$ and $v_n = e^{tc_n}$, where $0 < s < t$. If $c_n \to \infty$ and $c_n / \log n \to 0$, then (17.19) holds for $\lambda = \log t - \log s$ (use the middle member of (17.9)), and we obtain $\sum[\delta_p(m) \colon sc_n < \log p \le tc_n] \Rightarrow \zeta(\log t - \log s)$. To compare this Poisson case with the Poisson-Dirichlet case (Section 4), take $c_n = \log\log n$:

$$\sum\left[\delta_p(m) \colon s < \frac{\log p}{\log\log n} \le t\right] \Rightarrow_n \zeta(\log t - \log s). \tag{17.20}$$

This magnifies what goes on just to the right of 0 in Theorem 4.5. □

The Brownian Motion Limit

Suppose as in Example 17.1 that $f_n(p) \equiv 1$. For $1 \le m \le n$, define an element $X^n(m)$ of $D[0,1]$ by

$$X_t^n(m) = \frac{1}{\sqrt{\log\log n}} \sum\nolimits_{p \le e^{\log^t n}} \left(\delta_p(m) - \frac{1}{p}\right). \tag{17.21}$$

The reason for the range in the sum is this: Define a new $f_n(p)$ as 1 if $p \le e^{\log^t n}$ and 0 otherwise; then the corresponding μ_n and σ_n^2 of (17.5) are both $t \log\log n + O(1)$ by (17.9). The variance of X_t^n under P_n is therefore approximately t, the variance of W_t.

Theorem 17.2. *For the X^n of (17.21), $X^n \Rightarrow W$.*

PROOF. It follows immediately from (17.10) that $X_1^n \Rightarrow N$. Suppose that $0 < s < t < 1$, and apply Example 17.2 to

$$f_n(p) = \begin{cases} a & \text{if } p \le e^{\log^s n}, \\ b & \text{if } e^{\log^s n} < p \le e^{\log^t n}, \\ 0 & \text{otherwise.} \end{cases}$$

Then the σ_n^2 of (17.5) is $(a^2 s + b^2(t-s)) \log\log n + O(1)$, and (17.17) gives

$$aX_s^n + b(X_t^n - X_s^n) \Rightarrow (a^2 s + b^2(t-s))^{1/2} N.$$

Since the limit here has the same distribution as $aW_s + b(W_t - W_s)$, it follows by the Cramér-Wold argument that $(X_s^n, X_t^n - X_s^n)$ converges in distribution to $(W_s, W_t - W_s)$. An extension shows that all the finite-dimensional distributions of X^n converge weakly to those of W.

Define Y^n by (17.21) but with p further constrained by $p \le n^{1/4}$ in the sum. Then

$$\mathsf{P}_n[\|X^n - Y^n\| \ge \epsilon] \le \mathsf{P}_n\left[\sum\nolimits_{n^{1/4} < p \le n} \left(\delta_p + \frac{1}{p}\right) \ge \epsilon\sqrt{\log\log n}\right] \to 0$$

by Markov's inequality (first order) and (17.9). Therefore, it is enough to prove $Y^n \Rightarrow W$, and since the finite-dimensional distributions of Y^n also converge to those of W, it is enough to show that $\{Y^n\}$ is tight.

Let r and s be positive integers, let U and V be disjoint sets of primes, and consider the four sums

$$S_U = \sum\nolimits_{p \in U} \xi_p, \quad S_V = \sum\nolimits_{q \in V} \xi_q, \quad f_U = \sum\nolimits_{p \in U} \delta_p, \quad f_V = \sum\nolimits_{q \in V} \delta_q.$$

We have (see (17.14))

$$\mathsf{E}[S_U^r S_V^s] = \sum_{u=1}^{r} \sum_{v=1}^{s} \sum{}' \frac{r!}{r_1! \cdots r_u!} \frac{s!}{s_1! \cdots s_v!} \sum{}'' \frac{1}{p_1 \cdots p_u q_1 \cdots q_v},$$

where $\sum'$ extends over the u-tuples $\{r_i\}$ adding to r and the v-tuples $\{s_j\}$ adding to s, and $\sum''$ extends over the $\{p_i\} \subset U$ and $\{q_j\} \subset V$ satisfying $p_1 < \cdots < p_u$ and $q_1 < \cdots < q_v$. And $\mathsf{E}_n[f_U^r f_V^s]$ is the same thing with the inner summand replaced by $n^{-1}\lfloor n/p_1 \cdots p_u q_1 \cdots q_v \rfloor$.

Therefore, $|\mathsf{E}[S_U^r S_V^s] - \mathsf{E}_n[f_U^r f_V^s]| \le n^{-1}|U|^r|V|^s$, where $|U|$ and $|V|$ denote cardinality. This also holds if $r = 0$ or $s = 0$, and the proof is simpler. Let $a = \sum_{p\in U} p^{-1}$ and $b = \sum_{q \in V} q^{-1}$. By a double binomial expansion,

$$(17.22)\quad \begin{aligned} &|\mathsf{E}[(S_U - a)^r(S_V - b)^s] - \mathsf{E}_n[(f_U - a)^r(f_V - b)^s]| \\ &\qquad \le \frac{1}{n}\sum_{i=0}^{r}\sum_{j=0}^{s}\binom{r}{i}\binom{s}{j}|U|^i a^{r-i}|V|^j b^{s-j} \\ &\qquad = (|U| + a)^r(|V| + b)^s \le \frac{2^{r+s}}{n}|U|^r|V|^s. \end{aligned}$$

Take $r = s = 2$ and $W = U \cup V$. Since we have $\mathsf{E}[(S_U - a)^2] = \sum_{p\in U} p^{-1}(1 - p^{-1}) \le a$ and similarly for S_V, it follows by (17.22) that (use $xy \le (x + y)^2$ and Schwarz's inequality)

$$\begin{aligned} \mathsf{E}_n[f_U - a)^2(f_V - b)^2] &\le (a + b)^2 + \frac{16}{n}|W|^4 \\ &\le \Big[1 + \frac{16}{n}\Big(\sum\nolimits_{p\in W} p\Big)^2\Big]\Big(\sum\nolimits_{p\in W}\frac{1}{p}\Big)^2. \end{aligned}$$

If $\alpha < \beta \le n^{1/4}$ and W is contained in the set of primes in the range $\alpha < p \le \beta$, then

$$\begin{aligned} &\mathsf{P}_n[|f_U - a| \wedge |f_V - b| \ge \lambda] \\ &\le \frac{1}{\lambda^4}\Big[1 + \frac{16}{n}(\sum\nolimits_{\alpha<p\le\beta} p)^2\Big]\Big(\sum\nolimits_{\alpha\le p\le\beta}\frac{1}{p}\Big)^2 \le \frac{17}{\lambda^4}\Big(\sum\nolimits_{\alpha<p\le\beta}\frac{1}{p}\Big)^2. \end{aligned}$$

Suppose that $\alpha < r \le \beta \le n^{1/4}$, and let U and V consist of the primes in the ranges $\alpha < p \le r$ and $r < p \le \beta$, respectively. The preceding inequality gives

$$\mathsf{P}_n\Big[\Big|\sum_{\alpha<p\le r}\Big(\delta_p - \frac{1}{p}\Big)\Big| \wedge \Big|\sum_{r<p\le\beta}\Big(\delta_p - \frac{1}{p}\Big)\Big| \ge \lambda\Big] \le \frac{17}{\lambda^4}\Big(\sum_{\alpha<p\le\beta}\frac{1}{p}\Big)^2.$$

Suppose now that $\lambda > 4$, so that $\lambda/4$ exceeds the maximum absolute value of the summands $\delta_p - p^{-1}$. Apply Theorem 10.1 and then (10.6):

$$(17.23)\qquad \mathsf{P}_n\Big[\max_{\alpha<r\le\beta}\Big|\sum_{\alpha<p\le r}\Big(\delta_p - \frac{1}{p}\Big)\Big| \ge \lambda\Big] \le \frac{17K}{(\lambda/4)^4}\Big(\sum_{\alpha<p\le\beta}\frac{1}{p}\Big)^2.$$

Let $s_x^2 = \sum_{p\le x} p^{-1}$. Change the definition of Y^n: Normalize by s_n rather than $\sqrt{\log\log n}$. It is enough to prove tightness for the new random functions Y^n. By (17.23), there is a constant K', such that, if $\alpha < \beta \le n^{1/4}$ and n is large enough that $\epsilon s_n > 4$, then

$$\mathsf{P}_n\Big[\max_{\alpha<r\le\beta}\Big|\sum_{\alpha<p\le r}\Big(\delta_p - \frac{1}{p}\Big)\Big| \ge \epsilon s_n\Big] \le \frac{K'}{\epsilon^4}\Big(\frac{s_\beta^2}{s_n^2} - \frac{s_\alpha^2}{s_n^2}\Big)^2. \tag{17.24}$$

As x increases from 0 to n, s_x^2 increases in jumps from 0 to s_n^2, and no jump exceeds 1. Therefore, for a given δ, if n is large enough that $s_n^2 < \delta$, then there exist $\alpha_0, \ldots, \alpha_v$ such that, for $t_i = s_{\alpha_i}^2/s_n^2$, $0 = t_0 < t_1 < \cdots < t_v = 1$ and $\delta < t_i - t_{i-1} < 2\delta$ for each $i < v$. By (17.24) and (7.11),

$$\mathsf{P}_n[w(Y^n, \delta) \ge 3\epsilon] \le \sum_{i=1}^{v} \mathsf{P}_n\Big[\sup_{t_{i-1}\le s\le t_i} |Y_s^n - Y_{t_{i-1}}^n| \ge \epsilon\Big] \le \frac{K'}{\epsilon^4} 2\delta$$

for large n. This proves tightness. □

We can clarify the number-theoretic meaning of this theorem by transforming (17.21). First, $\mathsf{P}_n[m\colon \log\log m \le (1-\epsilon)\log\log n] \to 0$ by (17.9). Therefore, if P_n governs the choice of m,

$$\frac{\log\log m}{\log\log n} \Rightarrow_n 1. \tag{17.25}$$

Let $\gamma_x = \sum_{p<x} p^{-1}$; this is essentially s_x^2, but we now regard it as an approximate mean rather than variance. Define $\Phi_n(\cdot, m)$ by

$$\Phi_n(0,m) = 0; \quad \Phi_n(1,m) = 1; \tag{17.26}$$
$$\Phi_n\Big(\frac{\gamma_p}{\gamma_{m+1}}, m\Big) = \frac{\log\log p}{\log\log n}, \quad p \le m,$$

and interpolate linearly between these points. It follows by (17.25) and two more applications of (17.9) that

$$\sup_{0\le t\le 1} |\Phi_n(t,m) - t| \Rightarrow 0,$$

and therefore, if I is the identity function on $[0,1]$, then $\Phi_n \Rightarrow I$ in the sense of $D[0,1]$. By Theorem 3.9, $(X^n, \Phi_n) \Rightarrow (W, I)$, and by the lemma to Theorem 14.4, $X^n \circ \Phi_n \Rightarrow W \circ I = W$. But $X^n \circ \Phi_n$ is the random function Z^n defined by

$$Z_t^n(m) = \frac{1}{\sqrt{\log\log n}} \sum_{p:\gamma_p/\gamma_{m+1} \le t} \left(\delta_p(m) - \frac{1}{p}\right). \tag{17.27}$$

Thus $Z^n \Rightarrow W$. We could also divide by $\sqrt{\log\log m}$—the result is the same.

The random function Z^n can be described this way: If $t = \gamma_p/\gamma_{m+1}$, then

$$Z_t^n(m) = \frac{1}{\sqrt{\log\log n}} \sum_{q \le p} (\delta_q(m) - q^{-1}); \tag{17.28}$$

and if p' is the next prime after p, then $Z^n(m)$ is constant on the interval $[\gamma_p/\gamma_{m+1}, \gamma_{p'}/\gamma_{m+1}]$, which has length proportional to p^{-1}.

Suppose an idiot-savant successively checks the primes up to m in order to see which ones divide m, the amount of time he devotes to p being proportional to $1/p$—he calculates with ever-increasing fury. Call p "advantageous" with respect to m if $\sum_{q<p} \delta_q(m) > \gamma_p$; this means that, with respect to divisibility by the primes preceding p, m is "more composite" or "less prime-like" than the average integer. If q' is the product of the prime divisors of m that precede p, then p divides m if and only if it divides m/q', and the latter is easier to check than the former. The savant knows all this. If p is advantageous, then q' is likely to be large, which simplifies the computational task and hence irritates the idiot (computation being his life). The proportion $\tau_n(m)$ of time he spends on advantageous (and to him vexatious) primes is the Lebesgue measure of the set of t in $[0, 1]$ for which $Z_t^n(m) > 0$, and in the limit this follows the arc sine law (9.27). Since the corresponding density is U-shaped, τ_n is more likely to be near 0 or 1 than to be near the middle of $[0, 1]$. For example, for about 20% of the integers under n (n large), $\tau_n > .9$, for about 20%, $\tau_n < .1$, and for only about 6% do we have $.45 < \tau_n < .55$.

Let $\alpha_p(m)$ be the highest power of p that divides m. If we redefine Z^n by substituting $\alpha_p(m)$ for $\delta_p(m)$ in (17.27), then $Z^n \Rightarrow W$ still holds. Indeed, if $\beta_p(m) = \alpha_p(m) - \delta_p(m)$, then $\mathsf{P}_n[\beta_p \ge k] = n^{-1}\lfloor np^{-k-1}\rfloor \le p^{-k-1}$, and hence $\mathsf{E}_n[\beta_p] \le \sum_{k=1}^{\infty} p^{-k-1} = (p(p-1))^{-1}$. It follows that $\mathsf{E}_n[\sum_p \beta_p]$ is bounded, and therefore, by Markov's inequality, $\sum_p \beta_p/\log\log n \Rightarrow_n 0$: The new Z^n does converge in distribution to W.

We can now imagine that the idiot-savant finds for each successive p its exact power in the prime factorization of m. Now he can quit after he has encountered the largest prime divisor $p_1(m)$ of m. By Theorem 4.5, $\log p_1(m)/\log m =$

$\chi_n(m) \Rightarrow_n \chi$, where χ has the density $d_1(x;1)$ as defined by (4.16). It follows that $\log\log p_1(m)/\log\log m = 1 + \log\chi_n(m)/\log\log m \Rightarrow_n 1$. Therefore, if the idiot quits after encountering $p_1(m)$, he saves (to his satisfaction) very little time.

The skewing of the distribution of τ_n toward the ends of $[0,1]$, as described above, is even more pronounced if our idiot-savant spends the same amount of time on each p: Let A_δ be the set of x in $D[0,1]$ that are nonzero and have constant sign over $[1-\delta,1]$. For each ϵ there is a δ such that $\mathsf{P}[W \in A_\delta \cap C] > 1-\epsilon$ (see (9.28)). And by Theorem 2.1(iv) and the fact that $A_\delta^\circ \cap C = A_\delta \cap C$, it follows that $\liminf_n \mathsf{P}_n[Z^n \in A_\delta] \geq \mathsf{P}[W \in A_\delta^\circ] = \mathsf{P}[W \in A_\delta \cap C] > 1-\epsilon$. The points of discontinuity of $Z^n(m)$ are the ratios $\gamma_p/\gamma_{m+1} \approx \log\log p/\log\log m$, and the great majority of these lie to the right of $1-\delta$. In fact, if $\pi(x)$ is the number of primes up to x, then the proportion of the discontinuity points less than $1-\delta$ is about $\pi(\exp(\log^{1-\delta} m))/\pi(m)$, and this goes to 0 in probability because $\pi(x)$ is of the order $x/\log x$. Therefore, under the new regime for our idiot-savant, the limiting distribution of τ_n has mass $\frac{1}{2}$ at 0 and mass $\frac{1}{2}$ at 1.

The Poisson-Process Limit

There is a functional limit theorem corresponding to (17.20). First, transform the ratio there by taking its logarithm. If

$$u_n = \exp\exp(s + \log\log\log n), \quad v_n = \exp\exp(t + \log\log\log n),$$

then

$$\sum[\delta_p(m) \colon s < \log\log p - \log\log\log n \leq t] = \sum_{u_n < p \leq v_n} \delta_p(m) \Rightarrow_n \zeta(t-s). \tag{17.29}$$

For each prime divisor p of m, place a point at $\log\log p - \log\log\log n$. If P_n governs the choice of m, this gives a point process on $(-\infty,\infty)$. Let $X_t^n(m)$ be the number of points in $(0,t]$ for $t \geq 0$ and minus the number in $(t,0]$ for $t \leq 0$. Then $X_t^n(m) - X_s^n(m)$ is the sum in (17.29).

Now consider a Poisson process on (∞,∞) with a constant intensity of 1, and let X_t be the number of points in $(0,t]$ or minus the number in $(t,0]$ as $t \geq 0$ or $t \leq 0$. Then X has independent increments, and $X_t - X_s$ has the Poisson distribution with mean $t-s$. Thus (17.29) can be written as $X_t^n - X_s^n \Rightarrow X_t - X_s$.

We can extend this in the usual way by the Cramér-Wold method. Suppose that $r < s < t$, define $r_n = \exp\exp(r - \log\log\log n)$, and define u_n and v_n as before. Now define

$$S'_n = \sum_{r_n < p \le u_n} \xi_p, \quad S''_n = \sum_{u_n < p \le v_n} \xi_p,$$
$$f'_n(m) = \sum_{r_n < p \le u_n} \delta_p(m), \quad f''_n(m) = \sum_{u_n < p \le v_n} \delta_p(m).$$

By the analysis above, S'_n and f'_n converge in distribution to $X_s - X_r$, and S''_n and f''_n converge in distribution to $X_t - X_s$. By independence, Theorem 2.8(ii), and the mapping theorem, $S_n = aS'_n + bS''_n \Rightarrow a(X_s - X_r) + b(X_t - X_s) = \xi$. If $f_n = af'_n + bf''_n$, $\lambda' = \log s - \log r$, and $\lambda'' = \log t - \log s$, then the μ_n and σ_n^2 of (17.5) satisfy $\mu_n \to \mu = a\lambda' + b\lambda''$ and $\sigma_n^2 \to \sigma^2 = a^2\lambda' + b^2\lambda''$. But now, $(S_n - \mu_n)/\sigma_n \Rightarrow (\xi - \mu)/\sigma$, and it follows by Theorem 17.1 that $(f_n - \mu_n)/\sigma_n \Rightarrow (\xi - \mu)/\sigma$. But then it follows further that $f_n \Rightarrow \xi$. This means that $af'_n + bf''_n \Rightarrow a(X_s - X_r) + b(X_t - X_s)$ for all a and b, and therefore, $(X_s^n - X_r^n, X_t^n - X_s^n) = (f'_n, f''_n) \Rightarrow (X_s - X_t, X_t - X_s)$. The argument extends: The finite dimensional distributions of the process X^n converge to those of X.

The sample paths of X^n and X are elements of the space $D(-\infty, \infty)$ of cadlag functions over $(-\infty, \infty)$. If we restrict X^n and X to $[s, t]$, then, by Theorem 12.6, there is convergence in distribution in the sense of the space $D[s, t]$. It is possible to adapt the reasoning of Section 16 to the space $D(-\infty, \infty)$. Replace the g_m of (16.2) by the function that is 1 on $[-m+1, m-1]$ and 0 outside $[-m, m]$, and is linear on $[-m, -m+1]$ and $[m-1, m]$. Define $x^m(t)$ by (16.3), but for all t on the line, and define a metric on $D(-\infty, \infty)$ by (16.4), where now d_m° is the metric for the Skorohod topology on $D[-m, m]$. One can derive for $D(-\infty, \infty)$ the analogues of Theorems 16.1 through 16.7. If $r_{s,t}x$ is the restriction of x to $[s, t]$, then, as observed above, the distributions P_n and P of X^n and X satisfy $P_n r_{s,t}^{-1} \Rightarrow P r_{s,t}^{-1}$ for all s and t, and by the analogue of Theorem 16.7, $P_n \Rightarrow P$:

Theorem 17.3. *We have $X^n \Rightarrow X$ in the sense of the space $D(-\infty, \infty)$.*

For negative [nonnegative] i, let β_i be the successive X-points (points where X jumps) preceding [following] 0: $\cdots < \beta_{-2} < \beta_{-1} < 0 < \beta_0 < \beta_1 < \cdots$ (with probability 1, 0 is not an X-point). It is a standard property of the Poisson process that the interarrival intervals $\beta_i - \beta_{i-1}$ have the exponential density $\phi_1(x) = e^{-x}$ for $i \neq 0$, while $\beta_0 - \beta_{-1}$ has density $\phi_2(x) = (\phi_1 * \phi_1)(x) = xe^{-x}$.

Thus the interarrival intervals have mean 1, except for the one that covers 0, which has mean 2: A long interval has a better chance of covering 0 than a short one has. In fact, for every fixed t, the interval that covers it has density ϕ_2 and mean 2. This seems paradoxical, because an arbitrary interval can be made to cover t simply by shifting t the right amount. But this is like trying to place your bet after the roulette wheel has come to rest, which casino operators vigorously discourage.

For negative [nonnegative] i, let $p_i(m)$ be the prime divisors of m preceding [following] $\log n$: $\cdots < p_{-2}(m) < p_{-1}(m) < \log n < p_0(m) < p_1(m) < \cdots$ ($p = \log n$ is impossible because e is transcendental). Then the X^n-points corresponding to the β_i are the points $\beta_i^n(m) = \log\log p_i(m) - \log\log\log n$. Let D_c be the set of functions in $D(-\infty, \infty)$ that are nondecreasing, take integers as values, and have jumps of exactly 1 at the points of discontinuity. The corresponding set for $D[0,1]$ is described after (12.38).

Suppose that x_n and x lie in D_c and that $x_n \to x$ in the topology of $D(-\infty, \infty)$. Assume that $s < 0 < t$ and that x is continuous at s and t. Then x has a finite number M of jumps in $[s, t]$, and it is not hard to show (see the discussion following the proof of Theorem 12.6) that, for large n, x_n has exactly M jumps in $[s, t]$ and that the positions of these jumps converge to the positions of the jumps in x. Since X^n and X have their values in D_c, it follows by Example 3.1 and the mapping theorem that $(\beta_u^n, \ldots, \beta_v^n) \Rightarrow (\beta_u, \ldots, \beta_v)$ for all u and v. Therefore, $\log\log p_i(m) - \log\log p_{i-1}(m) \Rightarrow_n \beta_i - \beta_{i-1}$, and from this it follows that $\log p_i(m)/\log p_{i-1}(m) \Rightarrow_n \exp(\beta_i - \beta_{i-1})$ and (for $y > 1$)

$$(17.30) \qquad \lim_n \mathsf{P}_n[m: p_i(m) \geq p_{i-1}^y(m)] = \begin{cases} y^{-1}(1 + \log y) & \text{if } i = 0, \\ y^{-1} & \text{if } i \neq 0. \end{cases}$$

Also, $\mathsf{E}[\exp(\beta_i - \beta_{i-1})] = \infty$ (for $i = 0$ as well as for $i \neq 0$), and since $p_i - p_{i-1} \geq \log p_i / \log p_{i-1}$, Theorem 3.4 implies that $\mathsf{E}_n[p_i - p_{i-1}] \to \infty$.

Because of (17.25), $\Delta_{mn} = \log\log\log m - \log\log\log n \Rightarrow_n 0$ if P_n governs the choice of m. The points corresponding to the process X^n are the $\log\log p - \log\log\log n$. If instead we use the points $\log\log p - \log\log\log m$, we have a different process Y^n. Since this is just a matter of translating the time scale by the amount Δ_{mn}, the distance in $D(-\infty, \infty)$ between X^n and Y^n goes to 0 in probability, and therefore, by Theorem 3.1, $Y^n \Rightarrow X$. For example, (17.30) also holds if we redefine the $p_i(m)$ so that $\cdots < p_{-2}(m) < p_{-1}(m) < \log m < p_0(m) < p_1(m) < \cdots$.

SECTION 18. MARTINGALES

This section makes use of the basic facts of ergodic theory and discrete-time martingale theory. Here is a statistical example to which the limit theory of the section applies.

Example 18.1. Let $\{\zeta_0, \zeta_1, \ldots\}$ be a stationary Markov chain with finite state space. Suppose the transition and stationary probabilities $p(i, j; \theta)$ and $p(i; \theta)$ are positive and depend on a parameter θ ranging over an open interval. Define (assume the derivatives exist)

$$\xi_0 = \frac{d}{d\theta} \log p(\zeta_0; \theta) = \frac{p'(\zeta_0; \theta)}{p(\zeta_0; \theta)};$$

$$\xi_k = \frac{d}{d\theta} \log p(\zeta_{k-1}, \zeta_k; \theta) = \frac{p'(\zeta_{k-1}, \zeta_k; \theta)}{p(\zeta_{k-1}, \zeta_k; \theta)}, \quad k \geq 1.$$

The log-likelihood of the observation $\zeta_0, \ldots, \zeta_n$ is $L_n(\theta) = \log p(\zeta_0; \theta) + \sum_{k=1}^n \log p(\zeta_{k-1}, \zeta_k; \theta)$, and its derivative is $L_n'(\theta) = \sum_{k=0}^n \xi_k$. It is easy to verify that $\mathsf{E}[\xi_0] = 0$ and $\mathsf{E}[\xi_k \| \zeta_0, \ldots, \zeta_{k-1}] = 0$, from which it follows that $\{\xi_n\}$ is a martingale difference and (equivalently) $\{L_n'(\theta)\}$ is a martingale.

Large-sample theory for Markov chains starts with a central limit theorem: If θ is the true value of the parameter, then $n^{-1/2} L_n'(\theta) \Rightarrow \sigma(\theta) N$, where $\sigma^2(\theta) = \mathsf{E}[\xi_1^2]$. But $\{\xi_1, \xi_2, \ldots\}$ is stationary (not so if ξ_0 is included) and ergodic, and it follows by Theorem 18.3 that $n^{-1/2} \sum_{k=1}^n \xi_k \Rightarrow \mathsf{E}[\xi_1^2] N$. And because of the norming factor $n^{-1/2}$, this still holds if we include ξ_0 in the sum, which gives the required limit theorem. These arguments can be made to cover chains of higher order as well as more general processes. □

Triangular Arrays

Suppose we have a triangular array of random variables. For each n, $\xi_{n1}, \xi_{n2}, \ldots$ is a martingale difference with respect to the σ-fields $\mathcal{F}_0^n, \mathcal{F}_1^n, \ldots$: ξ_{nk} is $\mathcal{F}_k^n$-measurable, and $\mathsf{E}[\xi_{nk} \| \mathcal{F}_{k-1}^n] = 0$. There may be a different probability space for each n. Suppose that the ξ_{nk} have second moments, and put $\sigma_{nk}^2 = \mathsf{E}[\xi_{nk}^2 \| \mathcal{F}_{k-1}^n]$. If the martingale is originally defined only for $1 \leq k \leq r_n$, take $\xi_{nk} = 0$ and $\mathcal{F}_k^n = \mathcal{F}_{r_n}^n$ for $k > r_n$. Assume that $\sum_{k=1}^\infty \xi_{nk}$ and $\sum_{k=1}^\infty \sigma_{nk}^2$ converge with probability 1. The development begins with the following theorem [PM.476].

Theorem 18.1. *Assume that*

$$\sum_{k=1}^\infty \sigma_{nk}^2 \Rightarrow \sigma^2, \tag{18.1}$$

where σ is a nonnegative constant,[†] *and that*

$$\sum_{k=0}^{\infty} \mathsf{E}[\xi_{nk}^2 I_{[|\xi_{nk}|\geq\epsilon]}] \to 0 \tag{18.2}$$

for each ϵ. Then $\sum_{k=1}^{\infty} \xi_{nk} \Rightarrow \sigma N$.

We turn now to the corresponding functional limit theorem. The ξ_{nk} are still a martingale difference, and the notation is the same as before.

Theorem 18.2. *Assume that*

$$\sum_{k\leq nt} \sigma_{nk}^2 \Rightarrow_n t \tag{18.3}$$

for every t and that

$$\sum_{k\leq nt} \mathsf{E}[\xi_{nk}^2 I_{[|\xi_{nk}|\geq\epsilon]}] \to_n 0 \tag{18.4}$$

for every t and ϵ. If $X_t^n = \sum_{k\leq nt} \xi_{nk}$, then $X^n \Rightarrow W$ in the sense of $D[0,\infty)$.

A simple way to construct a Brownian motion W on $[0,\infty)$ is to take a Brownian bridge W° on $[0,1]$ and define $W_t = (1+t)W^\circ_{t/(1+t)}$ for $t \in [0,\infty)$.

PROOF. Suppose that $s < t$, and define η_{nk} as $a\xi_{nk}$ for $k \leq \lfloor ns \rfloor$ and as $b\xi_{nk}$ for $\lfloor ns \rfloor < k \leq \lfloor nt \rfloor$. By (18.3), $\sum_{k\leq nt} \mathsf{E}[\eta_{nk}^2 \| \mathcal{F}_{k-1}^n] \Rightarrow a^2 s + b^2(t-s)$; therefore, Theorem 18.1 applies to $\{\eta_{nk}\}$, and it follows that $aX_s^n + b(X_t^n - X_s^n) \Rightarrow aW_s + b(W_t - W_s)$. By the Cramér-Wold argument, the two-dimensional distributions converge, and the same is true for higher dimensions.

For tightness, we use the corollary to Theorem 16.10. Since $\xi_{nk}^2 \leq \epsilon^2 + \xi_{nk}^2 I_{[|\xi_{nk}|\geq\epsilon]}$, (18.4) implies that $\mathsf{E}[\max_{k\leq nm} \xi_{nk}^2] \to_n 0$ for each m. Therefore, $\{\jmath_m(X^n)\}$ is tight on the line for each m, and of course $\{X_0^n\}$ is tight.

To verify Condition 1° of the theorem, we use the sequential version, (16.32). Assume then that τ_n are discrete X^n-stopping times bounded by m and that $\delta_n \to 0$. We must show that

$$X_{\tau_n+\delta_n}^n - X_{\tau_n}^n \Rightarrow_n 0. \tag{18.5}$$

[†] The theorem is stated [PM.476] for $\sigma > 0$, but the proof there covers the case $\sigma = 0$ as well.

If A_k^n is the event where $\lfloor n\tau_n \rfloor < k \le \lfloor n(\tau_n + \delta_n)\rfloor$, and if $\zeta_{nk} = I_{A_k^n}\xi_{nk}$, then the difference in (18.5) is $\sum_k \zeta_{nk}$. For large n, $\delta_n < 1$, and in that case ($\tau_n \le m$) $A_k^n \ne \emptyset$ implies $k \le n(m+1)$, and we can take the last sum as $\sum_{k \le n(m+1)} \zeta_{nk}$. Now apply Theorem 18.1 for the case where $\sigma^2 = 0$ in (18.1): (18.5) will follow if we prove, first, that the ζ_{nk} form a martingale difference with respect to the $\mathcal{F}_k^n$ and, second, that

$$\sum_{k \le n(m+1)} \mathsf{E}[\zeta_{nk}^2 \| \mathcal{F}_{k-1}^n] \Rightarrow_n 0 \tag{18.6}$$

and

$$\sum_{k \le n(m+1)} \mathsf{E}[\zeta_{nk}^2 I_{[|\zeta_{nk}| \ge \epsilon]}] \to_n 0. \tag{18.7}$$

Since $|\zeta_{nk}| \le |\xi_{nk}|$, (18.7) is a consequence of (18.4).

If r is the largest point in the (finite) range of τ_n for which $r < k/n$, then $[\lfloor n\tau_n \rfloor < k] = [n\tau_n < k] = [\tau_n \le r] \in \sigma[X_s^n : s \le r] \subset \mathcal{F}_{\lfloor nr \rfloor}^n \subset \mathcal{F}_{k-1}^n$. Since $\tau_n + \delta_n$ is another discrete X^n-stopping time, $[\lfloor n(\tau_n + \delta_n) \rfloor \ge k]$ also lies in $\mathcal{F}_{k-1}$, and hence $A_k^n \in \mathcal{F}_{k-1}^n$. From this it follows that $\mathsf{E}[\zeta_{nk} \| \mathcal{F}_{k-1}^n] = I_{A_k^n} \mathsf{E}[\xi_{nk} \| \mathcal{F}_{k-1}^n] = 0$, and so $\{\zeta_{nk}\}$ is indeed a martingale difference, as required.

Only (18.6) remains to be proved. For each integer q, let

$$B_q^n = \bigcap_{i=1}^{qm+1} \left[\left| \sum_{k \le ni/q} \sigma_{nk}^2 - \frac{i}{q} \right| \le \frac{1}{q} \right].$$

By (18.3), $\mathsf{P}B_q^n \to_n 1$ for each q. For large n, $\delta_n < 1/q$, and then $\tau_n + \delta_n < \tau_n + 1/q \le m + 1$, which implies that

$$\begin{aligned} \sum_{k \le n(m+1)} \mathsf{E}[\zeta_{nk}^2 \| \mathcal{F}_{k-1}^n] &= \sum_{k \le n(m+1)} I_{A_k^n} \sigma_{nk}^2 \\ &\le \sum_{k=\lfloor n\tau_n \rfloor + 1}^{\lfloor n(\tau_n + 1/q) \rfloor} \sigma_{nk}^2 \le \sup_{t \le m} \sum_{k = \lfloor nt \rfloor + 1}^{\lfloor n(t+1/q) \rfloor} \sigma_{nk}^2. \end{aligned}$$

Given t, choose j so that $(j-1)/q < t \le j/q$. If $t \le m$, then $j \le qm+1$, and so, on the set B_q^n,

$$\sum_{k=\lfloor nt \rfloor + 1}^{\lfloor n(t+1/q) \rfloor} \sigma_{nk}^2 \le \sum_{k = \lfloor n(j-1)/q \rfloor}^{\lfloor n(j+1)/q \rfloor} \sigma_{nk}^2 \le \frac{j+1}{q} - \frac{j-1}{q} + \frac{2}{q} = \frac{4}{q}.$$

Since $\mathsf{P}B_q^n \to_n 1$ for each q, (18.6) follows. □

Ergodic Martingale Differences

Now consider the stationary case. Let

$$\dots, \xi_{-1}, \xi_0, \xi_1 \dots \tag{18.8}$$

be a doubly infinite sequence of random variables. It is a martingale difference if ξ_n is $\mathcal{F}_n$-measurable and $\mathsf{E}[\xi_n \| \mathcal{F}_{n-1}] = 0$, where $\mathcal{F}_n = \sigma[\xi_k : k \le n]$.

Theorem 18.3. *Let* (18.8) *be a stationary, ergodic martingale difference for which* $\mathsf{E}[\xi_n^2] = \sigma^2$ *is positive and finite. If we take* $X_t^n = \sum_{k \le nt} \xi_k / \sigma\sqrt{n}$, *then* $X^n \Rightarrow W$ *in the sense of* $D[0, \infty)$.

PROOF. Represent $\{\xi_n\}$ as the coordinate variables on the space $R^{+\infty}$ of doubly infinite sequences [M22] and put $\sigma_n^2 = \mathsf{E}[\xi_n^2 \| \mathcal{F}_{n-1}]$. Then [M22] $\{\sigma_n^2\}$ is stationary and ergodic.

Now we use Theorem 18.2. Let $\xi_{nk} = \xi_k / \sigma\sqrt{n}$, $\mathcal{F}_k^n = \mathcal{F}_k$, and $\sigma_{nk}^2 := \mathsf{E}[\xi_{nk}^2 \| \mathcal{F}_{k-1}^n] = \sigma_k^2 / \sigma^2 n$. By stationarity,

$$\sum_{1 \le k \le nt} \mathsf{E}[\xi_{nk}^2 I_{[|\xi_{nk}| \ge \epsilon]}] = \lfloor nt \rfloor \mathsf{E}\left[\frac{1}{n\sigma^2} \xi_0^2 I_{[|\xi_0| \ge \epsilon\sigma\sqrt{n}]}\right] \to 0.$$

Hence (18.4). By the ergodic theorem,

$$\sum_{1 \le k \le nt} \sigma_{nk}^2 = \frac{1}{n\sigma^2} \sum_{1 \le k \le nt} \sigma_k^2 \Rightarrow t.$$

Hence (18.3). □

Theorem 18.3 also holds if (18.8) is replaced by a one-sided sequence $(\xi_1, \xi_2, \dots)$ for which $\mathsf{E}[\xi_n \| \xi_1, \dots, \xi_{n-1}] = 0$: Represent the process on $R^{+\infty}$ [M22].

SECTION 19. ERGODIC PROCESSES

The Basic Theorem

Consider a stationary, ergodic process

$$\dots, \xi_{-1}, \xi_0, \xi_1, \dots. \tag{19.1}$$

Represent the process on the space of doubly infinite sequences [M22], use $\|\cdot\|$ to denote the L^2 norm, let T be the shift, and put

$$\mathcal{F}_n = \sigma[\xi_k \colon k \le n], \qquad \mathcal{G}_n = \sigma[\xi_k \colon k \ge n]. \tag{19.2}$$

Theorem 19.1. *Assume that* (19.1) *is stationary and ergodic, that the ξ_n have finite second moments, and that*

$$\sum_{n=1}^{\infty} \|\mathsf{E}[\xi_n \| \mathcal{F}_0]\| < \infty. \tag{19.3}$$

Then

$$\mathsf{E}[\xi_n] = 0, \tag{19.4}$$

and the series

$$\sigma^2 = \mathsf{E}[\xi_0^2] + 2\sum_{n=1}^{\infty} \mathsf{E}[\xi_0 \xi_n] \tag{19.5}$$

converges absolutely. If $\sigma > 0$ and

$$S_n = \sum_{k=1}^{n} \xi_k, \qquad X_t^n = S_{\lfloor nt \rfloor}/\sigma\sqrt{n}, \tag{19.6}$$

then $X^n \Rightarrow W$ in the sense of $D[0,\infty)$.

That (19.4) holds is usually obvious a priori. The condition (19.3) can be replaced by

$$\sum_{n=1}^{\infty} \|\mathsf{E}[\xi_0 \| \mathcal{G}_n]\| < \infty. \tag{19.7}$$

This is simply a matter of reversing time: T^{-1} is ergodic if T is, $\sum_{k=1}^{n} \xi_k$ has the same distribution as $\sum_{k=1}^{n} \xi_{-k}$, and (19.7) is the same thing as $\sum_{n=1}^{\infty} \|\mathsf{E}[\xi_{-n} \| \mathcal{G}_0]\| < \infty$.

If we start with a one-sided sequence $(\xi_0, \xi_1, \ldots)$ that is stationary and ergodic and satisfies (19.7), we can construct the two-sided version [M22] and still conclude that $X^n \Rightarrow W$.

The proof uses Theorem 18.3, on martingale differences. Before proceeding to the proof itself, consider two special cases that explain the idea behind it. The first idea is to consider $\eta_k = \xi_k - \mathsf{E}[\xi_k \| \mathcal{F}_{k-1}]$. Now $\mathsf{E}[\eta_k \| \mathcal{F}_{k-1}] = 0$—$\{\eta_n\}$ is a martingale difference—and $M_k = \sum_{k=1}^{n} \eta_k$ will be approximately normally distributed by the theory of the preceding section. On the other hand, the difference $S_n - M_n$ is $\sum_{k=1}^{n} \mathsf{E}[\xi_k \| \mathcal{F}_{k-1}]$, and there is no a priori reason why this sum should be small (unless, for example, $\{\xi_n\}$ is a martingale difference to start with). But now suppose that the ξ_n are 1-dependent.† If we redefine η_k as

$$\eta_k = \xi_k + \mathsf{E}[\xi_{k+1} \| \mathcal{F}_k] - \mathsf{E}[\xi_k \| \mathcal{F}_{k-1}], \tag{19.8}$$

then, since $\mathsf{E}[\mathsf{E}[\xi_{k+1} \| \mathcal{F}_k] \| \mathcal{F}_{k-1}] = \mathsf{E}[\xi_{k+1} \| \mathcal{F}_{k-1}] = 0$, $\{\eta_n\}$ is again a martingale difference. This time, $S_n = M_n + \sum_{k=1}^{n}(\mathsf{E}[\xi_{k+1} \| \mathcal{F}_k] - \mathsf{E}[\xi_k \| \mathcal{F}_{k-1}])$, and the last sum telescopes to $\mathsf{E}[\xi_{n+1} \| \mathcal{F}_n] - \mathsf{E}[\xi_1 \| \mathcal{F}_0]$, which *is* small compared with S_n and M_n (we use a norming factor of the order $\sqrt{n}$). If $\{\xi_n\}$ is 2-dependent, the same argument works for

$$\eta_k = \xi_k + \mathsf{E}[\xi_{k+1} \| \mathcal{F}_k] + \mathsf{E}[\xi_{k+2} \| \mathcal{F}_k] - \mathsf{E}[\xi_k \| \mathcal{F}_{k-1}] - \mathsf{E}[\xi_{k+1} \| \mathcal{F}_{k-1}]. \tag{19.9}$$

The assumption (19.3) makes it possible to replace the "correction" terms in (19.8) and (19.9) by infinite series.

PROOF OF THE THEOREM. First [M22(40)],

$$\mathsf{E}[\xi_i \| \mathcal{F}_k] T^n = \mathsf{E}[\xi_{i+n} \| \mathcal{F}_{k+n}]. \tag{19.10}$$

By Schwarz's inequality, we have $|\mathsf{E}[\xi_0 \xi_n]| \le \mathsf{E}[|\xi_0| \cdot |\mathsf{E}[\xi_n \| \mathcal{F}_0]|] \le \|\xi_0\| \cdot \|\mathsf{E}[\xi_n \| \mathcal{F}_0]\|$, and it follows by (19.3) that the series (19.5) converges absolutely. If $\rho_k = \mathsf{E}[\xi_0 \xi_k]$, then, by stationarity, $\mathsf{E}[S_n^2] = n\rho_0 + 2\sum_{k=1}^{n-1}(n-k)\rho_k$. From

$$\left| \sigma^2 - \frac{1}{n}\mathsf{E}[S_n^2] \right| \le 2\sum_{k=n}^{\infty} |\rho_k| + \frac{2}{n}\sum_{i=1}^{n-1}\sum_{k=i}^{\infty} |\rho_k|$$

it now follows that

$$\frac{1}{n}\mathsf{E}[S_n^2] \to \sigma^2. \tag{19.11}$$

Thus $\sigma \ge 0$.‡ Since $|\mathsf{E}[\xi_0]| = |\mathsf{E}[\xi_n]| = |\mathsf{E}[\mathsf{E}[\xi_n \| \mathcal{F}_0]]| \le \mathsf{E}[|\mathsf{E}[\xi_n \| \mathcal{F}_0]|] \le \|\mathsf{E}[\xi_n \| \mathcal{F}_0]\|$ by Lyapounov's inequality, (19.4) follows by (19.3).

† The ξ_n are m-dependent if $(\xi_i, \ldots, \xi_k)$ and $(\xi_{k+n}, \ldots, \xi_j)$ are independent whenever $n > m$. An independent process is 0-dependent.

‡ If $\sigma = 0$, then, by Chebyshev's inequality, $S_n/\sqrt{n} \Rightarrow 0$.

By Lyapounov's inequality again, together with (19.3),

$$\mathsf{E}\Big[\sum_{i=1}^{\infty}|\mathsf{E}[\xi_{k+i}\|\mathcal{F}_k|\Big] \le \sum_{i=1}^{\infty}\|\mathsf{E}[\xi_{k+i}\|\mathcal{F}_k]\| < \infty. \tag{19.12}$$

Define

$$\theta_k = \sum_{i=1}^{\infty}\mathsf{E}[\xi_{k+i}\|\mathcal{F}_k]. \tag{19.13}$$

Because of (19.12), the series here converges with probability 1. Now define $\eta_k = \xi_k + \theta_k - \theta_{k-1}$ ((19.8) and (19.9) are special cases). Because of (19.12) again, the partial sums of the series in (19.13) are dominated by an integrable random variable, and so the sum can be interchanged with $\mathsf{E}[\cdot\|\mathcal{F}_{k-1}]$: $\mathsf{E}[\theta_k\|\mathcal{F}_{k-1}] = \sum_{i=1}^{\infty}\mathsf{E}[\xi_{k+i}\|\mathcal{F}_{k-1}]$. Therefore, $\mathsf{E}[\eta_k\|\mathcal{F}_{k-1}] = 0$, and $\{\eta_n\}$ is a martingale difference.

To apply the results of the preceding section, we must show that the ξ_n have finite second moments. If $\beta_i = \mathsf{E}[\xi_i\|\mathcal{F}_0]$, then $\mathsf{E}[\theta_0^2] \le \sum_{i,j=1}^{\infty}\mathsf{E}[|\beta_i|\cdot|\beta_j|] \le \sum_{i,j=1}^{\infty}\|\beta_i\|\cdot\|\beta_j\|$, and this is finite by (19.3). Therefore, $\tau^2 = \mathsf{E}[\eta_n^2]$ is finite, and it is a consequence of Theorem 18.3 that $n^{-1/2}M_n = n^{-1/2}\sum_{k=1}^{n}\eta_n \Rightarrow \tau N$. But $M_n = S_n + \theta_n - \theta_0$, and since $\|\theta_n\| = \|\theta_0\|$ is finite, $\|n^{-1/2}(M_n - S_n)\| \to 0$. This means in the first place that $n^{-1/2}S_n \Rightarrow \tau N$, and in the second place, that $\tau = \sigma$ because of (19.11) and the fact that the martingale property implies $\|n^{-1/2}M_n\|^2 = \tau^2$. Therefore, $S_n/\sigma\sqrt{n} \Rightarrow N$.

In most of the preceding sections, we have proved weak convergence by verifying tightness and the convergence of the finite-dimensional distributions. But here we can prove it directly by comparing X^n with the random function Y^n defined by $Y_t^n = M_{\lfloor nt\rfloor}/\sigma\sqrt{n}$, and we need only tighten the preceding argument and use the full force of Theorem 18.3. By Theorem 16.7, it is enough to show that, for each m,

$$\|Y^n - X^n\|_m = \sup_{t\le m}|Y_t^n - X_t^n| = \frac{1}{\sigma\sqrt{n}}\max_{k\le mn}|\theta_k - \theta_0| \Rightarrow_n 0. \tag{19.14}$$

Since $\mathsf{E}[\theta_0^2] < \infty$, we have $\sum_n \mathsf{P}[\theta_n^2/n \ge \epsilon] = \sum_n \mathsf{P}[\theta_0^2 \ge n\epsilon] < \infty$, and hence, by the Borel-Cantelli lemma, $\theta_n^2/n \to 0$ with probability 1. But

$$\frac{1}{n}\max_{k\le mn}\theta_k^2 \le \Big(\max_{k\le n_0}\frac{\theta_k^2}{n}\Big) \vee \Big(\sup_{k>n_0}\frac{\theta_k^2}{k}\Big),$$

and (19.14) follows from this. □

Example 19.1. Suppose that (19.1) is m-dependent. If $\mathsf{E}[\xi_n] = 0$, then $\|\mathsf{E}[\xi_n \| \mathcal{F}_0]\| = 0$ for $n > m$, and so (19.3) holds. If $\xi_n = \zeta_n - \zeta_{n+1}$, where the ζ_n are independent with mean 0 and variance 1, then $\{\xi_n\}$ is 1-dependent. In this case, $\mathsf{E}[\xi_n^2] = 2$, but the σ^2 of (19.5) is 0. □

Example 19.2. Define an autoregressive process as a sum $\xi_n = \sum_{i=0}^{\infty} \beta^i \zeta_{n-i}$, where $|\beta| < 1$ and the ζ_i are independent and identically distributed with mean 0 and variance 1. The σ-field $\mathcal{F}_0$ is generated by the ξ_k for $k \le 0$ and hence also by the ζ_k for $k \le 0$. Since $\zeta_1 \ldots \zeta_n$ are independent of $\mathcal{F}_0$, $\mathsf{E}[\xi_n \| \mathcal{F}_0] = \sum_{i=n}^{\infty} \beta^i \zeta_{n-i}$ and $\|\mathsf{E}[\xi_n \| \mathcal{F}_0]\|^2 = \beta^{2n}/(1-\beta^2)$. Therefore, (19.3) holds, and since $\{\xi_n\}$ is ergodic [PM.495], Theorem 19.1 applies. □

Uniform Mixing

Consider three measures of dependence between the σ-fields $\mathcal{F}_k$ and $\mathcal{G}_{k+n}$ of (19.2) ($\xi \in \mathcal{F}_k$ means ξ is $\mathcal{F}_k$-measurable):

$$\alpha_n = \sup[|\mathsf{P}(A \cap B) - \mathsf{P}A \cdot \mathsf{P}B| \colon A \in \mathcal{F}_k,\ B \in \mathcal{G}_{k+n}], \tag{19.15}$$

$$\rho_n = \sup[|\mathsf{E}[\xi\eta]| \colon \xi \in \mathcal{F}_k,\ \mathsf{E}[\xi] = 0,\ \|\xi\| \le 1,\ \eta \in \mathcal{G}_{k+n},\ \mathsf{E}[\eta] = 0,\ \|\eta\| \le 1], \tag{19.16}$$

$$\varphi_n = \sup[|\mathsf{P}(B|A) - \mathsf{P}B| \colon A \in \mathcal{F}_k,\ \mathsf{P}A > 0,\ B \in \mathcal{G}_{k+n}]. \tag{19.17}$$

These are independent of k if $\{\xi_n\}$ is stationary. The idea is that, if one or another of these quantities is small, then $\mathcal{F}_k$ and $\mathcal{G}_{k+n}$ are "almost" independent. If $\lim_n \alpha_n = 0$, the sequence $\{\xi_n\}$ is defined to be α-mixing, and the notions of ρ-mixing and φ-mixing are defined in the same way.

The three measures of the rate of mixing are related by the inequalities [M23(46)]

$$\alpha_n \le \rho_n \le 2\sqrt{\varphi_n}. \tag{19.18}$$

Theorem 19.2. *Assume that* (19.1) *is stationary, that* $\mathsf{E}[\xi_n] = 0$ *and the* ξ_n *have second moments, and that*

$$\sum_n \rho_n < \infty. \tag{19.19}$$

Then the conclusions of Theorem 19.1 *follow.*

PROOF. First [M23(48)], $\|\mathsf{E}[\xi_n \| \mathcal{F}_0]\| \le \rho_n \|\xi_0\|$. Therefore, (19.19) implies (19.3). We need only prove ergodicity, and for this it suffices [M22] to show that the shift satisfies $\mathsf{P}(A \cap T^{-n}B) \to \mathsf{P}A \cdot \mathsf{P}B$ for cylinders A and B. But since $\alpha_n \le \rho_n$, from $A \in \mathcal{F}_i$ and $B \in \mathcal{G}_j$ it follows that $|\mathsf{P}(A \cap T^{-n}B) - \mathsf{P}A \cdot \mathsf{P}B| \le \alpha_{j-i+n} \le \rho_{j-i+n} \to 0$. □

Example 19.3. Let $\{\zeta_n\}$ be a stationary Markov chain with finite state space, and let $\xi_n = f(\zeta_n)$, where f is a real function over the states. Let $p_{uv}^{(n)}$ be the nth-order transition probabilities and let p_v be the stationary probabilities. Suppose that the chain is irreducible and aperiodic, so that the p_v are all positive and [PM.131]

$$\left| \frac{p_{uv}^{(n)}}{p_v} - 1 \right| \le K\theta^n, \quad \theta < 1 \tag{19.20}$$

for all u, v, n. Then for cylinders $A = [\zeta_{k-i} = u_i, 0 \le i \le l]$ and $B = [\zeta_{k+n+i} = v_i, 0 \le i \le m]$, we have

$$|\mathsf{P}(A \cap B) - \mathsf{P}A \cdot \mathsf{P}B| \le K\theta^n \cdot \mathsf{P}A \cdot \mathsf{P}B. \tag{19.21}$$

For fixed A, the class of B satisfying (19.21) is a λ-system, and for fixed B, the class of A satisfying (19.21) is also a λ-system. Since the class of cylinders A and the class of cylinders B are π-systems, (19.21) holds for $A \in \sigma[\zeta_i, i \le k]$ and $B \in \sigma[\zeta_i, i \ge k+n]$. Since these last σ-fields are larger than $\mathcal{F}_k$ and $\mathcal{G}_{k+n}$, we have $\varphi_n \le K\theta^n$. And now (19.19) follows by (19.18), so that Theorem 19.2 applies. □

Functions of Mixing Processes

Let f be a measurable mapping from the space of doubly infinite sequences of real numbers to the real line: $f(\ldots, x_{-1}, x_0, x_1, \ldots) \in R^1$. Let $\ldots \eta_{-1}, \eta_0, \eta_1 \ldots$ be a function of the process (19.1), in the sense that

$$\eta_n = f(\ldots, \xi_{n-1}, \xi_n, \xi_{n+1} \ldots), \quad n = 0, \pm 1, \pm 2, \ldots, \tag{19.22}$$

where ξ_n occupies the 0th place in the argument of f. (Although the η_n are real, the ξ_n could now take values in a general measurable space.)

Let f_k be a measurable map from the space of left-infinite sequences to R^1: $f(\ldots, x_{-1}, x_0) \in R^1$. Put

$$\eta_{kn} = f_k(\ldots, \xi_{n+k-1}, \xi_{n+k}). \tag{19.23}$$

We want to show that $\{\eta_n\}$ satisfies the conclusions of Theorem 19.1 if $\{\xi_n\}$ is ρ-mixing and there exist functions f_k for which

$$\sum_k \|\eta_0 - \eta_{k0}\| < \infty. \tag{19.24}$$

Theorem 19.3. *Suppose that* (19.1) *is stationary and* $\sum_n \rho_n < \infty$. *Suppose the* η_n *of* (19.22) *have mean* 0 *and finite variance and that the* η_{kn} *of* (19.23) *have second moments and satisfy* (19.24). *Then* $\ldots, \eta_{-1}, \eta_0, \eta_1, \ldots$ *satisfies the conclusions of Theorem* 19.1.

The η_n of (19.22) involves all the ξ_i and hence extends its influence into the future of the η-process. But η_{kn} does not involve the ξ_i for $i > n + k$, and hence its influence on the future is muted. But (19.24) controls the influence of η_n itself by ensuring that it is near η_{kn} for large k. There is an alternative version of the theorem: Reverse time and replace (19.23) by $\eta_{kn} = f_k(\xi_{n-k}, \xi_{n-k+1}, \ldots)$.

PROOF. If η has a second moment and σ-fields $\mathcal{M}$ and $\mathcal{N}$ satisfy $\mathcal{M} \subset \mathcal{N}$, then it follows by Jensen's inequality that $\|\mathsf{E}[\eta\|\mathcal{M}]\|^2 = \mathsf{E}[(\mathsf{E}[\mathsf{E}[\eta\|\mathcal{N}]\|\mathcal{M}])^2] \le \mathsf{E}[\mathsf{E}[(\mathsf{E}[\eta\|\mathcal{N}])^2\|\mathcal{M}]] = \|\mathsf{E}[\eta\|\mathcal{N}]\|^2$. Therefore,

$$\|\mathsf{E}[\eta\|\mathcal{M}]\| \le \|\mathsf{E}[\eta\|\mathcal{N}]\|, \qquad \|\mathsf{E}[\eta\|\mathcal{M}]\| \le \|\eta\|, \tag{19.25}$$

where the second inequality follows from the first.

Denote the kth term in (19.24) by β_k. Since η_{k0}, as defined by (19.23), is $\mathcal{F}_k$-measurable, it follows by (19.25) that $\|\eta_{k0} - \mathsf{E}[\eta_0\|\mathcal{F}_k]\| = \|\mathsf{E}[\eta_{k0} - \eta_0\|\mathcal{F}_k]\| \le \beta_k$. Therefore, $\|\eta_0 - \mathsf{E}[\eta_0\|\mathcal{F}_k]\| \le 2\beta_k$. This means that if we take

$$\eta_{kn} = \mathsf{E}[\eta_n\|\mathcal{F}_{n+k}], \tag{19.26}$$

then (19.24) still holds.[†] For each k, $\{\eta_{kn}\}$ is stationary. Suppose that $0 < k < n$. We have

$$\|\mathsf{E}[\eta_0\|\mathcal{G}_n]\| \le \|\mathsf{E}[\eta_0\|\mathcal{G}_n] - \mathsf{E}[\eta_{k0}\|\mathcal{G}_n]\| + \|\mathsf{E}[\eta_{k0}\|\mathcal{G}_n]\|.$$

By a second application of (19.25), the first term on the right is at most $\|\eta_0 - \eta_{k0}\|$. Since $\mathsf{E}[\eta_{k0}] = \mathsf{E}[\eta_0] = 0$ by the new definition (19.26) of η_{k0}, the second term on the right is at most [M23(48)] $\rho_{n-k}\|\eta_{k0}\|$, and by a third application of (19.25), $\|\eta_{k0}\| \le \|\eta_0\|$. This brings us to $\|\mathsf{E}[\eta_0\|\mathcal{G}_n]\| \le \|\eta_0 - \eta_{k0}\| + \rho_{n-k}\|\eta_0\|$. Take $k = \lfloor n/2 \rfloor$. From (19.24) and the assumption $\sum_n \rho_n < \infty$ it now follows that $\sum_n \|\mathsf{E}[\eta_0\|\mathcal{G}_n]\| < \infty$.

A final application of (19.25): If $\mathcal{H}_n = \sigma(\eta_n, \eta_{n+1}, \ldots)$, then $\mathcal{H}_n \subset \mathcal{G}_n$, and hence $\|\mathsf{E}[\eta_0\|\mathcal{H}_n]\| \le \|\mathsf{E}[\eta_0\|\mathcal{G}_n]\|$. Therefore, $\sum_n \|\mathsf{E}[\eta_0\|\mathcal{H}_n]\| < \infty$. Since $\{\xi_n\}$ is ergodic (being ρ-mixing), $\{\eta_n\}$ is also ergodic and hence satisfies the hypotheses of Theorem 19.1 (see (19.7)). □

[†] The point is that this new η_{kn} is $\mathcal{F}_{n+k}$-measurable, not that it has the specific form (19.23) (although it does in fact have this form: see PM.186, Problem 13.3 and the note).

Example 19.4. Suppose that the ξ_n are independent and identically distributed with mean 0 and variance 1 and define

$$\eta_n = \sum_{i=-\infty}^{\infty} a_i \xi_{n+i},$$

where we assume that $\sum_i a_i^2 < \infty$. If for (19.23) we take

$$\eta_{kn} = \sum_{i=-\infty}^{k} a_i \xi_{n+i},$$

then the requirement (19.24) becomes

$$\sum_{k=1}^{\infty} \Big(\sum_{i=k+1}^{\infty} a_i^2 \Big)^{1/2} < \infty. \qquad \square$$

For the one-sided version of Theorem 19.3, suppose that $\xi_1, \xi_2, \ldots$ is stationary and define

$$\eta_n = f(\xi_n, \xi_{n+1}, \ldots) \tag{19.27}$$

and

$$\eta_{kn} = f_k(\xi_n, \ldots, \xi_{n+k}). \tag{19.28}$$

Take $\mathcal{F}_n = \sigma(\xi_1, \ldots, \xi_n)$ and $\mathcal{G}_n = \sigma(\xi_n, \xi_{n+1}, \ldots)$, and modify (19.15), (19.16), and (19.17) by inserting to the left of each supremum another supremum extending over positive k. Again assume that $\sum_n \rho_n < \infty$, the η_n have mean 0 and finite variance, and the η_{kn} have second moments; and assume that $\sum_k \|\eta_1 - \eta_{k1}\| < \infty$. Then $\eta_1, \eta_2, \ldots$ satisfies the conclusions of Theorem 19.3.

Example 19.5. Let P be Lebesgue measure on the unit interval, and let $\xi_1(x), \xi_2(x), \ldots$ be the digits of the dyadic expansion of x. This is a stationary, independent process under P, and hence it is ergodic. A random variable η_n of the form (19.27) can be regarded as a function $\eta_n(x) = f(T^{n-1}x)$, where f is a function on the unit interval and T is the transformation $Tx = 2x (\text{mod } 1)$. Suppose that f is square-integrable and that $\sum_k \beta_k < \infty$, where $\beta_k^2 = \int_0^1 (f(x) - f_k(x))^2\, dx$ and each f_k depends on x only through the first k digits of its dyadic expansion. It then follows by Theorem 19.3 (one-sided version) that, if $\mu = \int_0^1 f(x)\,dx$, then

$$\frac{1}{\sqrt{n}}\Big(\sum_{i=1}^{n} f(T^{i-1}x) - \mu n\Big) \Rightarrow \sigma N \tag{19.29}$$

for the appropriate asymptotic variance σ^2. And there is the corresponding functional limit theorem. If there exist such functions f_k at all, they may be taken to be

$$f_k(x) = 2^k \int_{(i-1)/2^k}^{i/2^k} f(s)\,ds \quad \text{for } x \in \Big(\frac{i-1}{2^k}, \frac{i}{2^k}\Big], \tag{19.30}$$

since then $f_k(.\xi_1, \dots \xi_k) = \mathsf{E}[\eta_1 \| \mathcal{F}_k]$.

If $f = I_{(0,t)}$ and f_k is defined by (19.30), then $\beta_k \le 2^{-k}$, so that (19.29) holds ($\mu = t$). If t is a dyadic rational, then $f = f_k$ for some k; but if t is not dyadic, then $f(x)$ involves the entire expansion of x.

If f is continuous and f_k is defined by (19.30), then $\beta_k^2 \le w_f^2(2^{-k})$, where w_f is the modulus of continuity. Therefore, (19.29) holds if $\sum_k w_f(2^{-k}) < \infty$, a condition which follows if f satisfies a uniform Hölder condition of some positive order: $|f(x) - f(x')| \le K|x - x'|^\theta$, $\theta > 0$. For example, we can take $f(x) \equiv x$. In this last case, $\{\eta_n\}$ is not φ-mixing (or even α-mixing) at all—to know $T^n x$ is to know $T^{n+1}x, T^{n+2}x \dots$ exactly. □

Diophantine Approximation

Suppose that Ω consists of the irrationals in $[0,1]$, $\mathcal{F}$ consists of the linear Borel sets contained in Ω, and P is Gauss's measure:

$$\mathsf{P}A = \frac{1}{\log 2}\int_A \frac{dx}{1+x}, \quad A \in \mathcal{F}. \tag{19.31}$$

Let T be the continued-fraction transformation: $Tx = x^{-1} - \lfloor x^{-1} \rfloor$. And let $a_k(x)$ be the kth partial quotient in the continued-fraction expansion of x, so that

$$T^{n-1}x = 1\rfloor a_n(x) + 1\rfloor a_{n+1}(x) + \cdots; \tag{19.32}$$

$(a_1, a_2, \dots)$ will play the role of the $(\xi_1, \xi_2, \dots)$ in Theorem 19.3.

We need several facts about continued fractions [PM.319-324]. The nth convergent to x is

$$1\rfloor\overline{a_1(x)} + \cdots + 1\rfloor\overline{a_n(x)} = \frac{p_n(x)}{q_n(x)}, \tag{19.33}$$

where the fraction is in lowest terms, and

$$\frac{1}{q_n(x)(q_n(x)+q_{n+1}(x))} < \left| x - \frac{p_n(x)}{q_n(x)} \right| < \frac{1}{q_n(x)q_{n+1}(x)}. \tag{19.34}$$

Further,

$$\left| \log x - \log \frac{p_n(x)}{q_n(x)} \right| \le \frac{4}{2^{n/2}}, \tag{19.35}$$

and

$$\left| \sum_{k=1}^{n} \log T^{k-1}x + \log q_n(x) \right| < 10. \tag{19.36}$$

Finally,

$$\frac{1}{\log 2} \int_0^1 \frac{-\log x}{1+x} dx = \frac{\pi^2}{12 \log 2}. \tag{19.37}$$

The basic fact [M24] is that $\{a_n\}$ is φ-mixing, where

$$\varphi_n \le K\theta^n, \quad \theta < 1. \tag{19.38}$$

Since $\sum_n \varphi^{1/2} < \infty$, we have $\sum_n \rho_n < \infty$, as required in Theorem 19.3. Let $\eta_n(x) = -\log T^{n-1}x - \mu$, where $\mu = \pi^2/12\log 2$; this in effect has the form (19.27) ($\xi_i = a_i$). And take

$$\eta_{kn}(x) = -\log\left[1\rfloor\overline{a_n(x)} + \cdots + 1\rfloor\overline{a_{n+k}(x)} \right] - \mu.$$

Then, by (19.35), $\sum_k \|\eta_1 - \eta_{k1}\| < \infty$. If X^n is the random element of $D[0,1]$ defined by

$$X_t^n = \frac{1}{\sigma\sqrt{n}} \sum_{k=1}^{\lfloor nt \rfloor} \eta_k \tag{19.39}$$

for the appropriate σ,[†] then $X^n \Rightarrow W$.

By (19.36), $\sum_{k=0}^{n-1} \eta_k - (\log q_n - n\mu)$, is bounded, from which it follows that $(\log q_n - n\mu)/\sigma\sqrt{n} \Rightarrow N$. This leads to theorems connected with Diophantine approximation. A fraction p/q is a best approximation to x if it minimizes the form $|q' \cdot x - p'|$ over fractions p'/q' with denominators q' not exceeding q. The successive best approximations to x are[‡] just the convergents $p_n(x)/q_n(x)$, and so the value of the form

[†] Which is positive; see Remark 5 on p. 482 of Gordin's paper [32].

[‡] See, for example, Khinchine [42] or Rockett & Szüsz [58].

for the nth in the series of best approximations is

$$d_n(x) = |q_n(x) \cdot x - p_n(x)|.$$

Since $|-\log d_n(x) - \log q_{n+1}(x)| \le \log 2$ by (19.34), we arrive at

$$\frac{1}{\sigma\sqrt{n}}\left(-\log d_n(x) - \frac{n\pi^2}{12\log 2}\right) \Rightarrow N. \tag{19.40}$$

There is the corresponding functional limit theorem. If we define Z^n by

$$Z_t^n(x) = \frac{1}{\sigma\sqrt{n}}\left(-\log d_{\lfloor nt\rfloor}(x) - \frac{\lfloor nt\rfloor\pi^2}{12\log 2}\right), \tag{19.41}$$

then $Z^n \Rightarrow W$ in the sense of $D[0,1]$. By (19.40), $k^{-1}\log d_k \Rightarrow_k -\mu$, so that the discrepency $d_k(x)$ has normal order $e^{-k\mu}$. Call the kth best approximation $p_k(x)/q_k(x)$ "superior" if

$$d_k(x) < e^{-k\pi^2/12\log 2}$$

and "inferior" otherwise. If $\tau_n(x)$ is the fraction of superior ones among the convergents

$$\frac{p_1(x)}{q_1(x)}, \ldots, \frac{p_n(x)}{q_n(x)},$$

then, by (9.28),

$$\mathsf{P}[\tau_n \le u] \to \frac{2}{\pi}\arcsin\sqrt{u}, \quad 0 < u < 1. \tag{19.42}$$

It is possible to prove Theorem 14.2 for X^n and Z^n as defined by (19.39) and (19.41). To prove that $X^n \Rightarrow W$ still holds if P (defined by (19.22)) is replaced by a probability measure P_0 absolutely continuous with respect to it, define $\bar{X}^n$ by (14.3) with η_i in place of ξ_i. If E lies in $\mathcal{F}_k$, then

$$|\mathsf{P}([\bar{X}^n \in A] \cap E) - \mathsf{P}[\bar{X}^n \in A] \cdot \mathsf{P}E| \le \varphi_{p_n - k} \to_n 0,$$

and so (14.7) holds just as before. Since $\mathcal{H} = \bigcup_k \mathcal{F}_k$ is a field and each $[\bar{X}^n \in A]$ lies in $\sigma(\mathcal{H})$, we have [M21] $\mathsf{P}_0[\bar{X}^n \in A] \to W(A)$. The rest of the proof is the same as that for Theorem 14.2. And Z^n is treated the same way. It follows, for example, that the P in (19.42) can be taken to be Lebesgue measure instead of Gauss's measure.

CHAPTER 5.

OTHER MODES OF CONVERGENCE

SECTION 20. CONVERGENCE IN PROBABILITY

There are four major modes of convergence for random variables:

1°: Convergence in distribution.
2°: Convergence in probability.
3°: Convergence with probability 1 (or almost-sure convergence).
4°: Convergence in L^2.

Neither of 3° and 4° implies the other, but each implies 2°, and 2° implies 1°. Each of these concepts extends to random elements of other metric spaces; 4° for the space C would be $\mathsf{E}[\|X^n - Y^n\|^2] \to 0$, where X^n and Y^n are random functions defined on the same probability space. And the relations between the four carry over as well. That 2° implies 1°, for example, is the corollary to Theorem 3.1.

Donsker's theorem and most of the other results of the preceding chapters concern convergence in distribution. Instances of convergence in distribution, even for random variables, often cannot be strengthened as they stand to convergence in probability.† But sometimes, if $\eta_n \Rightarrow \xi$ cannot be replaced by convergence in probability, it is possible to prove (on a new probability space) that $\eta_n - \xi_n \Rightarrow 0$, where each ξ_n has the distribution of ξ. In this section and in the next, Donsker's theorem will be strengthened in this way. And Strassen's theorem of Section 22 concerns the convergence with probability 1, in C, of the random paths (8.5) with new norming constants; it is a greatly generalized version of the law of the iterated logarithm.

In connection with the law of large numbers, 2° is called the "weak" law, and 3° is called the "strong" law. For convergence of random functions, the terminology in the literature is sometimes confusing. A theorem of type 1° concerns "weak convergence" as defined at the

† Problem 3.8.

beginning of this book, the invidious "weak" being an inheritance from functional analysis. A theorem of type 2° for random functions is often, for contrast, called "strong," a term best reserved for theorems of type 3°. Probably the simplest way out is to use the phrases in 1° through 4°, either as they stand or converted into adjectives.

As explained at the end of Section 8, the term "invariance principle" (IP) was first applied to Donsker's theorem (and its earlier forms) to indicate that various limit distributions are invariant under changes in the distribution common to the summands ξ_i (provided the mean and variance do not change). But now "invariance principle" has become a catchall phrase applied to a large miscellany of approximation theorems in probability; there are IPs in distribution, IPs in probability, almost-sure IPs, and L^2-IPs.

Section 22 depends on Theorem 20.1 but not on anything else in Sections 20 and 21.

A Convergence-in-Probability Version of Donsker's Theorem

Let $\xi_1, \xi_2, \ldots$ be independent and identically distributed random variables on $(\Omega, \mathcal{F}, \mathsf{P})$; suppose they have mean 0 and variance 1. Let $S_k = \sum_{i \le k} \xi_i$, and let X^n be the random polygonal function (8.5) (for $\sigma = 1$): $X^n_{i/n} = S_i/\sqrt{n}$, and X^n is linear between the points i/n, $i \le n$. According to Donsker's theorem, $X^n \Rightarrow W$ in the sense of $C = C[0,1]$. By Skorohod's representation theorem—Theorem 6.7—there exist on some (new) common probability space random elements $\bar{X}^n$ and $\bar{W}$ of C such that $\mathcal{L}(\bar{X}^n) = \mathcal{L}(X^n)$, $\mathcal{L}(\bar{W}) = \mathcal{L}(W)$, and $\bar{X}^n(\omega) \to \bar{W}(\omega)$ for each ω. But in this construction, the relations between the various X^n are not carried over to the $\bar{X}^n$: $\mathcal{L}(\bar{X}^n, \bar{X}^m)$ has no connection with $\mathcal{L}(X^n, X^m)$, for example.

Here we use an entirely different construction, also due to Skorohod. Let $(\Omega', \mathcal{F}', \mathsf{P}')$ be a space on which is defined a standard Brownian motion $[B(t)\colon 0 \le t < \infty]$: The increments are independent, the paths are continuous, and $B(t)$ is normally distributed with mean 0 and variance t. (We reserve W for Brownian motion as a random element of $C[0,1]$). This is Skorohod's theorem [PM.519]:

Theorem 20.1. *On $(\Omega', \mathcal{F}', \mathsf{P}')$ there exists a nondecreasing sequence $\tau_0 = 0, \tau_1, \tau_2, \ldots$ of random variables such that the differences $\zeta_n = B(\tau_n) - B(\tau_{n-1})$ are independent and have the distribution com-*

mon to the ξ_n *(above) and the differences* $\tau_n - \tau_{n-1}$ *are independent and identically distributed with mean* 1.[†]

Let $T_k = \sum_{i \le k} \zeta_i = B(\tau_n)$, and define a random element Y^n of C by linear interpolation between the values $Y^n_{i/n} = T_i/\sqrt{n}$ at the points i/n. Then the system $\{\zeta_i, T_k, Y^n\}$ on Ω' is an exact probabilistic replica of the system $\{\xi_i, S_k, X^n\}$ on Ω. Define $B^n(t) = B(nt)/\sqrt{n}$ for $0 \le t \le 1$. Then B^n is a random element of C, and it has there the same distribution as W (check the means and variances). Let $\|\cdot\|$ be the supremum norm on C: $\|x - y\| = \sup_t |x(t) - y(t)|$.

Theorem 20.2. *With these definitions and assumptions, we have*

$$\|Y^n - B^n\| \Rightarrow 0. \tag{20.1}$$

Since $\mathcal{L}(B^n) = \mathcal{L}(W)$ and $\mathcal{L}(Y^n) = \mathcal{L}(X^n)$, it follows by the corollary to Theorem 3.1 that (20.1) implies Donsker's theorem: $X^n \Rightarrow W$. No weak-convergence theory is required for the statement and proof of Theorem 20.2 itself.

PROOF. Define a random element Z^n of C by linear interpolation between the values $Z^n_{i/n} = B^n(i/n)$ at the i/n. A summary of the notation:

$$\begin{cases} X^n : X^n_{i/n} = n^{-1/2} S_i = n^{-1/2} \sum_{h \le i} \xi_h \ (\text{on } \Omega), \\ Y^n : Y^n_{i/n} = n^{-1/2} T_i = n^{-1/2} \sum_{h \le i} \zeta_h = n^{-1/2} B(\tau_i) \ (\text{on } \Omega'), \\ Z^n : Z^n_{i/n} = B^n(i/n) = n^{-1/2} B(i) \ (\text{on } \Omega'). \end{cases} \tag{20.2}$$

These random functions are linear on the subintervals $[(i-1)/n, i/n]$.

We prove (20.1) in two steps:

$$\|B^n - Z^n\| \Rightarrow 0 \tag{20.3}$$

and

$$\|Y^n - Z^n\| \Rightarrow 0. \tag{20.4}$$

First [M25(58)],

$$\mathsf{P}'[\sup_{s \le t} |B(s)| \ge \alpha] \le 4e^{-\alpha^2/2t}, \quad \alpha \ge 0. \tag{20.5}$$

[†] If the ξ_i had variance σ^2, the $\tau_n - \tau_{n-1}$ would have mean σ^2.

Since $Z^n(\cdot)$ is $B^n(\cdot)$ made linear between the i/n, $\|B^n - Z^n\|$ is at most the maximum over $i \leq n$ of the supremum over $t \leq 1/n$ of $2|B^n(t + i/n) - B^n(i/n)|$. Since the increments of Brownian motion are stationary, (20.5) gives

$$\text{(20.6)}\quad \mathsf{P}'[\|B^n - Z^n\| \geq \epsilon] \leq n\mathsf{P}'[2\sup_{t\leq 1/n}|B^n(t)| \geq \epsilon]$$
$$= n\mathsf{P}'[\sup_{s\leq 1}|B(s)| \geq \epsilon n^{1/2}/2] \leq 4ne^{-\epsilon^2 n/8} \to_n 0.$$

Hence (20.3).

Becaue of the polygonal form of Y^n and Z^n,

$$\text{(20.7)}\qquad \|Y^n - Z^n\| = n^{-1/2}\max_{i\leq n}|B(\tau_i) - B(i)|$$
$$= \max_{i\leq n}|B^n(\tau_i/n) - B^n(i/n)|.$$

Since the differences $\tau_i - \tau_{i-1}$ are independent and identically distributed with mean 1, it follows by the strong law of large numbers that $\tau_i/i \to_i 1$ with P'-probability 1. This means that τ_i/n is near i/n (i large), and we can use path continuity to show that $B^n(\tau_i/n)$ is near $B^n(i/n)$.

Let ϵ be given. If we define $B^n(t)$ as $B(nt)/\sqrt{n}$ for all $t \geq 0$ (rather than for $0 \leq t \leq 1$ only), then it is another standard Brownian motion, and so its paths are uniformly continuous over $[0, 2]$. Therefore, there is a δ small enough that

$$\text{(20.8)}\qquad \mathsf{P}'[\sup_{s\leq 1, |t-s|\leq\delta}|B^n(t) - B^n(s)| \geq \epsilon] < \epsilon.$$

This holds for every n. Since $\tau_i/i \to_i 1$ with probability 1, there is an m such that

$$\text{(20.9)}\qquad \mathsf{P}'\Big[\max_{m\leq i\leq n}\Big|\frac{\tau_i}{n} - \frac{i}{n}\Big| \geq \delta\Big] \leq \mathsf{P}'\Big[\sup_{i\geq m}\Big|\frac{\tau_i}{i} - 1\Big| \geq \delta\Big] < \epsilon.$$

Fix the pair δ, m. If $|B^n(t) - B^n(s)| < \epsilon$ for $s \leq 1$ and $|t - s| \leq \delta$, and if $|\tau_i/n - i/n| < \delta$ for $m \leq i \leq n$, then $|B^n(\tau_i/n) - B^n(i/n)| < \epsilon$ for $m \leq i \leq n$. Therefore, by (20.8) and (20.9),

$$\text{(20.10)}\qquad \mathsf{P}'\Big[\max_{m\leq i\leq n}\Big|B^n\Big(\frac{\tau_i}{n}\Big) - B^n\Big(\frac{i}{n}\Big)\Big| \geq \epsilon\Big] < 2\epsilon.$$

This holds for all n. But obviously there is an n_0 such that

$$\text{(20.11)}\qquad \mathsf{P}'\Big[\max_{i\leq m}\Big|B^n\Big(\frac{\tau_i}{n}\Big) - B^n\Big(\frac{i}{n}\Big)\Big| \geq \epsilon\Big] < \epsilon, \qquad n > n_0.$$

And now (20.4) follows from (20.7), (20.10), and (20.11). □

SECTION 21. APPROXIMATION BY INDEPENDENT NORMAL SEQUENCES

The approximation theorem is this:

Theorem 21.1. *Let $\{\xi_i\}$ be an independent, identically distributed sequence of random variables on $(\Omega, \mathcal{F}, \mathsf{P})$, and assume that $\mathsf{E}[\xi_i] = 0$ and $\mathsf{E}[\xi_i^2] = 1$. Suppose there is on $(\Omega, \mathcal{F}, \mathsf{P})$ a random variable U that is independent of $\{\xi_i\}$ and is uniformly distrubuted over $[0,1]$. Then there exists on $(\Omega, \mathcal{F}, \mathsf{P})$ a sequence $\{\eta_i\}$ of standard normal random variables such that*

$$\frac{1}{\sqrt{n}} \max_{k \le n} \left| \sum\nolimits_{i \le k} \xi_i - \sum\nolimits_{i \le k} \eta_i \right| \Rightarrow_n 0. \tag{21.1}$$

For an example, take the probability space to be the unit square with Lebesgue measure, and define $U(\omega_1, \omega_2) = \omega_2$ and $\xi_i(\omega_1, \omega_2) = r_i(\omega_1)$, where the r_i are the Rademacher functions. Like Theorem 20.2, this theorem implies that of Donsker.

It is impossible to strengthen convergence in probability to convergence with probability 1 in (21.1),† although this is possible if the norming factor $\sqrt{n}$ is replaced by $\sqrt{n \log\log n}$ (see (22.25) in the next section). The sequences $\{\xi_i\}$ and $\{\eta_i\}$ cannot be independent of one another, since this would contradict the central limit theorem theorem for $\{\xi_i - \eta_i\}$.

We give two proofs of this theorem, and each of them requires some preliminary definitions and results. Let S and T be metric spaces with Borel σ-fields $\mathcal{S}$ and $\mathcal{T}$. Call S and T Borel isomorphic, and write $S \sim T$, if there is a one-to-one map φ of S onto T such that φ is measurable $\mathcal{S}/\mathcal{T}$ and φ^{-1} is measurable $\mathcal{T}/\mathcal{S}$. We need the following isomorphism relations:

$$R^1 \sim (0,1) \sim [0,1] \sim [0,1]^\infty \sim R^\infty. \tag{21.2}$$

The first one is easy: There are many continuous, increasing maps φ of R onto $(0,1)$. For the second, let x_n be points of $(0, \frac{1}{4})$ that increase to $\frac{1}{4}$, let y_n be points of $(\frac{3}{4}, 1)$ that decrease to $\frac{3}{4}$, let φ map $x_1 \to 0$, $x_{n+1} \to x_n$, $y_1 \to 1$, $y_{n+1} \to y_n$, and take φ to be the identity elsewhere.

For the third isomorphism, we construct a map $\varphi: I^\infty \to I$, where $I = [0,1]$. Associate with each point of I a unique binary expansion:

† Major [45].

Take $0 = .00\cdots$ and $1 = .11\cdots$, and use the nonterminating expansion (say) for points of $(0,1)$. Let $\mathcal{N} = \{1,2,\ldots\}$, take f to be some one-to-one map of $\mathcal{N}$ onto $\mathcal{N}\times\mathcal{N}$, and let $g = f^{-1}$. For $y \in I$ write $y = .y_1y_2\cdots$, and for $x = (x_1, x_2, \ldots) \in I^\infty$ write $x_i = .x_{i1}x_{i2}\cdots$. Define a one-to-one map φ of I^∞ onto I by requiring of $y = \varphi(x)$ that $y_k = x_{f(k)}$ for all k: Intertwine the expansions of the coordinates of x to arrive at the expansion of $\varphi(x)$. Then $\varphi^{-1}(y) = x$ if and only if $x_{ij} = y_{g(ij)}$ for all i, j. Let $\mathcal{I}$ be the Borel σ-field in I. If J is the set of points of I whose expansions start with the digits $d_1, \ldots, d_m$, then $\varphi^{-1}J$ is the set of x such that $x_{f(k)} = d_k$ for $1 \le k \le m$, a set that lies in $\mathcal{I}^\infty$. Since these sets J generate $\mathcal{I}$, φ is measurable $\mathcal{I}^\infty/\mathcal{I}$. On the other hand, if J_i is the set of points of I whose expansions start with the digits $d_{i1}, \ldots, d_{im_i}$, then $\varphi(J_1 \times \cdots \times J_l \times I \times \cdots)$ consists of those y such that $y_{g(ij)} = d_{ij}$ for $1 \le i \le l$ and $1 \le j \le m_i$. This set lies in $\mathcal{I}$, and it follows that φ^{-1} is measurable $\mathcal{I}/\mathcal{I}^\infty$.

To construct the final isomorphism, take an isomorphism from I onto R^1 and apply it to each coordinate of the points in I^∞.

In the following lemma, R, S, and T can be any metric spaces that are Borel isomorphic to $[0,1]$. In fact, every uncountable, separable, complete metric space is Borel isomorphic to $[0,1]$,[†] but we have proved this result—and need it—only for the spaces in (21.2). In Lemma 2 below, ρ is a random element on $(\Omega, \mathcal{F}, \mathsf{P})$ with values in R. In both lemmas, σ, τ, and u are random elements on the same $(\Omega, \mathcal{F}, \mathsf{P})$, with values in S, T, $[0,1]$, respectively, ν is a probability measure on $S \times T$ with marginal measure μ on S, and $\mathcal{L}(\sigma) = \mu$:

$$\rho \in R, \quad \sigma \in S, \quad u \in [0,1], \quad \mu(\,\cdot\,) = \nu(\,\cdot \times T), \quad \mathcal{L}(\sigma) = \mu; \quad \tau \in T.$$

Lemma 1. *Assume that S and T are Borel isomorphic to $[0,1]$, that $\mathcal{L}(\sigma) = \mu$, that u is uniformly distributed over $[0,1]$, and that σ and u are independent. Then there is on $(\Omega, \mathcal{F}, \mathsf{P})$ a random element τ, a function of (σ, u), such that $\mathcal{L}(\sigma, \tau) = \nu$.*

Lemma 2. *Assume that R, S, T are Borel isomorphic to $[0,1]$, that $\mathcal{L}(\sigma) = \mu$, that u is uniformly distributed over $[0,1]$, and that ρ, σ, and u are independent. Then there is on $(\Omega, \mathcal{F}, \mathsf{P})$ a random element τ, a function of (σ, u), such that $\mathcal{L}(\sigma, \tau) = \nu$ and ρ is independent of (σ, τ).*

[†] Parthasarathy [48], p. 14.

PROOFS. We prove the second lemma first. Assume the result holds in the special case $R = S = T = [0,1] = I$, and let φ_R, φ_S, φ_T be isomorphisms of R, S, T onto I. Let $\varphi_{ST}(s,t) = (\varphi_S(s), \varphi_T(t))$, so that φ_{ST} is an isomorphism of $S \times T$ onto $I \times I$; take $\nu' = \nu\varphi_{ST}^{-1}$, $\mu' = \mu\varphi_S^{-1}$, $\rho' = \varphi_R\rho$, and $\sigma' = \varphi_S\sigma$. Then $\mathcal{L}(\sigma') = \mu'$ (the first marginal of ν'), and ρ', σ', u are independent. By the special case, there is a random element τ' of I, a function of (σ', u), such that $\mathcal{L}(\sigma', \tau') = \nu'$ and ρ' is independent of (σ', u). If $\tau = \varphi_T^{-1}\tau'$, then τ is a function of (σ, u), $\mathcal{L}(\sigma, \tau) = \nu$ and ρ' is independent of (σ, u). The point is that isomorphic spaces are measure-theoretically indistinguishable, and the lemma has to do with measure theory only (no topology).

In the special case, there is for ν a conditional probability distribution over the second component of $I \times I$ given the first. That is, there exists a function $p(s, B)$, defined for $s \in I$ and $B \in \mathcal{I}$, such that, for fixed s, $p(s, \cdot\,)$ is a probability measure on $\mathcal{I}$, and for $A, B \in \mathcal{I}$, $p(\,\cdot\,, B)$ is $\mathcal{I}$-measurable and $\nu(A \times B) = \int_A p(s,B)\mu(ds)$. Define a distribution function $F(s, \cdot\,)$ by $F(s,z) = p(s, [0,z])$, and let $\phi(s, \cdot\,)$ be the corresponding inverse, or quantile function: $\phi(s,t) = \inf[z: t \le F(s,z)]$, so that $\phi(s,t) \le z$ if and only if $t \le F(s,z)$. Then $[(s,t): \phi(s,t) \le z] = \bigcap_r([s: r \le F(s,z)] \times [r, 1])$, where the intersection extends over rational r. Thus ϕ is measurable $\mathcal{I} \times \mathcal{I}$, and $\tau(\omega) := \phi(\sigma(\omega), u(\omega))$ is a random variable. If λ is Lebesgue measure on I, then $\lambda[t: \phi(s,t) \le z] = \lambda[t: t \le F(s,z)] = F(s,z) = p(s,[0,z])$, and it follows that $\lambda[t: \phi(s,t) \in B] = p(s,B)$ for $B \in \mathcal{I}$. Since $\mathcal{L}(\sigma) = \mu$ and $\mathcal{L}(u) = \lambda$, and since σ and u are independent, $\mathcal{L}(\sigma, u) = \mu \times \lambda$. Therefore,

$$\begin{aligned}
\nu(A \times B) &= \int_A p(s,B)\mu(ds) = \int_A \lambda[t: \phi(s,t) \in B]\mu(ds) \\
&= (\mu \times \lambda)[(s,t): s \in A, \phi(s,t) \in B] \\
&= \mathsf{P}[\sigma \in A, \phi(\sigma, u) \in B] = \mathsf{P}[\sigma \in A, \tau \in B]:
\end{aligned}$$

(σ, τ) has distribtion ν.

In Lemma 2, ρ is assumed independent of (σ, u), and so it is independent of $(\sigma, \phi(\sigma, u)) = (\sigma, \tau)$ as well.

For the proof of Lemma 1, simply remove from the preceding argument all reference to R and ρ (or introduce a dummy space R and take $\rho \equiv r$ for some r in R). □

FIRST PROOF OF THEOREM 21.1. The first proof depends on Theorem 20.2. Changing the notation there, write ξ_i' for ζ_i and η_i' for

$B(i) - B(i-1)$; these are defined on the space Ω'. Then $\mathcal{L}(\{\xi'_i\}) = \mathcal{L}(\{\xi_i\})$, $\sum_{i \le j} \xi'_i = B(\tau_j)$, and the η_i are independent standard normal variables. And by (20.4), (21.1) holds if we replace ξ_i and η_i (variables on Ω) by ξ'_i and η'_i (variables on Ω').

But we can use Lemma 1 to pull everything back to Ω. Take $S = T = R^\infty$ and $\nu = \mathcal{L}(\{\xi'_i\}, \{\eta'_i\})$, the first marginal of which is $\mathcal{L}(\{\xi'_i\}) = \mathcal{L}(\{\xi_i\})$. Take $\sigma = \{\xi_i\}$ on $(\Omega, \mathcal{F}, \mathsf{P})$, and use U in the role of u. By the lemma, there is on $(\Omega, \mathcal{F}, \mathsf{P})$ a $\tau = \{\eta_i\}$ such that $\mathcal{L}(\{\xi_i\}, \{\eta_i\}) = \mathcal{L}(\{\xi'_i\}, \{\eta'_i\})$. And this $\{\eta_i\}$ is exactly the sequence we want. □

This first proof depends on Theorem 20.1. Our second proof is intricate but has the advantage that it avoids the Skorohod representation theorem.[†]

SECOND PROOF OF THEOREM 21.1. Suppose ϵ given. Define

$$t_k = \lfloor (1+\epsilon^4)^k \rfloor, \qquad n_k = t_{k+1} - t_k. \tag{21.3}$$

Next define

$$X_k = n_k^{-1/2} \sum\nolimits_{t_k < i \le t_{k+1}} \xi_i. \tag{21.4}$$

Let $S_j = \sum_{i \le j} \xi_i$. By the central limit theorem, there is a $j_0(\epsilon)$ such that

$$\pi(\mathcal{L}(j^{-1/2} S_j), \mathcal{L}(N)) < \epsilon^6, \qquad j > j_0(\epsilon), \tag{21.5}$$

where π is the Prohorov distance. And since $n_k \to \infty$, it follows that $\pi(\mathcal{L}(X_k), \mathcal{L}(N)) \le \epsilon^6$ for k exceeding some $k_0(\epsilon)$. By Theorem 6.9, there exists a probability measure ν_k on $R^1 \times R^1$ that has first and second marginal measures $\mathcal{L}(X_k)$ and $\mathcal{L}(N)$, respectively, and satisfies

$$\nu_k[(s,t) \colon |s-t| > \epsilon^6] < \epsilon^6, \qquad k > k_0(\epsilon). \tag{21.6}$$

It is possible [PM.265] to define independent random variables $U_0, U_1, U_2, \ldots$, each uniformly distributed over $[0,1]$ and each a function of U, so that, by the hypothesis of the theorem, $\{U_i\}$ is independent of $\{\xi_i\}$ and hence of $\{X_i\}$. The next step is to construct on

[†] Which does not extend in a simple way to vector-valued processes, for example; see Monrad & Philipp [46].

$(\Omega, \mathcal{F}, \mathsf{P})$ an independent sequence $Y_1, Y_2, \ldots$ of standard normal variables, each independent of U_0, such that

$$\mathsf{P}[|X_k - Y_k| > \epsilon^6] < \epsilon^6, \qquad k > k_0(\epsilon). \tag{21.7}$$

First apply Lemma 1 for $S = T = R^1$, $\sigma = X_1$, $u = U_1$, and $\nu = \nu_1$. Since X_1 and U_1 are independent, there is a random variable $\tau = Y_1$, a function of (X_1, U_1), such that $\mathcal{L}(X_1, Y_1) = \nu_1$. Since $\mathcal{L}(Y_1)$ is the second marginal of ν_1, namely $\mathcal{L}(N)$, Y_1 is a standard normal variable. Since Y_1 is a function of (X_1, U_1) and hence of $(\{\xi_i\}, U_1)$, it is independent of U_0.

Suppose that $Y_1, \ldots, Y_k$ have been defined: They are independent, standard normal variables, Y_i is a function of (X_i, U_i), and $\mathcal{L}(X_i, Y_i) = \nu_i$ $(i \leq k)$. Apply Lemma 2 for $R = R^k$, $S = T = R^1$, $\rho = (Y_1, \ldots, Y_k)$, $\sigma = X_{k+1}$, $u = U_{k+1}$, and $\nu = \nu_{k+1}$. Since ρ is a function of $(X_1, \ldots, X_k, U_1, \ldots, U_k)$, the three random elements ρ, σ, u are independent. Therefore, there is a random variable $\tau = Y_{k+1}$, a function of (X_{k+1}, U_{k+1}), such that $\mathcal{L}(X_{k+1}, Y_{k+1}) = \nu_{k+1}$, and $(Y_1, \ldots, Y_k)$ and (X_{k+1}, Y_{k+1}) are independent. Again Y_{k+1} is a standard normal variable: $\mathcal{L}(Y_{k+1})$ is $\mathcal{L}(N)$, the second marginal of ν_{k+1}. And Y_{k+1}, being a function of $(\{\xi_i\}, U_{k+1})$, is independent of U_0. This gives the sequence we want: (21.7) holds because of (21.6).

Next, let $\zeta_1, \zeta_2, \ldots$ be an independent sequence of standard normal variables on some (new) probability space and set

$$W_k = n_k^{-1/2} \sum_{t_k < i \leq t_{k+1}} \zeta_i. \tag{21.8}$$

Apply Lemma 1 again, this time with $S = T = R^\infty$, $\sigma = (Y_1, Y_2, \ldots)$, $u = U_0$, and $\nu = \mathcal{L}(\{W_k\}, \{\zeta_i\})$. Since σ and u are independent, there exists (on the original $(\Omega, \mathcal{F}, \mathsf{P})$) a random element $\tau = (\eta_1, \eta_2, \ldots)$ of R^∞ such that $\mathcal{L}(\{Y_k\}, \{\eta_i\}) = \nu$. This means in the first place that the η_i (like the ζ_i) are independent standard normal variables, and in the second place that the joint law of $\{Y_k\}$ and $\{\eta_i\}$ is the same as the joint law of $\{W_k\}$ and $\{\zeta_i\}$. But then (21.8) implies that

$$Y_k = n_k^{-1/2} \sum_{t_k < i \leq t_{k+1}} \eta_i \tag{21.9}$$

holds with probability 1 on $(\Omega, \mathcal{F}, \mathsf{P})$.

To sum up: We have constructed independent standard normal variables η_i such that the Y_k given by (21.9) satisfy (21.7). Write

$S_j = \sum_{i \le j} \xi_i$ and $T_j = \sum_{i \le j} \eta_i$. To prove (21.1) is to prove that there exists an absolute constant K with the property that for each ϵ there is an $n_0(\epsilon)$ such that

$$\mathsf{P}[n^{-1/2} \max_{j \le n} |S_j - T_j| \ge \epsilon] \le K\epsilon, \qquad n > n_0(\epsilon). \tag{21.10}$$

The following argument does prove (21.10), but it leaves a major problem: The sequence $\{\eta_i\}$ we have constructed depends on ϵ itself. Nonetheless, we proceed to prove (21.10), leaving to the end the resolution of this problem.

Define M, s, m (functions of n and ϵ) by

$$t_{M-1} \le n < t_M, \quad s = \lfloor \epsilon^{-5} \rfloor, \quad m = M - s. \tag{21.11}$$

By the definitions (21.3), for large n we have $n \ge t_M/4$, and so it is enough to find an upper estimate for $\mathsf{P}[\max_{j \le n} |S_j - T_j| \ge 7\epsilon t_M^{1/2}]$. Define

$$A_n = \max_{j \le t_m} |S_j|, \quad B_n = \max_{j \le t_m} |T_j|, \quad C_n = \max_{m \le k < M} \max_{i < n_k} |S_{t_k + i} - S_{t_k}|,$$
$$D_n = \max_{m \le k < M} \max_{i < n_k} |T_{t_k + i} - T_{t_k}|.$$

Then

$$\begin{aligned}\max_{j \le n} |S_j - T_j| &\le A_n + B_n + \max_{t_m < j \le n} |S_j - T_j| \\ &\le 2A_n + 2B_n + \max_{t_m < j \le n} |(S_j - S_{t_m}) - (T_j - T_{t_m})| \\ &\le 2A_n + 2B_n + C_n + D_n + \max_{m < k \le M} |(S_{t_k} - S_{t_m}) - (T_{t_k} - T_{t_m})|.\end{aligned}$$

And for $m < k \le M$, each term in this last maximum is at most $\sum_{i=m}^{M-1} n_i^{1/2} |X_i - Y_i|$. Therefore, it will be enough to bound the terms in

$$\begin{aligned}(21.12)\quad \mathsf{P}[\max_{j \le n} |S_j - T_j| \ge 7\epsilon t_M^{1/2}] &\\ \le \mathsf{P}[\max_{j \le t_m} |S_j| \ge \epsilon t_M^{1/2}] &+ \mathsf{P}[\max_{j \le t_m} |T_j| \ge \epsilon t_M^{1/2}] \\ &+ \mathsf{P}[\max_{m \le k < M} \max_{i < n_k} |S_{t_k + i} - S_{t_k}| \ge \epsilon t_M^{1/2}] \\ &+ \mathsf{P}[\max_{m \le k < M} \max_{i < n_k} |T_{t_k + i} - T_{t_k}| \ge \epsilon t_M^{1/2}] \\ &+ \mathsf{P}\Big[\sum_{m \le k < M} n_k^{1/2} |X_k - Y_k| \ge \epsilon t_M^{1/2}\Big] \\ &= \mathrm{I} + \mathrm{II} + \mathrm{III} + \mathrm{IV} + \mathrm{V}.\end{aligned}$$

We estimate the terms separately. In several places we use the fact that $\lfloor x \rfloor \geq x/2$ if $x \geq 2$. We can use Kolmogorov's inequality on the first two terms. By (21.3) and (21.11),

$$\text{(21.13)}\quad \mathrm{I} \leq \epsilon^{-2} t_M^{-1} \operatorname{Var}[S_{t_m}] \leq 2\epsilon^{-2}(1+\epsilon^4)^{-s}$$
$$\leq 2\epsilon^{-2}\exp\Big(-\frac{1}{2}\epsilon^{-5}\log(1+\epsilon^4)\Big) \leq 2\epsilon^{-2}\exp\Big(-\frac{1}{4}\epsilon^{-1}\Big) < K_1\epsilon,$$

where K_1 is a constant large enough that the last inequality holds for $0 < \epsilon < 1$. This and the same argument for T_{t_m} give

$$\text{(21.14)}\qquad \mathrm{I} < K_1\epsilon, \qquad \mathrm{II} < K_1\epsilon.$$

By Etemadi's inequality [M19],

$$\text{(21.15)}\qquad \mathrm{III} \leq \sum\nolimits_{m \leq k < M} \mathsf{P}[\max\nolimits_{i < n_k} |S_{t_k+i} - S_{t_k}| \geq \epsilon t_M^{1/2}]$$
$$\leq 3 \sum\nolimits_{m \leq k < M} \max_{i < n_k} \mathsf{P}[|S_i| \geq \epsilon t_M^{1/2}/3].$$

The number of summands here is $M - m = s \leq \epsilon^{-5}$. To estimate III, consider separately the terms for $i \leq n^{1/2}$ and $i > n^{1/2}$. By Chebyshev's inequality, the terms for $i \leq n^{1/2}$ contribute to III at most (use (21.11))

$$\text{(21.16)}\qquad 3 \cdot \epsilon^{-5}(\epsilon t_M^{1/2}/3)^{-2} n^{1/2} = 27\epsilon^{-7} t_M^{-1} n^{1/2} \leq 27\epsilon^{-7} n^{-1/2} < \epsilon,$$

where the last inequality holds for n sufficiently large.

If $k < M$ and $i < n_k$, then $i < n_M = t_{M+1} - t_M$ and hence

$$\frac{i}{t_M} < \frac{t_{M+1} - t_M}{t_M} \leq 2\frac{(1+\epsilon^4)^{M+1} - (1+\epsilon^4)^M + 1}{(1+\epsilon^4)^M} \leq 2\epsilon^4 + \frac{2}{(1+\epsilon^4)^M}.$$

Since M goes to infinity along with n, $i/t_M < 4\epsilon^4$ for large n. But also, for large n, $i > n^{1/2}$ implies that i exceeds the $k_0(\epsilon)$ of (21.7), and (if $\epsilon^7 < 1/12$) it follows that [M25(55)]

$$\mathsf{P}\Big[|S_i| \geq \frac{1}{3}\epsilon t_M^{1/2}\Big] = \mathsf{P}\Big[i^{-1/2}|S_i| \geq \frac{1}{3}\epsilon\Big(\frac{t_M}{i}\Big)^{1/2}\Big] \leq \mathsf{P}\Big[i^{-1/2}|S_i| \geq \frac{1}{6\epsilon}\Big]$$
$$\leq \mathsf{P}\Big[|N| \geq \frac{1}{6\epsilon} - \epsilon^6\Big] + \epsilon^6 \leq \mathsf{P}\Big[|N| \geq \frac{1}{12\epsilon}\Big] + \epsilon^6$$
$$\leq 2\exp\Big(-\frac{1}{2(12\epsilon)^2}\Big) + \epsilon^6 < K_2\epsilon^6,$$

where K_2 is a constant large enough that the last inequality holds for $0 < \epsilon < 1$. Since the number of summands in (21.15) is at most ϵ^{-5}, the terms for $i > n_m$ contribute to III at most $3\epsilon^{-5}K_2\epsilon^6 = 3K_2\epsilon$. Therefore

$$\text{III} \leq \epsilon + 3K_2\epsilon. \tag{21.17}$$

And IV satisfies the same inequality.

To estimate V, note that, by Schwarz's inequality,

$$\sum\nolimits_{m\leq k<M} n_k^{1/2} \leq \Big(\sum\nolimits_{m\leq k<M} n_k\Big)^{1/2}(M-n)^{1/2} \leq t_M^{1/2}\epsilon^{-5/2}.$$

It follows by (21.7) that

$$\begin{aligned} \text{V} &\leq \mathsf{P}\Big[\sum\nolimits_{m\leq k<M} n_k^{1/2}|X_k - Y_k| \geq \epsilon^{7/2}\sum\nolimits_{m\leq k<M} n_k^{1/2}\Big] \\ &\leq \sum\nolimits_{m\leq k<M} \mathsf{P}[|X_k - Y_k| \geq \epsilon^{7/2}] \leq (M-m)\epsilon^6 \leq \epsilon. \end{aligned} \tag{21.18}$$

Putting the five estimates into (21.12) leads to (21.10) with $K = 2K_1 + 6K_2 + 3$. We must still find a single sequence $\{\eta_i\}$ that works for every ϵ—satisfies (21.10) for every value of ϵ.

There exists a (new and eventually irrelevant) probability space on which are defined random variables $\xi_i^{(p)}$, $i, p \geq 1$, and $U^{(p)}$, $p \geq 1$, such that all are independent of one another, each $\xi_i^{(p)}$ is distributed like the original ξ_i, and each $U^{(p)}$ is uniformly distributed over $[0,1]$. Let $S_k^{(p)} = \sum_{i\leq k}\xi_i^{(p)}$. The argument leading to (21.10) shows that for each p there exists on this probability space an independent sequence $(\eta_1^{(p)}, \eta_2^{(p)}, \ldots)$ of standard normal variables, a function of $(U^{(p)}, \xi_1^{(p)}, \xi_2^{(p)}, \ldots)$, such that, if $T_k^{(p)} = \sum_{i\leq k}\eta_i^{(p)}$, then there is an $n_0(p)$ such that

$$\mathsf{P}[\max\nolimits_{k\leq n}|n^{-1/2}S_k^{(p)} - n^{-1/2}T_k^{(p)}| \geq 2^{-p}] \leq 2^{-p}, \quad \text{for } n \geq n_0(p). \tag{21.19}$$

By the construction, the sets $\{\xi_i^{(p)}, \eta_i^{(p)} : i = 1, 2, \ldots\}$ for different values of p are independent of one another (but $S_k^{(p)}$ and $T_k^{(p)}$ are not independent).

Put $r(v) = \sum_{p\leq v} n_0(p)$, and define

$$\xi_i' = \xi_{i-r(p)}^{(p)}, \quad \eta_i' = \eta_{i-r(p)}^{(p)}, \qquad \text{for } r(p) < i \leq r(p+1). \tag{21.20}$$

Then $\{\xi_i'\}$ is an independent sequence of random variables, each having the distribution of the original ξ_i, and $\{\eta_i'\}$ is an independent sequence of standard normal variables. Define $S_k' = \sum_{i\le k}\xi_i'$ and $T_k' = \sum_{i\le k}\eta_i'$. Given ϵ, choose p_0 so that $2^{-p_0} < \epsilon$. And choose n^* large enough that

$$\mathsf{P}[n^{-1/2}\max\nolimits_{k\le r(p_0)}|S_k' - T_k'| \ge \epsilon] \le \epsilon \quad \text{for } n \ge n^*. \tag{21.21}$$

Suppose that $n > n^* \vee r(p_0)$, and choose v so that $r(v) < n \le r(v+1)$. Write

$$V_a^b := \Big| n^{-1/2} \sum_{a<i\le b} (\xi_i' - \eta_i') \Big|, \quad M_a^b := \max_{a<i\le b} V_a^i.$$

If $k \le r(p_0)$, then

$$V_0^k \le M_0^{r(p_0)}.$$

If $r(q) < k \le r(q+1)$ and $q \ge p_0$, then

$$V_0^k \le V_0^{r(p_0)} + \sum_{p_0\le p<q} V_{r(p)}^{r(p+1)} + V_{r(q)}^k \le M_0^{r(p_0)} + \sum_{p_0\le p<q} M_{r(p)}^{r(p+1)} + M_{r(q)}^k.$$

And now, since $r(v) < n \le r(v+1)$ (which, together with $n > r(p_0)$, implies $v \ge p_0$),

$$M_0^n \le M_0^{r(p_0)} + \sum_{p_0\le p<v} M_{r(p)}^{r(p+1)} + M_{r(v)}^n.$$

By (21.20), if $r(p) < k \le r(p+1)$, then

$$M_{r(p)}^k = \max_{i\le k-r(p)} |n^{-1/2}(S_i^{(p)} - T_i^{(p)})|.$$

Put all this together:

$$\begin{aligned} M_0^n := \max_{i\le n}|n^{-1/2}(S_i' - T_i')| &\le \max_{i\le r(p_0)} |n^{-1/2}(S_i - T_i)| \\ &+ \sum_{p_0\le p<v} \max_{i\le n_0(p+1)} |n^{-1/2}(S_i^{(p)} - T_i^{(p)})| \\ &+ \max_{i\le n}|n^{-1/2}(S_i^{(v)} - T_i^{(v)})| = A + \sum_{p_0\le p<v} B_p + C. \end{aligned}$$

By (21.21), $\mathsf{P}[A \ge \epsilon] \le \epsilon$. If $p_0 \le p < v$, then $n > r(v) \ge n_0(p+1)$, and hence, by (21.19), $\mathsf{P}[B_p \ge 2^{-p}] \le 2^{-p}$. And $n > r(v) \ge n_0(v)$, so that, by (21.19) again, $\mathsf{P}[C \ge 2^{-v}] \le 2^{-v}$. It follows that $\mathsf{P}[M_0^n \ge 2\epsilon] \le \epsilon + \sum_{p_0\le p\le v} 2^{-p} < 2\epsilon$.

Thus $\{\eta_i'\}$ stands in the desired relation to $\{\xi_i'\}$. What we want, however, is a sequence $\{\eta_i\}$ on the original $(\Omega, \mathcal{F}, \mathsf{P})$ that stands in this relation to the original sequence $\{\xi_i\}$. But now we can apply Lemma 1 again, just as in the first proof of the theorem. □

SECTION 22. STRASSEN'S THEOREM

The Theorem

Let $\xi_1, \xi_2, \ldots$ be an independent and identically distributed sequence of random variables on $(\Omega, \mathcal{F}, \mathsf{P})$; suppose the ξ_i have mean 0 and variance 1. Let $S_n = \sum_{i \le n} \xi_i$ $(S_0 = 0)$, and let X^n be the random element on Ω, with values in $C = C[0,1]$, defined by

$$X_t^n(\omega) = \frac{1}{\sqrt{2n \log \log n}} [S_{\lfloor nt \rfloor}(\omega) + (nt - \lfloor nt \rfloor)\xi_{\lfloor nt \rfloor + 1}(\omega)], \tag{22.1}$$

for $0 \le t \le 1$. This is the random function (8.5) of Donsker's theorem, but with a very different norming constant. (We consider (22.1) only for $n \ge 3$, so that the square root is well defined.) Strassen's theorem has to do not with the asymptotic distribution of X^n (the approximate behavior of $X^n(\omega)$ for fixed large n and ω varying over Ω), but with the properties of the entire sequence

$$(X^3(\omega), X^4(\omega), \ldots) \tag{22.2}$$

for each individual ω in a set of probability 1.

Consider elements x of C that are absolutely continuous in the sense of having a derivative x' outside a set of Lebesgue measure 0, a derivative that is integrable and in fact integrates back to x, so that $x(t) - x(0) = \int_0^t x'(s)\,ds$ for $0 \le t \le 1$. Let K be the set of absolutely continuous functions x in C for which $x(0) = 0$ and the associated x' satisfies

$$\int_0^1 (x'(t))^2\,dt \le 1. \tag{22.3}$$

Lemma 1. *The set K is compact.*

PROOF. If $x \in K$ and $s \le t$, then by Schwarz's inequality,

$$\begin{aligned} |x(t) - x(s)| &\le \int_s^t |x'(u)|\,du \\ &\le \left[(t-s)\int_s^t (x'(u))^2 du\right]^{1/2} \le \sqrt{t-s}. \end{aligned} \tag{22.4}$$

It follows by the Arzelà-Ascoli theorem that K is relatively compact. But K is also closed: Suppose that points x_n of K converge to a point

x of C. Then there exists [PM.246] a sequence $\{n_i\}$ of integers and a function x' on $[0,1]$ such that $\int_0^1 (x'(s))^2 ds \le 1$ and $\int_0^1 x'_{n_i}(s)z(s)\,ds \to_i \int_0^1 x'(s)z(s)\,ds$ for all square-integrable z on $[0,1]$. (This is the weak compactness of the unit ball in L^2.) But then we have $x(t) - x(0) = \lim_i (x_{n_i}(t) - x_{n_i}(0)) = \lim_i \int_0^t x'_{n_i}(s)\,ds = \int_0^t x'(s)\,ds$. □

It is not hard to see that K is convex as well as compact. But since the set is nowhere dense (Example 1.3), it bears no resemblence to, say, a closed disk in the plane—it is more like a closed line-segment in the plane. Strassen's theorem:

Theorem 22.1. *With probability* 1, *the sequence* (22.2) *is relatively compact and the set of its limit points coincides with* K.

In this section, to say that a property holds with probability 1 means that, if E is the set of ω having the property, then there is an $\mathcal{F}$-set F such that $F \subset E$ and $\mathsf{P}F = 1$. If $(\Omega, \mathcal{F}, \mathsf{P})$ is complete, it follows that $E \in \mathcal{F}$ and $\mathsf{P}E = 1$. Theorem 22.1 implies the ordinary law of the iterated logarithm, and this and other consequences are discussed at the end of the section.

Preliminaries on Brownian Motion

To simplify the notation, write

$$L_n = \sqrt{2n \log\log n}. \tag{22.5}$$

Let $[B(t): t \ge 0]$ be a standard Brownian motion on some $(\Omega, \mathcal{F}, \mathsf{P})$, define a random element B^n of C by

$$B^n_t(\omega) = L_n^{-1} B_{nt}(\omega), \qquad 0 \le t \le 1, \tag{22.6}$$

and consider the sequence

$$(B^3(\omega), B^4(\omega), \ldots). \tag{22.7}$$

Before proving Theorem 22.1 for the X^n, we prove the analogous result for the B^n.

Theorem 22.2. *With probability* 1, *the sequence* (22.7) *is relatively compact and the set of its limit points coincides with* K.

For ω in a set of probability 1, the path $B(\cdot, \omega)$ is nowhere differentiable [PM.505], and for no such ω can $B^n(\omega)$ lie in K; the reason why subsequences of (22.7) can approach the points of K is that K is nowhere dense.

PROOF. The proof is in two parts. In the first we show that, with probability 1, (22.7) is relatively compact and all its limit points lie in K. In the second part we show that each point of K is such a limit point.

First part. It is enough to show that, with probability 1, for each (rational) ϵ we have $B^m(\omega) \in K^\epsilon$ for all sufficiently large n. For then there is probability 1 that $\text{dist}(B^n(\omega), K) \to 0$, and there are points $y_n(\omega)$ of K such that $\|B^n(\omega) - y_n(\omega)\| \to 0$, which implies that (22.7), like $\{y_n(\omega)\}$, is relatively compact and all its limit points lie in K.

Given ϵ, fix a large positive integer m and a real number r just greater than 1, each to be specified later. We have

$$(22.8)\quad \mathsf{P}[B^n \notin K^\epsilon] \le \mathsf{P}\Big[m\sum_{i=1}^{m}\Big(B^n\Big(\frac{i}{m}\Big) - B^n\Big(\frac{i-1}{m}\Big)\Big)^2 \ge r^2\Big]$$
$$+\mathsf{P}\Big[B^n \notin K^\epsilon,\ m\sum_{i=1}^{m}\Big(B^n\Big(\frac{i}{m}\Big) - B^n\Big(\frac{i-1}{m}\Big)\Big)^2 < r^2\Big] = \mathrm{I} + \mathrm{II}.$$

Let χ_m^2 have the χ^2-distribution with m degrees of freedom. Since $\sqrt{m}(B^n(i/m) - B^n((i-1)/m))$ has variance $nL_n^{-2} = 1/(2\log\log n)$,

$$(22.9)\quad \mathrm{I} = \mathsf{P}[\chi_m^2 \ge 2r^2 \log\log n]$$
$$= \frac{1}{2^{m/2}\Gamma(m/2)}\int_{2r^2\log\log n}^{\infty} t^{(m/2)-1}e^{-t/2}dt$$
$$\sim \frac{1}{\Gamma(m/2)}(r^2\log\log n)^{(m/2)-1}e^{-r^2\log\log n}$$

as $n \to \infty$ (l'Hôpital).

Let Z^n be the random function obtained by linear interpolation between its values $Z^n(i/m) = B^n(i/m)$ at the points i/m, $0 \le i \le m$. The paths of Z^n are absolutely continuous, and the derivative is $(Z^n)'(t) = m(B^n(i/m) - B^n((i-1)/m))$ for t in $((i-1)/m, i/m)$. The second condition in II is exactly the requirement that $\int_0^1((Z^n)'(t))^2dt < r^2$, which implies $r^{-1}Z^n \in K$. Therefore,

$$\mathrm{II} \le \mathsf{P}[r^{-1}Z^n \in K,\ B^n \notin K^\epsilon] \le \mathsf{P}[r^{-1}Z^n \in K,\ \|r^{-1}Z^n - B^n\| \ge \epsilon].$$

Now $\|r^{-1}Z^n - B^n\| \le \|r^{-1}Z^n - Z^n\| + \|Z^n - B^n\|$, and the first term on the right, $(r-1)\|r^{-1}Z^n\|$, is by (22.4) at most $r-1$ if $r^{-1}Z^n \in K$.

Choose r close enough to 1 that $r - 1 < \epsilon/2$. Then

$$\mathrm{II} \le \mathsf{P}[\|Z^n - B^n\| \ge \epsilon/2].$$

Since

$$\|Z^n - B^n\| = \max_{1 \le i \le m} \sup_{\frac{i-1}{m} \le t \le \frac{i}{m}} |Z^n(t) - B^n(t)|$$

and the suprema here all have the same distribution, it follows by the definition of Z^n that

$$\begin{aligned} \text{(22.10)} \qquad \mathrm{II} &\le m\mathsf{P}[\sup_{t \le 1/m} |Z^n(t) - B^n(t)| \ge \epsilon/2] \\ &= m\mathsf{P}[\sup_{t \le 1/m} |mtB^n(1/m) - B^n(t)| \ge \epsilon/2]. \end{aligned}$$

If $t \le 1/m$, then $|mtB^n(1/m) - B^n(t)| \le |B^n(1/m)| + |B^n(t)|$, and so

$$\sup_{t \le 1/m} |mtB^n(1/m) - B^n(t)| \le 2 \sup_{t \le 1/m} |B^n(t)|.$$

And now, by(22.10) and [M25(58)],

$$\begin{aligned} \text{(22.11)} \quad \mathrm{II} &\le m\mathsf{P}[\sup_{t \le 1/m} |B^n(t)| \ge \epsilon/4] \\ &= m\mathsf{P}[\sup_{s \le n/m} |B(s)| \ge \epsilon L_n/4] \le 4me^{-(\epsilon^2 m/16)\log\log n}. \end{aligned}$$

The r of (22.8) has already been chosen so that $0 < r - 1 < \epsilon/2$. Choose m large enough and then choose c close enough to 1 that

$$\text{(22.12)} \qquad \epsilon^2 m/16 > 1, \quad 1 < c < r^2, \quad c < \epsilon^2 m/16.$$

By (22.8), (22.9), and (22.11), we have, for all sufficiently large n,

$$\text{(22.13)} \qquad \mathsf{P}[B^n \notin K^\epsilon] \le 5me^{-c\log\log n}.$$

Take $n_k = \lfloor c^k \rfloor$. Then $n_k \ge c^{k-1}$ for large k, and

$$\mathsf{P}[B^{n_k} \notin K^\epsilon] \le 5me^{-c\log\log c^{k-1}} = \frac{5m}{((k-1)\log c)^c},$$

and therefore $\sum_k \mathsf{P}[B^{n_k} \notin K^\epsilon] < \infty$. By the Borel-Cantelli lemma, there is probability 1 that $B^{n_k} \in K^\epsilon$ for all large enough k. This is what we want to prove, but we must also account for n's not of the form n_k. To do this, we must control the size of

$$\text{(22.14)} \qquad M_k = \max_{n_{k-1} \le n \le n_k} \|B^{n_k} - B^n\|.$$

Since

$$\|B^{n_k} - B^n\| \le \sup_{u\le 1} L_{n_k}^{-1}|B(n_k u) - B(nu)| + \sup_{u\le 1}|L_{n_k}^{-1} - L_n^{-1}|\cdot|B(nu)|,$$

we can control M_k by separately controlling

$$A_k = L_{n_k}^{-1} \max_{n_{k-1}\le n\le n_k} \sup_{t\le c^k}|B(n_k t/n) - B(t)| \tag{22.15}$$

and

$$B_k = \max_{n_{k-1}\le n\le n_k} |L_{n_k}^{-1} - L_n^{-1}| \sup_{t\le c^k}|B(t)|. \tag{22.16}$$

For large k, $n_{k-1} \ge c^{k-2}$, so that $n_{k-1} \le n \le n_k$ implies $t \le n_k t/n \le n_k t/n_{k-1} \le c^2 t$. Thus

$$A_k \le L_{n_k}^{-1} \sup_{\substack{t\le c^k\\ t\le s\le c^2 t}} |B(s) - B(t)| \le A_k' \vee A_k'',$$

where

$$A_k' = L_{n_k}^{-1} \sup_{\substack{t\le 1\\ t\le s\le c^2}} |B(s) - B(t)| \le 2L_{n_k}^{-1}\sup_{s\le c^2}|B(s)|$$

and

$$\begin{aligned} A_k'' &= L_{n_k}^{-1} \max_{1\le i\le k} \sup_{\substack{c^{i-1}\le t\le c^i\\ t\le s\le c^2 t}} |B(s) - B(t)| \\ &\le 2L_{n_k}^{-1} \max_{1\le i\le k} \sup_{c^{i-1}\le s\le c^{i+2}} |B(s) - B(c^{i-1})|. \end{aligned}$$

Now $n_k \ge c^{k-1}$ for large k, and then [M25(58)]

$$\begin{aligned} \mathsf{P}[A_k' \ge \epsilon] &\le \mathsf{P}\Big[\sup_{s\le c^2}|B(s)| \ge \epsilon L_{c^{k-1}}/2\Big] \\ &\le 4\exp\Big[-\frac{\epsilon^2}{8c^2}L_{c^{k-1}}^2\Big] \le 4\exp\Big[-\frac{\epsilon^2}{4}c^{k-3}\Big]. \end{aligned}$$

Therefore, $\sum_k \mathsf{P}[A_k' \ge \epsilon] < \infty$. Also, for large k [M25(57)],

$$\begin{aligned} \mathsf{P}[A_k'' \ge \epsilon] &\le \sum_{i=1}^k \mathsf{P}\Big[\sup_{c^{i-1}\le s\le c^{i+2}} |B(s) - B(c^{i-1})| \ge \epsilon L_{c^{k-1}}/2\Big] \\ &\le \sum_{i=1}^k \frac{8}{\sqrt{2\pi}} \frac{\sqrt{c^{i-1}(c^3-1)}}{\epsilon L_{c^{k-1}}} \exp\Big[-\frac{\epsilon^2 L_{c^{k-1}}^2}{8c^{k-1}(c^3-1)}\Big]. \end{aligned}$$

Since $\sum_{i=1}^{k} c^{(i-1)/2} = (c^{k/2}-1)/(c^{1/2}-1)$, we have

$$\mathsf{P}[A_k'' \geq \epsilon] \leq a \exp\Big[-\frac{\epsilon^2}{4(c^3-1)} \log\log c^{k-1}\Big], \tag{22.17}$$

where a is a constant. Now c was chosen to satisfy (22.12); if we move it closer to 1, then the factor in front of the iterated logarithm in (22.17) will exceed 1, in which case, $\sum_k \mathsf{P}[A_k'' \geq \epsilon] < \infty$. Therefore, $\sum_k \mathsf{P}[A_k \geq \epsilon] < \infty$.

The maximum preceding the supremum in (22.16) is at most

$$\Delta_k := L_{c^{k-2}}^{-1} - L_{c^k}^{-1} = L_{c^{k-2}}^{-1}\left(1 - c^{-1}\sqrt{\frac{\log\log c^{k-2}}{\log\log c^k}}\right) = L_{c^{k-2}}^{-1}(1-c^{-1}\theta_k),$$

where $\theta_k \to 1$. Therefore [M25(58)],

$$\begin{aligned} \mathsf{P}[B_k \geq \epsilon] &\leq \mathsf{P}[\sup_{t \leq c^k} |B(t)| \geq \epsilon/\Delta_k] \\ &\leq 4\exp\Big[-\frac{\epsilon^2}{2c^k\Delta_k^2}\Big] = 4\exp\Big[-\frac{\epsilon^2}{c^2(1-c^{-1}\theta_k)^2}\log\log c^{k-2}\Big]. \end{aligned} \tag{22.18}$$

By moving c still closer to 1, we can arrange that $\epsilon^2/c^2(1-c^{-1})^2 > \alpha > 1$. Then, for large k, the factor in front of the iterated logarithm in (22.18) will exceed α, so that $\sum_k \mathsf{P}[B_k \geq \epsilon] < \infty$.

Since we have controlled A_k and B_k, the M_k of (22.14) satisfies $\sum_k \mathsf{P}[M_k \leq \epsilon] < \infty$ for each ϵ, and there is probability 1 that $M_k \to_k 0$. As we already know, there is probability 1 that $B^{n_k} \in K^\epsilon$ for all large k, and hence $B^n \in K^{2\epsilon}$ for all large n. This completes the first part of the proof.

Second part. Suppose that, for each x in K, there is probability 1 that some subsequence of (22.7) converges to x, or

$$\liminf_n \|B^n(\omega) - x\| = 0. \tag{22.19}$$

There is a countable, dense subset D of K, and it follows that, for ω in a set of probability 1, (22.19) holds for every x in D. Fix such an ω. For y in K and ϵ arbitrary, there is an x in D such that $\|y - x\| < \epsilon$. Since (22.19) holds, for infinitely many n we have $\|B^n(\omega) - x\| < \epsilon$ and hence $\|B^n(\omega) - y\| < 2\epsilon$. For each y in K, this holds for every ϵ, and so a subsequence of (22.7) converges to y. It is therefore enough to deal with (22.19) for a single x in K, and we can even assume that

$$\int_0^1 (x'(u))^2 du < 1, \tag{22.20}$$

since the x with this property are dense in K.

Fix such an x and let ϵ be given. First choose m and then choose δ so that

$$m\epsilon^2 > 1, \quad \sqrt{1/m} < \epsilon, \quad x(1/m) < \epsilon, \quad m\delta < \epsilon; \tag{22.21}$$

the third condition is possible because $x(0) = 0$. Let $A_n = A_n(x, m, \delta)$ be the ω-set where

$$\left| \left(B^n\left(\frac{i}{m}\right) - B^n\left(\frac{i-1}{m}\right)\right) - \left(x\left(\frac{i}{m}\right) - x\left(\frac{i-1}{m}\right)\right) \right| < \delta, \quad \text{for} \quad 2 \le i \le m; \tag{22.22}$$

note that i starts at 2 rather than 1. Write

$$d_i = |x(i/x) - x((i-1)/m)|, \quad L_{n,m} = \sqrt{2m \log\log n}.$$

Since the distribution of N is symmetric about 0, it follows by the definition (22.6) of B^n that

$$\mathsf{P}A_n \ge \prod\nolimits_{i=2}^m \mathsf{P}[d_i \le L_{n,m}^{-1} N \le d_i + \delta].$$

By (22.4), $(x(t) - x(s))^2 \le (t-s)\int_s^t (x'(u))^2 du$; apply this with $s = (i-1)/m$ and $t = i/m$, add, and use (22.20): $m\sum_{i=1}^m d_i^2 \le \int_0^1 (x'(u))^2 du < 1$. By reducing δ still further, we can arrange that $m\sum_{i=1}^m (d_i + \delta)^2 < 1$. If n is large enough that $\delta L_{n,m} \ge \sqrt{2\pi}$, then [M25(56)]

$$\begin{aligned}\mathsf{P}[d_i \le L_{n,m}^{-1} N \le d_i + \delta] &\ge \exp[-L_{n,m}^2 (d_i + \delta)^2/2] \\ &= \exp[-m(d_i + \delta)^2 \log\log n],\end{aligned}$$

and it follows that

$$\mathsf{P}A_n \ge \exp[-m\sum_{i=1}^m (d_i + \delta)^2 \log\log n] \ge \exp[-\log\log n] = \frac{1}{\log n}.$$

Let $n_k = m^k$. If $\mathcal{F}(s,t)$ is the σ-field generated by the differences $B(v) - B(u)$ for $s \le u \le v \le t$, then $A_n \in \mathcal{F}(n/m, n)$, and $A_{n_k} \in \mathcal{F}(m^{k-1}, m^k)$. Therefore the A_{n_k} are independent. (If we had started i at 1 instead of at 2 in the definition of A_n, then A_{n_k} would lie in $\mathcal{F}(0, m^k)$, and we could not have drawn the conclusion that the A_{n_k}

were independent.) Since $\mathsf{P}A_n \geq 1/\log n$, $\sum_k \mathsf{P}A_{n_k} = \infty$, and by the Borel-Cantelli lemma, there is probability 1 that infinitely many of the A_{n_k} occur. For large n we have [M25(55)]

$$\mathsf{P}[|B^n(1/m)| \geq \epsilon] = \mathsf{P}[|N| \geq \epsilon L_{n,m}] \leq 2(\log n)^{-m\epsilon^2}.$$

Since $m\epsilon^2 > 1$ by (22.21), $\sum_k \mathsf{P}[|B^{n_k}(1/m)| \geq \epsilon] < \infty$, and hence, with probability 1, $|B^{n_k}(1/m)| \leq \epsilon$ for all large k.

We have now shown that, if (22.21) holds, then there is probability 1 that (22.22) and $|B^n(1/m)| \leq \epsilon$ both hold for infinitely many values of n. Since $|x(1/m)| < \epsilon$ by (22.21), it follows for such an n that $|B^n(1/m)-x(1/m)| \leq 2\epsilon$. But now, this and (22.22) imply ($B^n(0) = 0$)

$$|B^n(i/m) - x(i/m)| \leq m\delta + 2\epsilon, \quad 0 \leq i \leq m. \tag{22.23}$$

We know from the first part of the proof that, with probability 1 and for large n, $\|B^n - y\| \leq \epsilon$ for some $y \in K$, so that, by (22.4) for y, $|B^n(t) - B^n((i-1)/m)| \leq 2\epsilon + \sqrt{1/m} < 3\epsilon$ for $(i-1)/m \leq t \leq i/m$. Again by (22.4), $|x(t) - x((i-1)/m)| \leq \sqrt{1/m} < \epsilon$ for $(i-1)/m \leq t \leq i/m$. We conclude from (22.21) and (22.23) that $\|B^n - x\| \leq m\delta + 6\epsilon < 7\epsilon$. □

Proof of Strassen's Theorem

We are now in a position to complete the proof of Strassen's theorem; the argument remaining is similar to that for Theorem 20.2. We start with the sequence $\{\xi_n\}$ on $(\Omega, \mathcal{F}, \mathsf{P})$. It is independent, and the ξ_n all have the same distribution, with mean 0 and variance 1. Now consider the Brownian motion $[B(t): t \geq 0]$ and the stopping times τ_n of Theorem 20.1. They are defined on a new space $(\Omega', \mathcal{F}', \mathsf{P}')$. Of course, Theorem 22.2 applies to this Brownian motion. Define random variables ζ_i and T_k as for Theorem 20.2, and define random elements X^n, Y^n, and Z^n of C by linear interpolation between these values at the points i/n:

$$\begin{cases} X^n\colon & X^n_{i/n} = L_n^{-1} S_i = L_n^{-1} \sum_{h \leq i} \xi_h \quad (\text{on } \Omega), \\ Y^n\colon & Y^n_{i/n} = L_n^{-1} T_i = L_n^{-1} \sum_{h \leq i} \zeta_h = L_n^{-1} B(\tau_i) \quad (\text{on } \Omega'), \\ Z^n\colon & Z^n_{i/n} = B^n(i/n) = L_n^{-1} B(i) \quad (\text{on } \Omega'). \end{cases} \tag{22.24}$$

This is the same as (20.2), except for the new norming constant, and X^n coincides with the random function of (22.1). Since the ξ_i have the same joint distribution as the ζ_i, the X^n will satisfy Theorem 22.1 on Ω if the Y^n satisfy the same theorem on Ω'. The B^n are now defined on Ω' rather than on Ω, but they do satisfy Theorem 22.2.

Suppose we can prove that

$$\|Y^n - B^n\| \to 0 \tag{22.25}$$

holds with P'-probability 1. By Theorem 22.2, there is P'-probability 1 that $(B^3, B^4 \ldots)$ is relatively compact and the set of its limit points is K. But then the sequence $(Y^3, Y^4, \ldots)$ will have the same property by (22.25), which will complete the proof.

We prove (22.25) in two steps:

$$\|B^n - Z^n\| \to 0 \tag{22.26}$$

and

$$\|Y^n - Z^n\| \to 0 \tag{22.27}$$

hold with P'-probability 1. For large n we have [M25(58)]

$$\begin{aligned} \mathsf{P}'[\|B^n - Z^n\| \geq \epsilon] &\leq n\mathsf{P}'[\sup_{s \leq 1} |B(s)| \geq \epsilon L_n/2] \\ &\leq 4n \exp[-\epsilon^2 L_n^2/8] = \mathrm{o}(1/n^2). \end{aligned}$$

Now (22.26) follows by the Borel-Cantelli lemma.

As for (22.27), $\|Y^n - Z^n\| = L_n^{-1} \max_{i \leq n} |B(\tau_i) - B(i)|$ by the definitions, and so it is enough to prove that

$$L_n^{-1}|B(\tau_n) - B(n)| \to 0 \tag{22.28}$$

with P'-probability 1. Let ϵ be given. By the strong law of large numbers (recall that the $\tau_n - \tau_{n-1}$ are independent and identically distributed with mean 1), there is an n_0 such that, with probability exceeding $1 - \epsilon$, we have $|\tau_n - n| < n\epsilon$ for $n > n_0$, in which case $|B(\tau_n) - B(n)| \leq \sup |B(t) - B(n)|$, where the supremum extends over $(1-\epsilon)n \leq t \leq (1+\epsilon)n$. Define n', a function of n, by $n' = \lceil (1+\epsilon)n \rceil$. For large n, $(1-\epsilon)n/n' > 1 - 2\epsilon$, and this implies

$$\begin{aligned} |B(\tau_n) - B(n)| &\leq 2 \sup_{(1-\epsilon)n \leq t \leq n'} |B(t) - B(n')| \\ &\leq 2 \sup_{1-2\epsilon \leq s \leq 1} |B(sn') - B(n')|. \end{aligned}$$

For large n we have $L_{n'}/L_n < 2$, and then

$$L_n^{-1}|B(\tau_n) - B(n)| \leq 4 \sup_{1-2\epsilon \leq s \leq 1} L_{n'}^{-1}|B(sn') - B(n')| \tag{22.29}$$

$$= 4 \sup_{1-2\epsilon \leq s \leq 1} |B^{n'}(s) - B^{n'}(1)|.$$

With probability exceeding $1-\epsilon$, (22.29) holds for all large enough n. By Theorem 22.2, there is probability exceeding $1 - 2\epsilon$ that both (22.29) and $B^{n'} \in K^\epsilon$ hold for large n. But then, by (22.4), we have $|B^{n'}(s) - B^{n'}(1)| \leq 2\epsilon + \sqrt{2\epsilon}$ for $1 - 2\epsilon \leq s \leq 1$. Therefore, with probability exceeding $1 - 2\epsilon$,

$$L_n^{-1}|B(\tau_n) - B(n)| \leq 4(2\epsilon + \sqrt{2\epsilon})$$

for all sufficiently large values of n. This shows that (22.28) holds with probability 1 and completes the proof of Theorem 22.1. □

Applications

The applications use this result:

Lemma 2. *If ϕ is a continuous map from C to R^1, then, for ω in a set of probability* 1, *the sequence*

$$(\phi(X^3(\omega)), \phi(X^4(\omega)), \ldots) \tag{22.30}$$

is relatively compact and the set of its limit points coincides with the closed interval $\phi(K)$.

PROOF. If x and y are points of C, then the function $f(t) = \phi((1-t)x + ty)$ on $[0, 1]$ is continuous and assumes the values $\phi(x)$ and $\phi(y)$ at the endpoints and hence must assume all values between the two. Since K is convex, $\phi(K)$ is a closed interval. It is easy to see that, if (22.2) is relatively compact and the set of its limit points is K, then (22.30) is also relatively compact and the set of its limit points is $\phi(K)$. The result follows by Theorem 22.1. □

Example 22.1 Take $\phi(x) = x(1)$. By (22.4), $|x(1)| \leq 1$ for $x \in K$, and since $\phi(x)$ is $+1$ and -1 for $x(t) \equiv t$ and $x(t) \equiv -t$, we have $\phi(K) = [-1, +1]$. And

$$\phi(X^n) = (2n \log\log n)^{-1/2} S_n. \tag{22.31}$$

By the lemma, the limit points of the sequence defined by (22.31) exactly fill out the interval $[-1, +1]$. This is the Hartman-Wintner theorem. □

Example 22.2 If $\phi(x) = \sup_t x(t)$, then $\phi(K) = [0, 1]$, and

$$\phi(X^n) = (2n \log\log n)^{-1/2} \max_{i \le n} S_i. \tag{22.32}$$

The limit points exactly fill out $[0, 1]$. □

Example 22.3 Let f have bounded derivative on $[0, 1]$ and take $\phi(x) = \int_0^1 x(t)f(t)\,dt$. If $F(t) = \int_t^1 f(s)\,ds$ and $x \in K$, then partial integration gives $\phi(x) = \int_0^1 F(t)x'(t)\,dt$. This is an inner product (F, x') in $L^2[0, 1]$, and under the constraint $\|x'\|_2^2 = \int_0^1 (x'(t))^2 dt \le 1$, it is maximal for $x' = F/\|F\|_2$ and the maximum is $m = \|F\|_2 = [\int_0^1 F^2(t)\,dt]^{1/2}$. Therefore, $\phi(K) = [-m, m]$.

We can approximate $\phi(X^n) = \int_0^1 X_t^n f(t)\,dt$ by

$$U_n = (2n^3 \log\log n)^{-1/2} \sum_{i=1}^{n} f(i/n)S_i. \tag{22.33}$$

To see this, note first that

$$|\phi(X^n) - U_n| \le \sum_{i=1}^{n} \int_{(i-1)/n}^{i/n} |X_t^n f(t) - L_n^{-1} f(i/n)S_i|\,dt.$$

If A bounds $|f|$ and B bounds $|f'|$, then the integrand is bounded by $A|X_t^n - L_n^{-1}S_i| + Bn^{-1}L_n^{-1}|S_i|$, and we have

$$|\phi(X^n) - U_n| \le AL_n^{-1}n^{-1} \sum_{i=1}^{n} |\xi_i| + BL_n^{-1}n^{-1} \sum_{i=1}^{n} |S_i/i|.$$

By the strong law of large numbers, this goes to 0 with probability 1. Therefore, the limit points of $\{U_n\}$ fill out $[-m, m]$.

If $f(t) \equiv c$, then $m = |c|/\sqrt{3}$. If $f(t) = t^\alpha$, $\alpha > 1$, then $m = ((\alpha + 2)(\alpha + 3/2))^{-1/2}$. □

Example 22.4. Because of Example 22.1, it is interesting to investigate the frequency of the integers i for which $L_i^{-1}S_i$ exceeds a fixed c. Define random variables γ_i by

$$\gamma_i = \begin{cases} 1 & \text{if } S_i > c(2i \log\log i)^{1/2}, \\ 0 & \text{otherwise.} \end{cases} \tag{22.34}$$

For $0 < c < 1$, there is probability 1 that

$$\limsup_n \frac{1}{n} \sum_{i=3}^{n} \gamma_i = 1 - \exp\Big(-4\Big(\frac{1}{c^2} - 1\Big)\Big). \tag{22.35}$$

The proof of this falls into two distinct parts. Define

$$\nu_c(x) := \lambda[t: x(t) \geq c\sqrt{t}], \quad s^*(c) := \sup_{x \in K} \nu_c(x), \tag{22.36}$$

where λ is Lebesgue measure and t is restricted to $[0, 1]$. We show first that

$$s^*(c) = 1 - \exp(-4(c^{-2} - 1)) \tag{22.37}$$

and second that

$$\limsup_n \frac{1}{n} \sum_{i=1}^{n} \gamma_i = s^*(c) \tag{22.38}$$

with probability 1.

To prove (22.37), we show first that the supremum $s^*(c)$ is achieved. Suppose that $x_m \to x$ (in the sense of C) and consider the inequalities $\lambda[t: x_m(t) \geq c\sqrt{t}] \leq \lambda[t: x(t) \geq c\sqrt{t} - k^{-1}] < \lambda[t: x(t) \geq c\sqrt{t}] + \epsilon$. Given ϵ, choose k so that the second inequality holds; for large m, the first holds as well, which proves upper semicontinuity: $\limsup_m \nu_c(x_m) \leq \nu_c(x)$. Now choose $\{x_m\}$ in K so that $\nu_c(x_m)$ converges to $s^*(c)$, and by passing to subsequence, arrange that $x_m \to x_0$. Then $x_0 \in K$, and by upper semicontinuity, x_0 is a maximal point of K: It achieves the supremum $s^*(c)$. It will turn out that there is only one of them, but for now let x_0 be any maximal point.

If $s < t$ and $x \in K$, it follows by Jensen's inequality that

$$\int_s^t (x'(u))^2 du \geq \frac{(x(t) - x(s))^2}{t - s}, \tag{22.39}$$

and there is equality if and only if x is linear over $[s, t]$. If $x(t) \geq c\sqrt{t}$ and $t > 0$, then (22.39) gives $\int_0^t (x'(u))^2 du \geq (x(t))^2/t \geq c^2$. Therefore, since the integral goes to 0 with t, we must have $x(t) < c\sqrt{t}$ in some interval $(0, s_0)$. This is true of each x in K, and it implies $\nu_c(x) > 0$.

If x_0 is maximal and $\beta^2 = \int_0^1 (x_0'(u))^2 du < 1$, then $x_1 = \beta^{-1} x_0$ lies in K and $\nu_c(x_1)$ exceeds $\nu_c(x_0)$ (since the latter must be positive). Therefore, every maximal x_0 satisfies

$$\int_0^1 (x_0'(u))^2 du = 1. \tag{22.40}$$

Next, the set $A = [t: x_0(t) > c\sqrt{t}]$ is empty: If $1 \in A$, then there is an $a > c$ and a t' such that $x_0(t') = a$ and $x_0(t) > a$ on $(t', 1]$. Redefine x_0 so that $x_0(t) = a$ on $[t', 1]$. This leaves $\nu_c(x_0)$ unchanged but decreases $\int_0^1 (x_0'(t))^2 dt$ by (22.39), which contradicts (22.40). It follows that, if $A \neq \emptyset$, then there are points t_1 and t_2 such that $0 < t_1 < t_2 < 1$, $x_0(t_1) = c\sqrt{t_1}$, $x_0(t_2) = c\sqrt{t_2}$, and $x_0(t) > c\sqrt{t}$ on (t_1, t_2). Since $p(t) = c\sqrt{t}$ is strictly concave, x_0 cannot be linear over $[t_1, t_2]$. If each point of (t_1, t_2) can be enclosed in an open interval on which x_0 is linear, then by compactness, x_0 is linear over $[t_1 + \epsilon, t_2 - \epsilon]$ for each ϵ and hence is linear over $[t_1, t_2]$. Therefore, there is a t_0 in (t_1, t_2) that is contained in no open interval over which x_0 is linear. There is some η small enough that the line segment L connecting $(t_0 - \eta, x_0(t_0 - \eta))$ to $(t_0 + \eta, x_0(t_0 + \eta))$ lies above the graph of p. Redefine x_0 over $[t_0 - \eta, t_0 + \eta]$ so that its graph coincides with L there. This leaves $\nu_c(x_0)$ unchanged, and again we have a contradiction of (22.40). Therefore, A is indeed empty.

We know that $x_0(t) \le c\sqrt{t}$ for all t, and $x_0(t) < c\sqrt{t}$ on $(0, s_0)$. Take s_0 to be the infimum of those positive t for which $x_0(t) = c\sqrt{t}$, with $s_0 = 1$ if there are no such t. Then $x_0(t) = tc/\sqrt{s_0}$ on $[0, s_0]$, for if x_0 is not linear there, then (22.39) (strict inequality) for $x = x_0$, $s = 0$, $t = s_0$ contradicts (22.40) once again. We prove below that $x_0(t) = c\sqrt{t}$ on $[s_0, 1]$, so that

$$x_0(t) = \begin{cases} tc/\sqrt{s_0} & \text{on } [0, s_0], \\ c\sqrt{t} & \text{on } [s_0, 1]. \end{cases} \tag{22.41}$$

Let us assume this and complete the proof of (22.37). If (22.41) holds for the maximizing x_0, then, by (22.40), $c^2 + \int_{s_0}^1 (c/2\sqrt{t})^2 dt = 1$, and this gives $s_0 = s_0(c)$, where $s_0(c) := \exp(-4(1 - c^{-2}))$. This proves (22.37), as well as the fact that $0 < s_0 < 1$.

To prove (22.41), we must eliminate two possibilities. First, suppose there exist s_1 and s_2 such that $s_0 \le s_1 < s_2 \le 1$, $x_0(s_1) = c\sqrt{s_1}$, $x_0(s_2) = c\sqrt{s_2}$, and $x_0(t) < c\sqrt{t}$ on (s_1, s_2). Let $\delta = s_2 - s_1$ and define a continuous y_0 by

(22.42)
$$y_0(t) = \begin{cases} tc/\sqrt{s_0+\delta} & \text{on } [0, s_0+\delta], \\ c\sqrt{s_0+\delta} + x_0(t-\delta) - x_0(s_0) & \text{on } [s_0+\delta, s_2], \\ c\sqrt{s_0+\delta} + x_0(s_1) - x_0(s_0) + x_0(t) - x_0(s_2) & \text{on } [s_2, 1]. \end{cases}$$

If t is a point of $[s_0, s_1]$ and $x_0(t) \geq c\sqrt{t}$, then $t+\delta$ is point of $[s_0+\delta, s_2]$ and $y_0(t+\delta) \geq c\sqrt{t+\delta}$ (algebra); if t is a point of $[s_2, 1]$, then $y_0(t) \geq x_0(t)$ (algebra). Therefore, $\nu_c(y_0) \geq \nu_c(x_0)$ and y_0 is a maximal point. An easy computation, together with the facts that $y_0'(t) = x_0'(t-\delta)$ on $[s_0+\delta, s_2]$ and $y_0'(t) = x_0'(t)$ on $[s_2, 1]$, shows that

$$\int_0^{s_0} (x_0'(t))^2 dt = \int_0^{s_0+\delta} (y_0'(t))^2 dt = c^2,$$

$$\int_{s_0}^{s_1} (x_0'(t))^2 dt = \int_{s_0+\delta}^{s_2} (y_0'(t))^2 dt, \quad \int_{s_2}^{1} (x_0'(t))^2 dt = \int_{s_2}^{1} (y_0'(t))^2 dt.$$

Therefore, by (22.39),

$$\int_0^1 (y_0'(t))^2 dt = 1 - \int_{s_1}^{s_2} (x_0'(t))^2 dt \leq 1 - \frac{(c\sqrt{s_2} - c\sqrt{s_1})^2}{s_2 - s_1} < 1,$$

which contradicts (22.40) and rules out the first possibility.

The second possibility is that there exists an s_1 such that $s_0 \leq s_1 < 1$, $x_0(t) = c\sqrt{t}$ on $[s_0, s_1]$, and $x_0(t) < c\sqrt{t}$ on $(s_1, 1]$. Redefine x_0 by taking $x_0(t) = c\sqrt{s_1}$ on $[s_1, 1]$. Since x_0 is still maximal, (22.40) gives $s_0 = s_1 \exp(-4(c^{-2} - 1))$. But from $s_1 < 1$ follows $s_1 - s_0 < 1 - \exp(-4(c^{-2} - 1))$. Since this last quantity is acheived by (22.41), the redefined x_0 cannot be maximal, which rules out the second possibility.

This completes the proof of (22.37). It remains to prove (22.38). Suppose we can show that, for $0 < c' < c < c'' < 1$, there is probability 1 that

$$\limsup_n \frac{1}{n} \sum_{i=3}^{n} \gamma_i \leq s^*(c') \tag{22.43}$$

and

$$\limsup_n \frac{1}{n} \sum_{i=3}^{n} \gamma_i \geq s^*(c''). \tag{22.44}$$

This will be enough, since (22.37) shows that $s^*(c)$ is continuous and decreasing.

To prove (22.43), first restrict the range of summation to

$$\lceil \alpha n \rceil + 1 \le i \le n, \tag{22.45}$$

where $0 < \alpha < 1$. Since this decreases the average[†] by at most α, if we can show that (22.43) holds for the new average, then (22.43) itself will follow: Let α decrease to 0 through the rationals. Fix the α and take ϵ small enough that $c(1-\epsilon) - 2\epsilon/\sqrt{\alpha} > c'$. Suppose that i is in the range (22.45), that $t \in I_i^n = [(i-1)/n, i/n]$, and that n is large enough that $(\log\log \alpha n/\log\log n)^{1/2} \ge 1-\epsilon$. Then $\gamma_i = 1$ implies

$$X_{i/n}^n = L_n^{-1} S_i \ge c(1-\epsilon)\sqrt{i/n} \ge c(1-\epsilon)\sqrt{t}. \tag{22.46}$$

With probability 1, for large n there is an x_n in K for which $\|X^n - x_n\| < \epsilon$, and then (22.46) implies $x_n(i/n) \ge c(1-\epsilon)\sqrt{t} - \epsilon$, which in turn implies $x_n(t) \ge c(1-\epsilon)\sqrt{t} - 2\epsilon$ for n large enough that $\sqrt{1/n} < \epsilon$ (use (22.4)). And finally, (22.45) and $t \in I_i^n$ further imply $x_n(t) \ge (c(1-\epsilon) - 2\epsilon/\sqrt{\alpha})\sqrt{t} > c'\sqrt{t}$. To sum up, there is probability 1 that, for n large and i in the range (22.45), $\gamma_i = 1$ implies that $x_n(t) \ge c'\sqrt{t}$ on I_i^n. This means that the average in (22.43) for the new range (22.45) is at most $\lambda[t: x_n(t) \ge c'\sqrt{t}]$, which proves the original (22.43).

As for (22.44), we can restrict the sum by

$$\lfloor \alpha n \rfloor - 1 \le i < n, \tag{22.47}$$

since this only decreases the average. Take x_0 to be the maximal point of K for c'', and take α less than $s_0(c'')$ as defined after (22.41). Then

$$\lambda[t: \alpha \le t \le 1,\ x_0(t) \ge c''\sqrt{t}] = \lambda[t: x_0(t) \ge c''\sqrt{t}] = s^*(c'').$$

Now $X_{i/n}^n > c\sqrt{i/n}$ implies $S_i > c\sqrt{i/n}L_n \ge cL_i$ and hence $\gamma_i = 1$. With probability 1, there are infinitely many n such that, for i in the range (22.47), $\|X^n - x_0\| < (c''-c)\sqrt{\alpha/2} \le (c''-c)\sqrt{i/n}$. Suppose that, for such an n and i, we have $x_0(t) \ge c''\sqrt{t}$ for some t in I_{i+1}^n. Then

$$X_{i/n}^n > x_0\Big(\frac{1}{n}\Big) - (c''-c)\sqrt{\frac{i}{n}} \ge c''\sqrt{t} - (c''-c)\sqrt{\frac{i}{n}} \ge c\sqrt{\frac{i}{n}},$$

[†] These are not quite averages, since the number of summands is less than n.

and hence $\gamma_i = 1$. Since $[\alpha, 1]$ is covered by the I_{i+1}^n for i in the range (22.47), it follows that $n^{-1}\sum_{i=\lfloor \alpha n \rfloor - 1}^{n-1} \gamma_i$ is at least

$$\lambda[t\colon \alpha \le t \le 1,\ x_0(t) \ge c''\sqrt{t}].$$

This proves (22.44).

As Strassen points out, $.99999 < s^*(1/2) < .999999$. Therefore, for $c = 1/2$ there is probability 1 that $n^{-1}\sum_{i \le n} \gamma_i$ exceeds .99999 infinitely often but exceeds .999999 only finitely often. □

APPENDIX M

METRIC SPACES

We review here a few properties of metric spaces, taking as known the very first definitions and facts.

M1. Some Notation. We denote the space on which the metric is defined by S and the metric itself by $\rho(x, y)$; the metric space proper is the pair (S, ρ). For subsets A of S, denote the closure by A^-, the interior by A°, and the boundary by $\partial A = A^- - A^\circ$. The distance from x to A is $\rho(x, A) = \inf[\rho(x, y): y \in A]$; from $\rho(x, A) \le \rho(x, y) + \rho(y, A)$ it follows that $\rho(\cdot, A)$ is uniformly continuous. Denote by $B(x, r)$ the open r-ball $[y: \rho(x, y) < r]$; "ball" will mean "open ball," and closed balls will be denoted $B(x, r)^-$. The ϵ-*neighborhood* of a set A is the open set $A^\epsilon = [x: \rho(x, A) < \epsilon]$.

M2. Comparing Metrics. Suppose ρ and ρ' are two metrics on the same space S. To say that the ρ'-topology is *larger* than the ρ-topology is to say that the corresponding classes $\mathcal{O}$ and $\mathcal{O}'$ of open sets stand in the relation

$$\mathcal{O} \subset \mathcal{O}'. \tag{1}$$

This holds if and only if for every x and r, there is an r' such that $B'(x, r') \subset B(x, r)$, and so in this case the ρ'-topology is also said to be *finer* than the ρ-topology. (The term *stronger* is sometimes confusing.) Regard the identity map i on S as a map from (S, ρ') to (S, ρ). Then i is continuous if and only if $G \in \mathcal{O}$ implies $G = i^{-1}G \in \mathcal{O}'$—that is, if and only if (1) holds. But also, i is continuous in this sense if and only if

$$\rho'(x_n, x) \to 0 \quad \text{implies} \quad \rho(x_n, x) \to 0.$$

This is another way of saying that the ρ'-topology is finer than the ρ-topology. The metric ρ is *discrete* if $\rho(x, y) = 1$ for $x \ne y$; this gives to S the finest (largest) topology possible.

Two metrics and the corresponding topologies are *equivalent* if each is finer than the other: (S, ρ) and (S, ρ') are homeomorphic. If ρ' is finer than ρ, then the two may be equivalent; in other words, "finer" does not mean "strictly finer."

M3. Separability. The space S is *separable* if it contains a countable, dense subset. A *base* for S is a class of open sets with the property that each open set is a union of sets in the class. An *open cover* of A is a class of open sets whose union contains A.

Theorem. *These three conditions are equivalent:*

(i) *S is separable.*
(ii) *S has a countable base.*
(iii) *Each open cover of each subset of S has a countable subcover.*

PROOF. *Proof that* (i) $\rightarrow$ (ii). Let D be countable and dense, and take $\mathcal{V}$ to be the class of balls $B(d, r)$ for d in D and r rational. Let G be open; to prove that $\mathcal{V}$ is a base, we must show that, if G_1 is the union of those elements of $\mathcal{V}$ that are contained in G, then $G = G_1$. Clearly, $G_1 \subset G$, and to prove $G \subset G_1$ it is enough to find, for a given x in G, a d in D and a rational r such that $x \in B(d, r) \subset G$. But if $x \in G$, then $B(x, \epsilon) \subset G$ for some positive ϵ. Since D is dense, there is a d in D such that $\rho(x, d) < \epsilon/2$. Take a rational r satisfying $\rho(x, d) < r < \epsilon/2$: $x \in B(d, r) \subset B(x, \epsilon)$.

Proof that (ii) $\rightarrow$ (iii). Let $\{V_1, V_2, \ldots\}$ be a countable base, and suppose that $\{G_\alpha\}$ is an open cover of A (α ranges over an arbitrary index set). For each V_k for which there exists a G_α satisfying $V_k \subset G_\alpha$, let G_{α_k} be some one of these G_α containing it. Then $A \subset \bigcup_k G_{\alpha_k}$.

Proof that (iii) $\rightarrow$ (i). For each n, $[B(x, n^{-1}): x \in S]$ is an open cover of S. If (iii) holds, there is a countable subcover $[B(x_{nk}, n^{-1}): k = 1, 2, \ldots]$. The countable set $[x_{nk}: n, k = 1, 2, \ldots]$ is dense in S. □

A *subset* M of S is separable if there is a countable set D that is dense in M ($M \subset D^-$). Although D need not be a subset of M, this can easily be arranged: Suppose that $\{d_k\}$ is dense in M, and take x_{kn} to be a point common to $B(d_k, n^{-1})$ and M, if there is one. Given an x in M and a positive $\epsilon > 0$ choose n and then d_k so that $\rho(x, d_k) < n^{-1} < \epsilon/2$. Since $B(d_k, n^{-1})$ contains the point x of M, it contains x_{kn}, and $\rho(x, x_{kn}) < \epsilon$. The x_{kn} therefore form a countable, dense subset of M.

Theorem. *Suppose the subset M of S is separable.*

(i) *There is a countable class* $\mathcal{A}$ *of open sets with the property that, if* $x \in G \cap M$ *and* G *is open, then* $x \in A \subset A^- \subset G$ *for some* A *in* $\mathcal{A}$.

(ii) *Every open cover of* M *has a countable subcover.*

PROOF. *Proof of* (i). Let D be a countable, dense subset of M and take $\mathcal{A}$ to consist of the balls $B(d, r)$ for d in D and r rational. If $x \in G \cap M$ and G is open, choose ϵ so that $B(x, \epsilon) \subset G$, then choose d in D so that $\rho(x, d) < \epsilon/2$, and finally, choose a rational r so that $\rho(x, d) < r < \epsilon/2$. It follows that $x \in B(d, r) \subset B(d, r)^- \subset B(x, \epsilon) \subset G$.

Proof of (ii). Let $\mathcal{A} = \{A_1, A_2, \ldots\}$ be the class of part (i). Given an open cover $\{G_\alpha\}$ of M, choose for each A_k a G_{α_k} containing it (if there is one). Then $M \subset \bigcup_k G_{\alpha_k}$. □

These arguments are essentially those of the preceding proof. Part (ii) is the *Lindelöf property.*

Separability is a topological property: If ρ and ρ' are equivalent metrics, then M is ρ-separable if and only if it is ρ'-separable.

M4. Completeness. A sequence $\{x_n\}$ is *fundamental*, or has the *Cauchy property*, if

$$\sup_{i,j \geq n} \rho(x_i, x_j) \to_n 0.$$

A set M is *complete* if every fundamental sequence in M has a limit lying in M. A complete set is obviously closed. Usually, the question is whether S itself is complete. A fundamental sequence converges if it contains a convergent subsequence, which provides a convenient way of checking completeness.

Completeness is not a topological property: $S = [1, \infty)$ is complete under the usual metric ($\rho'(x, y) = |x - y|$) but not under the equivalent metric $\rho(x, y) = |x^{-1} - y^{-1}|$ (or, to put it another way, $[1, \infty)$ and $(0, 1]$ are homeomorphic, although the first is complete and the second is not). A metric space (S, ρ) is *topologically* complete if, as in this example, there is a metric equivalent to ρ under which it is complete.

Given a metric ρ on S, define

$$b(x, y) = 1 \wedge \rho(x, y). \tag{2}$$

Since $\phi(t) = 1 \wedge t$ is nondecreasing and satisfies $\phi(s + t) \leq \phi(s) + \phi(t)$ for $s, t \geq 0$, b is a metric; it is clearly equivalent to ρ. Further, since $\phi(t) \leq t$ for $t \geq 0$ and $\phi(t) = t$ for $0 \leq t \leq 1$, a sequence is b-fundamental if and only if it is ρ-fundamental; this means that S is ρ-complete if and only if it is b-complete. The metric b has the advantage that it is bounded: $b(x, y) \leq 1$.

M5. Compactness. The set A is by definition *compact* if each open cover of A has a finite subcover. An *ϵ-net* for A is a set of points $\{x_k\}$ with the property that for each x in A there is an x_k such that $\rho(x, x_k) < \epsilon$; A is *totally bounded* if for each positive ϵ it has a finite ϵ-net (the points of which are not required to lie in A).

Theorem. *These three conditions are equivalent:*

(i) *A^- is compact.*

(ii) *Each sequence in A has a convergent subsequence (the limit of which necessarily lies in A^-).*

(iii) *A is totally bounded and A^- is complete.*

PROOF. It is easy to see that (ii) holds if and only if each sequence in A^- has a subsequence converging to a point in A^- and that A is totally bounded if and only if A^- is. Therefore, we may assume in the proof that $A = A^-$ is closed.

The proof is clearer if we put three more properties between (i) and (ii):

(i_1) *Each countable open cover of A has a finite subcover.*

(i_2) *If $A \subset \bigcup_n G_n$, where the G_n are open and $G_1 \subset G_2 \subset \cdots$, then $A \subset G_n$ for some n.*

(i_3) *If $A \supset F_1 \supset F_2 \supset \cdots$, where the F_n are closed and nonempty, then $\bigcap_n F_n$ is nonempty.*

We first prove that (i_1), (i_2), (i_3), (ii), (iii) are all equivalent.

Proof that (i_1) $\leftrightarrow$ (i_2). Obviously (i_1) implies (i_2). As for the converse, if $\{G_n\}$ covers A, simply replace G_n by $\bigcup_{k \leq n} G_k$.

Proof that (i_2) $\leftrightarrow$ (i_3). First, (i_2) says that $A \cap G_n \uparrow A$ implies that $A \cap G_n = A$ for some n. And (i_3) says that $A \cap F_n \downarrow \emptyset$ implies that $A \cap F_n = \emptyset$ for some n (here the F_n need not be contained in A). If $F_n = G_n^c$, the two statements say the same thing.

Proof that (i_3) $\leftrightarrow$ (ii). Assume that (i_3) holds. If $\{x_n\}$ is a sequence in A, take $B_n = \{x_n, x_{n+1}, \ldots\}$ and $F_n = B_n^-$. Each F_n is nonempty, and hence, if (i_3) holds, $\bigcap_n F_n$ contains some x. Since x is in the closure of B_n, there is an i_n such that $i_n \geq n$ and $\rho(x, x_{i_n}) < n^{-1}$; choose the i_n inductively so that $i_1 < i_2 < \cdots$. Then $\lim_n \rho(x, x_{i_n}) = 0$: (ii) holds. On the other hand, if F_n are decreasing, nonempty closed sets and (ii) holds, take $x_n \in F_n$ and let x be the limit of some subsequence; clearly, $x \in \bigcap_n F_n$: (i_3) holds.

Proof that (ii)$\rightarrow$(iii). If A is not totally bounded, then there exists a poitive ϵ and an infinite sequence $\{x_n\}$ in A such that $\rho(x_m, x_n) \geq \epsilon$

for $m \neq n$. But then $\{x_n\}$ contains no convergent subsequence, and so (ii) must imply total boundedness. And (ii) implies completeness, because, if $\{x_n\}$ is fundamental and has a subsequence converging to x, then the entire sequence converges to x.

Proof that (iii)→(ii). Use the the diagonal method [PM.538]. If A is totally bounded, it can for each n be covered by finitely many open balls $B_{n1}, \ldots, B_{nk_n}$ of radius n^{-1}. Given a sequence $\{x_m\}$ in A, first choose an increasing sequence of integers $m_{11}, m_{12}, \ldots$ in such a way that $x_{m_{11}}, x_{m_{12}}, \ldots$ all lie in the same B_{1k}, which is possible because there are only finitely many of these balls. Then choose a sequence $m_{21}, m_{22}, \ldots$, a subsequence of $m_{11}, m_{12}, \ldots$, in such a way that $x_{m_{21}}, x_{m_{22}}, \ldots$ all lie in the same B_{2k}. Continue. If $r_i = m_{ii}$, then $x_{r_n}, x_{r_{n+1}}, \ldots$ all lie in the same B_{nk}. It follows that $x_{r_1}, x_{r_2}, \ldots$ is fundamental and hence by completeness converges to some point of A.

Thus (i_1) through (iii) are equivalent. Since obviously (i) implies (i_1), we can complete the proof by showing that (i_1) and (iii) together imply (i). But if A is totally bounded, then it is clearly separable, and it follows by the Lindelöf property that an arbitrary open cover of A has a countable subcover. And now it follows by (i_1) that there is a further subcover that is finite. □

Compactness is a topological property, as follows by condition (ii) of the theorem. A set A is *bounded* if its diameter $\sup[\rho(x, y) : x, y \in A]$ is finite. The closure of a totally bounded set is obviously bounded in this sense; the converse is false, since, for example, the closed balls in C (Example 1.3) are not compact. On the other hand, a set in Euclidean k-space is totally bounded if and only if it is bounded, and so here the bounded sets are exacly the ones with compact closure.

A set A is *relatively compact* if A^- is compact. This is equivalent to the condition that every sequence in A contains a convergent subsequence, the limit of which may not lie in A.

A useful fact: *The continuous image of a compact set is compact.* For suppose that $f : S \to S'$ is continuous and A is compact in S. If $\{f(x_n)\}$ is a sequence in $f(A)$, choose $\{n_i\}$ so that $\{x_{n_i}\}$ converges to a point x of A. By continuity, $\{f(x_{n_i})\}$ converges to the point $f(x)$ of $f(A)$.

M6. Products of Metric Spaces. Suppose that (S_i, ρ_i), $i = 1, 2, \ldots$, are metric spaces and consider the infinite Cartesian product $S =$

$S_1 \times S_2 \times \cdots$. It is clear that

$$\rho(x,y) = \sum_{i=1}^{\infty} 2^{-i}(1 \wedge \rho_i(x_i, y_i)) \tag{3}$$

is a metric, the metric of coordinatewise convergence.

If each S_i is separable, then S is separable. For suppose that D_i is a countable set dense in S_i, and consider the countable set D in S consisting of the points of the form

$$x = (x_1, \ldots, x_k, x_{k+1}^\circ, x_{k+2}^\circ, \ldots), \tag{4}$$

where $k \geq 1$, x_i is a variable point of D_i for $i \leq k$, and x_i° is some fixed point of S_i for $i > k$. Given an ϵ and a point y of S, choose k so that $\sum_{i>k} 2^{-i} < \epsilon$ and then choose points x_i of the D_i so that $\rho_i(y_i, x_i) < \epsilon$. With this choice, the point (4) satisfies $\rho(y,x) < 2\epsilon$.

If each S_i is complete, then S is complete. Suppose that $x^n = (x_1^n, x_2^n, \ldots)$ are points of S forming a fundamental sequence. Then each sequence $x_i^1, x_i^2, \ldots$ is fundamental in S_i and hence $\rho_i(x_i^n, x_i) \to_n 0$ for some x_i in S_i. By the M-test, $\rho(x^n, x) \to 0$.

If A_i is compact in S_i, then $A_1 \times A_2 \times \cdots$ is compact in S. (This is a special case of Tihonov's theorem.) Given a sequence of points $x^n = (x_1^n, x_2^n, \ldots)$ in A, consider for each i the sequence $x_i^1, x_i^2, \ldots$ in A_i. Since A_i is compact, there is a sequence $n_1, n_2, \ldots$ of integers such that $x_i^{n_k} \to_k x_i$ for some x_i in A_i. But by the diagonal method, the sequence $\{n_k\}$ can be chosen so that $x_i^{n_k} \to_k x_i$ holds for all i at the same time. And then, $x^{n_k} \to_k (x_1, x_2, \ldots)$.

M7. Baire Category. A set A is dense in B if $B \subset A^-$. And A is everywhere dense (or simply dense) if $S = A^-$, which is true if and only if A is dense in every open ball B. And A is defined to be *nowhere dense* if there is *no* open ball B in which it is dense. The Cantor set is nowhere dense in the unit interval, for example, but a nowhere-dense set can be entirely ordinary: A line is nowhere dense in the plane.

To say that A is nowhere dense is to say that, for every open ball B, A fails to be dense in B, which in turn is to say that B contains an x such that, for some ϵ, the ball $B(x,\epsilon)$ fails to meet A: $B(x,\epsilon) \subset A^c$. But since B is open, $B(x,\epsilon) \subset B$ for small enough ϵ:

$$B(x,\epsilon) \subset B \cap A^c. \tag{5}$$

Thus A is nowhere dense if and only if every open ball B contains a ball $B(x, \epsilon)$ satisfying (5). This condition is usually taken as the definition, although the words "nowhere dense" then lose their direct meaning. By making ϵ smaller, we can strengthen (5) to $B(x, \epsilon)^- \subset B \cap A^c$. The *Baire category theorem:*

Theorem. *If S is complete, then it cannot be represented as a countable union of nowhere dense sets.*

PROOF. Suppose that each of $A_1, A_2, \ldots$ is nowhere dense. There is in S an x_1 such that $B(x_1, \epsilon_1)^- \subset S \cap A_1^c$ for some ϵ_1. And $B(x_1, \epsilon_1)$ contains an x_2 such that $B(x_2, \epsilon_2)^- \subset B(x_1, \epsilon_1) \cap A_2^c$ for some ϵ_2. Continue. The ϵ_n can be chosen in sequence so that $\epsilon_n < 2^{-n}$, and then, since $\rho(x_n, x_{n+1}) < 2^{-n}$, the sequence $\{x_n\}$ is fundamental and hence has a limit x. For each k, x lies in $B(x_k, \epsilon_k)^-$ and hence lies outside A_k: $S = \bigcup_k A_k$ is impossible. □

A set is defined to be of the *first category* if it can be represented as a countable union of nowhere-dense sets; otherwise it is of the *second* category. According to Baire's theorem, the space itself must be of the second category if it is complete.

M8. Upper Semicontinuity. A function f is upper semicontinuous at x if for each ϵ there is a δ such that $\rho(x, y) < \delta$ implies $f(y) < f(x) + \epsilon$. It is easy to see that f is everywhere upper semicontinuous if and only if, for each real α, $[x: f(x) < \alpha]$ is an open set. *Dini's theorem:*

Theorem. *If $f_n(x) \downarrow 0$ for each x, and if each f_n is everywhere upper semicontinuous, then the convergence is uniform on each compact set.*

PROOF. For each ϵ, the open sets $G_n = [x: f_n(x) < \epsilon]$ cover S. If K is compact, then $K \subset G_n$ for some n, and uniformity follows. □

M9. Lipschitz Functions. A function satisfying a Lipschitz condition on a subset A of S can be extended to the entire space.

Theorem. *Supose that f is a function on A that satisfies $|f(x) - f(y)| \leq K\rho(x, y)$ for x and y in A. There is an extension g of f to S that satisfies the same condition: $|g(x) - g(y)| \leq K\rho(x, y)$ for x and y in S. If f satisfies $|f| \leq a$ on A, then g can be taken to satisfy $|g| \leq a$ on S.*

PROOF. Fix an arbitrary point z of A. If $y \in A$, then for all $x \in S$,

$$\begin{aligned} f(y) + K\rho(x,y) &= f(z) + K\rho(x,y) + (f(y) - f(z))) \\ &\geq f(z) + K\rho(x,y) - K\rho(y,z) \geq f(z) - K\rho(x,z). \end{aligned}$$

Therefore, the function $g(x) = \inf_{y \in A}(f(y) + K\rho(x,y))$ is well defined on S. If x and y lie in A, then $f(y) + K\rho(x,y) \geq f(x)$, with equality for $y = x$: $g(x) = f(x)$ for x in A.

Let x and x' be points of S. Given ϵ, choose y in A so that $g(x) \geq f(y) + K\rho(x,y) - \epsilon$. Then

$$\begin{aligned} g(x') - g(x) &\leq f(y) + K\rho(x',y) - [f(y) + K\rho(x,y) - \epsilon] \\ &= K(\rho(x',y)) - \rho(x,y)) + \epsilon \leq K\rho(x',x) + \epsilon, \end{aligned}$$

and so $g(x') - g(x) \leq K\rho(x',x)$. Interchange x' and x to get the Lipschitz condition for g.

If a bounds $|f|$ on A, truncate g above at a and below at $-a$. □

M10. Topology and Measurability. The Borel σ-field $\mathcal{S}$ for (S,ρ) is the one generated by the open sets. Let (S',ρ') be a second metric space, with Borel σ-field $\mathcal{S}'$. *If* $h: S \to S'$ *is continuous, then it is measurable* $\mathcal{S}/\mathcal{S}'$ (in the sense that $A' \in \mathcal{S}'$ implies $A \in \mathcal{S}$). To prove this, it is enough [PM.182] to show that $h^{-1}G' \in \mathcal{S}$ if G' is an open set in S'; but of course $h^{-1}G'$ is open. In particular, a continuous real function on S is $\mathcal{S}$-measurable.

Let $(\Omega,\mathcal{F})$ be a measurable space, and let h_n and h be maps from Ω to S. *If each* h_n *is measurable* $\mathcal{F}/\mathcal{S}$, *and if* $\lim_n h_n x = hx$ *for every* x, *then* h *is also measurable* $\mathcal{F}/\mathcal{S}$. In fact, $h^{-1}F \subset \liminf_n h_n^{-1}F^\epsilon \subset h^{-1}F^{2\epsilon}$. Intersect over positive, rational ϵ: If F is closed, then $h^{-1}F = \bigcap_\epsilon \liminf_n h_n^{-1}F^\epsilon$, which lies in $\mathcal{F}$.

The set D_h *of points at which* h *is not continuous lies in* $\mathcal{S}$. This is true even if h is not measurable $\mathcal{S}/\mathcal{S}'$. To prove it, let $A_{\epsilon\delta}$ be the set of x in S for which there exist points y and z in S satisfying $\rho(x,y) < \delta$, $\rho(x,z) < \delta$, and $\rho'(hy,hz) \geq \epsilon$. Then $A_{\epsilon\delta}$ is open, and $D_h \in \mathcal{S}$ because $D_h = \bigcup_\epsilon \bigcap_\delta A_{\epsilon\delta}$ (ϵ and δ ranging over the positive rationals).

Subspaces. A subset S_0 of S is a metric space in its own right. If $\mathcal{O}$ and $\mathcal{O}_0$ are the classes of open sets in S and S_0, then $\mathcal{O}_0 = \mathcal{O} \cap S_0$ ($= [G \cap S_0: G \in \mathcal{O}]$), and it follows [PM.159] that the Borel σ-field in S_0 is

$$\mathcal{S}_0 = \mathcal{S} \cap S_0. \tag{6}$$

If S_0 lies in $\mathcal{S}$, (6) becomes

$$\mathcal{S}_0 = [A: A \subset S_0, A \in \mathcal{S}]. \tag{7}$$

Product Spaces. Let S' and S'' be metric spaces with metrics ρ' and ρ'' and Borel σ-fields $\mathcal{S}'$ and $\mathcal{S}''$, and consider the product space $T = S' \times S''$. The *product topology* in T may be specified by various metrics, for example,

$$t((x', x''), (y', y'')) = \sqrt{[\rho'(x', y')]^2 + [\rho''(x'', y'')]^2} \tag{8}$$

and

$$t((x', x''), (y', y'')) = \rho'(x', y') \vee \rho''(x'', y''). \tag{9}$$

Under either of these metrics there is convergence $(x'_n, x''_n) \to (x', x'')$ in T if and only if $x'_n \to x'$ in S' and $x''_n \to x''$ in S''. Under the metric (9) we have, in an obvious notation,

$$B_t((x', x''), r) = B_{\rho'}(x', r) \times B_{\rho''}(x'', r), \tag{10}$$

which is convenient.

Consider the projections $\pi' : T \to S'$ and $\pi'' : T \to S''$ defined by $\pi'(x', x'') = x'$ and $\pi''(x', x'') = x''$; each is continuous. If T_0 is countable and dense in T, then $\pi' T_0$ and $\pi'' T_0$ are countable and dense in S' and S''. On the other hand, if S'_0 and S''_0 are countable and dense in S' and S'', then $S'_0 \times S''_0$ is countable and dense in T. Therefore: *T is separable if and only if S' and S'' are both separable.*

Let $\mathcal{T}$ be the Borel σ-field in T. Consider also the product σ-field $\mathcal{S}' \times \mathcal{S}''$—the one generated by the measurable rectangles, the sets $A' \times A''$ for $A' \in \mathcal{S}'$ and $A'' \in \mathcal{S}''$. Now this rectangle is $(\pi')^{-1}A' \cap (\pi'')^{-1}A''$; since the two projections are continuous, they are measurable $\mathcal{T}/\mathcal{S}'$ and $\mathcal{T}/\mathcal{S}''$, respectively, and it follows that the rectangle lies in $\mathcal{T}$. Therefore, $\mathcal{S}' \times \mathcal{S}'' \subset \mathcal{T}$. On the other hand, if T is separable, then each open set in T is a countable union of the sets (10) and hence lies in $\mathcal{S}' \times \mathcal{S}''$. It follows that

$$\mathcal{S}' \times \mathcal{S}'' = \mathcal{T}, \tag{11}$$

if T is separable.†

† Without separability, (11) may fail: If $S' = S''$ is discrete and has power exceeding that of the coninuum, then the diagonal $[(x, y): x = y]$ lies in $\mathcal{T}$ but not in $\mathcal{S}' \times \mathcal{S}''$. See Problem 2 on p. 261 of Halmos [36].

ANALYSIS

M11. The Gamma and Beta Functions. The gamma function is defined by

$$\Gamma(\alpha) = \int_0^\infty x^{\alpha-1} e^{-x} dx \tag{12}$$

for positive α. Integration by parts shows that $\Gamma(\alpha) = (\alpha-1)\Gamma(\alpha-1)$ for $\alpha > 1$, from which it follows that $\Gamma(m) = (m-1)!$ for positive integers m.

The gamma-α distribution (with unit scale parameter) has density

$$g_\alpha(x) = \frac{1}{\Gamma(\alpha)} x^{\alpha-1} e^{-x} \tag{13}$$

over $(0, \infty)$. A calculation with the Laplace transform shows that g_α and g_β convolve to $g_{\alpha+\beta}$, from which it follows that $g_{\alpha+\beta}(1) = \int_0^1 g_\alpha(y) g_\beta(1-y)\, dy$, or

$$\int_0^1 y^{\alpha-1}(1-y)^{\beta-1} dy = \frac{\Gamma(\alpha)\Gamma(\beta)}{\Gamma(\alpha+\beta)}. \tag{14}$$

The left side here defines the beta function; the beta-(α, β) distribution has over (0,1) the density $y^{\alpha-1}(1-y)^{\beta-1}\Gamma(\alpha+\beta)/\Gamma(\alpha)\Gamma(\beta)$.

M12. Dirichlet's Formula. Let D_k be the R^k-set defined by the inequalities $t_1, \ldots, t_k > 0$ and $\sum_{i=1}^k t_i < 1$. Dirichlet's integral formula is

$$\int_{D_k} f(t_1 + \cdots + t_k) t_1^{\alpha_1-1} \cdots t_k^{\alpha_k-1} dt_1 \cdots dt_k = \frac{\Gamma(\alpha_1)\cdots\Gamma(\alpha_k)}{\Gamma(\alpha_1 + \cdots + \alpha_k)} \int_0^1 f(t) t^{\alpha-1} dt, \tag{15}$$

where $\alpha = \alpha_1 + \ldots + \alpha_k$ and the α_i are positive. To prove it, fix $t_3, \ldots, t_k$, write $a = t_3 + \cdots + t_k$ ($a = 0$ if $k = 2$), and consider the integral

$$\iint_{\substack{t_1, t_2 > 0 \\ t_1 + t_2 < 1-a}} f(t_1 + t_2 + a) t_1^{\alpha_1-1} t_2^{\alpha_2-1} dt_1 dt_2.$$

By (14), the change of variables $t_1 = (1-s_1)s_2$, $t_2 = s_1 s_2$ (with Jacobian s_2) reduces this to

$$\iint_{\substack{0<s_1<1 \\ 0<s_2<1-a}} f(s_2+a)(1-s_1)^{\alpha_1-1} s_1^{\alpha_2-1} s_2^{\alpha_1+\alpha_2-1} ds_1 ds_2 = \frac{\Gamma(\alpha_1)\Gamma(\alpha_2)}{\Gamma(\alpha_1+\alpha_2)} \int_0^{1-a} f(s_2+a) s_2^{\alpha_1+\alpha_2-1} ds_2.$$

And now (15) follows by induction.

M13. The Dirichlet Distribution. The random vector $X = (X_1, \ldots, X_k)$ has the Dirichlet distribution with (positive) parameters $\alpha_1, \ldots, \alpha_k$ if $X_k = 1 - \sum_{i=1}^{k-1} X_i$ and $(X_1, \ldots, X_{k-1})$ has density

$$\frac{\Gamma(\alpha_1 + \cdots + \alpha_k)}{\Gamma(\alpha_1) \cdots \Gamma(\alpha_k)} t_1^{\alpha_1 - 1} \cdots t_{k-1}^{\alpha_{k-1}-1}(1 - t_1 - \cdots - t_{k-1})^{\alpha_k - 1} \tag{16}$$

on the set where $t_1, \ldots, t_{k-1}$ are positive and add to at most 1. By (14) and (15), (16) does integrate to 1. A change of variables shows that, if $\alpha_i = \alpha_j$, then the distribution of X remains the same if X_i and X_j are interchanged. If the α_i are all the same, then X has the *symmetric* Dirichlet distribution; in this case, the distribution of X is invariant under all permutations of the components.

M14. Scheffé's Theorem. Suppose that y_m^n are nonnegative, that $\sum_m y_m^n = \sum_m y_m$ (finite) for all n, and that $y_m^n \to_n y_m$ for all m. Then, by the series form of Scheffé's theorem [PM.215],

$$\sum_m |y_m^n - y_m| \to_n 0. \tag{17}$$

If f is a bounded and continuous real function, then

$$\sum_m y_m^n f(y_m^n) \to_n \sum_m y_m f(y_m). \tag{18}$$

For, if M bounds f, then

$$\begin{aligned} \Big| \sum_m y_m^n f(y_m^n) - \sum_m y_m f(y_m) \Big| & \\ \le \sum_m |y_m^n - y_m| \cdot |f(y_m^n)| &+ \sum_m y_m |f(y_m^n) - f(y_m)| \\ \le M \sum_m |y_m^n - y_m| &+ \sum_m y_m |f(y_m^n) - f(y_m)|. \end{aligned}$$

To the first sum on the right apply (17) and to the second apply the bounded convergence theorem.

M15. Measurability of Some Mappings. Let $\mathcal{T}$ be the class of Borel subsets of $T = [0, 1]$. For each t, the projection π_t from C to R^1 (Example 1.3) is measurable $\mathcal{C}$. Since the mapping

$$(x, t) \to x(t) \tag{19}$$

from $C \times T$ to R^1 is continuous in the product topology, and since $\mathcal{C} \times \mathcal{T}$ is the σ-field of Borel sets for this topology (see (11)), (19) is measurable $\mathcal{C} \times \mathcal{T}/\mathcal{R}^1$.

For $x \in C$, let $h(x)$ be the Lebesgue measure of the set of t in T for which $x(t) > 0$. We want to prove that h is measurable $\mathcal{C}$, and we can derive this from a more general result. If v is the indicator of $(0, \infty)$, then

$$h(x) = \int_0^1 v(x(t))\, dt. \tag{20}$$

It will be enough to prove that, if v is Borel measurable, bounded, and continuous except on a set of Lebegue measure 0, then (20) is measurable $\mathcal{C}$; we prove at the same time that it is continuous except on a set of Wiener measure 0.

Since $v(x(t))$ is a bounded, Borel measurable function of t, (20) is well defined. And since (19) is measurable $\mathcal{C} \times \mathcal{T}$, the mapping $\psi : C \times T \to R^1$ defined by $\psi(x, t) = v(x(t))$ is also measurable. Since ψ is bounded, $h(x) = \int_0^1 \psi(x, t)\, dt$ is measurable $\mathcal{C}$ as a function of x [PM.233].

If D_v is the set of discontinuities of v, then, by assumption, $\lambda D_v = 0$, where λ is Lebesgue measure. Let E be the set of (x, t) for which $x(t) \in D_v$. If W is Wiener measure, then $W[x\colon (x, t) \in E] = 0$, and it follows by Fubini's theorem applied to the measure $W \times \lambda$ on $\mathcal{C} \times \mathcal{T}$ that $\lambda[t\colon (x, t) \in E] = 0$ for $x \notin A$, where A is a $\mathcal{C}$-set satisfying $WA = 0$. Suppose that $x_n \to x$ in the topology of C. If $x \notin A$, then $x(t) \notin D_v$ for almost all t and hence, since $x_n(t) \to x(t)$ for all t, $v(x_n(t)) \to v(x(t))$ for almost all t. It follows by the bounded convergence theorem that

$$\int_0^1 v(x_n(t))\, dt \to \int_0^1 v(x(t))\, dt. \tag{21}$$

Thus h is continuous except at points forming a set of W-measure 0.

The argument goes through if W is replaced by a P with the property that $P\pi_t^{-1}$ is absolutely continuous with respect to Lebesgue measure for almost all t. This is true of W°.

Except in two places, this argument also goes through word for word if C is replaced by D. First, if $x_n \to x$ in the topology of D, then $x_n(t) \to x(t)$ for all but countably many t (rather than for all t), which is still enough for (21). Second, the proof that (19) is measurable—in this case, that it is measurable $\mathcal{D} \times \mathcal{T}/\mathcal{R}^1$—requires modification.

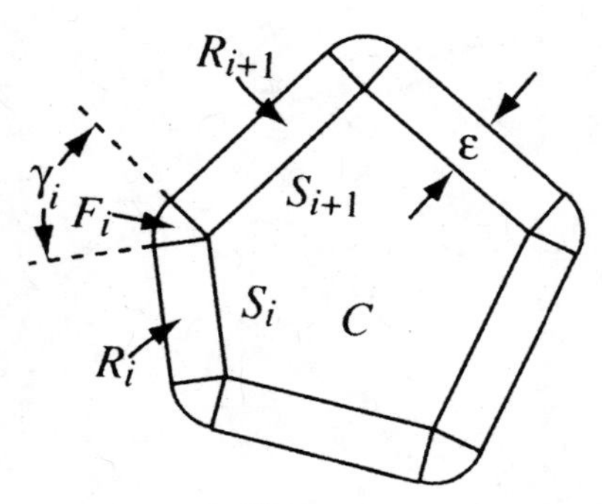

Figure 1

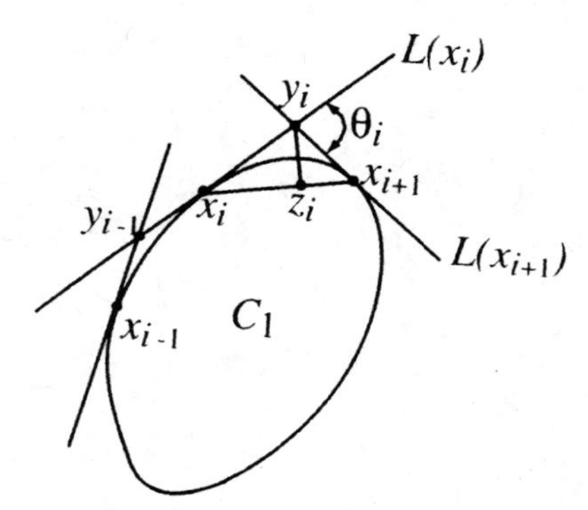

Figure 2

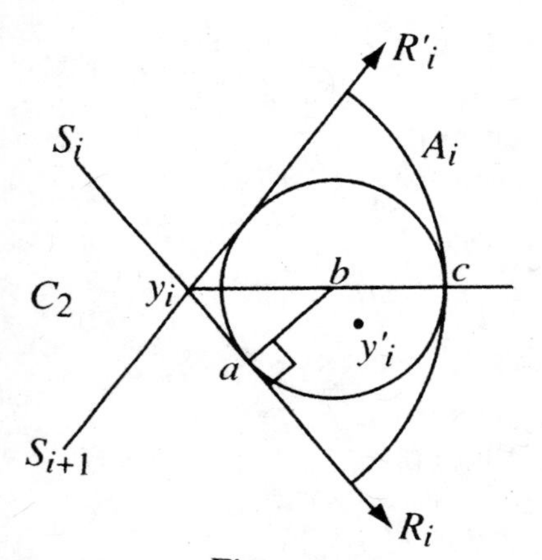

Figure 3

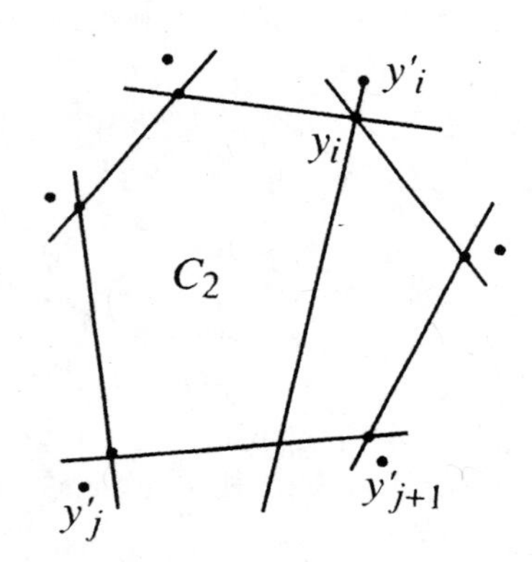

Figure 4

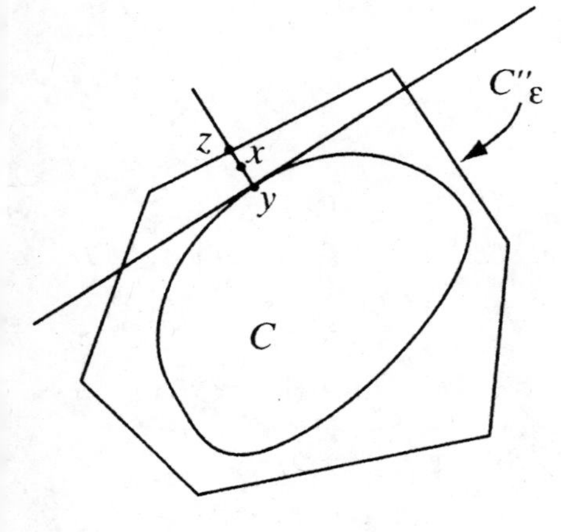

Figure 5

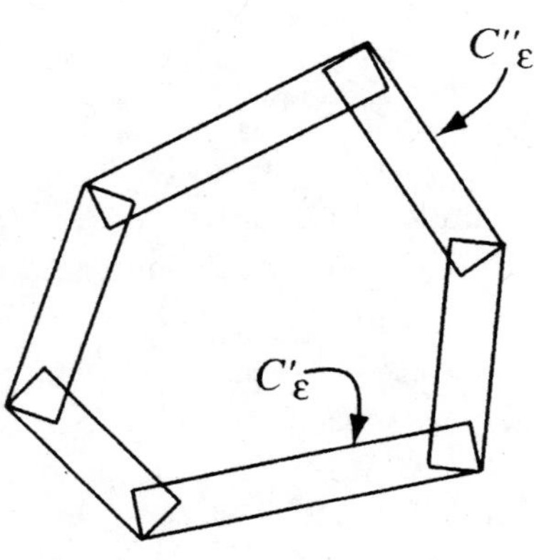

Figure 6

Define (see the proof in Section 12 that π_t is measurable) $h_\epsilon(x,t)$ as $\epsilon^{-1}\int_t^{t+\epsilon} x(s)\,ds$ for $t \le \frac{1}{2}$ and as $\epsilon^{-1}\int_{t-\epsilon}^{t} x(s)\,ds$ for $t > \frac{1}{2}$. Then h_ϵ is continuous on $D \times T$, and since $h_{m^{-1}}(x,t) \to_m x(t)$ for all x and t, measurability follows.

One more function on C remains to be analyzed, namely, the supremum $h(x)$ of those t in $[0,1]$ for which $x(t) = 0$. Since $[x: h(x) < \alpha]$ is open, h is measurable. If h is discontinuous at x, then $h(t)$ must keep to one side of 0 in $(h(x), 1)$ and keep to the same side of 0 in $(h(x) - \epsilon, h(x))$ for some ϵ. That h is continuous except on a set of Wiener measure 0 will therefore follow if we show that, for each t_0, the supremum and infimum of W over $[t_0, 1]$ have continuous distributions. Since $W_t - W_{t_0}$ for t ranging over $[t_0, 1]$ is distributed as a Wiener path with a linearly transformed time scale, $-W_{t_0} + \sup_{t \ge t_0} W_t$ has a continuous distribution (see (8.20)). This last random variable and W_{t_0} are independent, and hence their sum also has a continuous distribution. The infimum is treated the same way.

M16. More Measurability. In Section 14 we defined the mapping $\psi: D \times D_0 \to D$ by $\psi(x,\phi) = x \circ \phi$ (see (14.11)), and we are to prove it measurable $\mathcal{D} \times \mathcal{D}_0/\mathcal{D}$. Since the finite-dimensional sets in D generate $\mathcal{D}$, it is enough to prove that, for each t, the mapping

$$(x,\phi) \to \pi_t(x \circ \phi) = x(\phi(t)) \tag{22}$$

is measurable $\mathcal{D} \times \mathcal{D}_0$. If $\phi_k(t) = \lceil k\phi(t) \rceil / k$, then $\phi_k(t) \downarrow \phi(t)$ for each t. Hence the mapping

$$(x,\phi) \to x(\phi_k(t)) \tag{23}$$

converges pointwise to (22), and it is enough to prove this latter mapping measurable $\mathcal{D} \times \mathcal{D}_0$. Now $[(x,\phi): x(\phi_k(t)) \le \alpha]$ is the union of

$$[(x,\phi): \phi(t) = 0] \cap [(x,\phi): x(0) \le \alpha] \tag{24}$$

with the sets

$$\left[(x,\phi): \frac{i-1}{k} < \phi(t) \le \frac{i}{k}\right] \cap \left[(x,\phi): x\left(\frac{i}{k}\right) \le \alpha\right], \quad i = 1, \ldots, k. \tag{25}$$

If $H \in \mathcal{R}^1$, then $[\phi \in D_0: \phi(t) \in H] = D_0 \cap \pi_t^{-1} H$ lies in $\mathcal{D}_0$, and therefore $[(x,\phi): \phi(t) \in H] \in \mathcal{D} \times \mathcal{D}_0$. Similarly, $[(x,\phi): x(t) \in H] \in \mathcal{D} \times \mathcal{D}_0$. Thus the sets (24) and (25) all lie in $\mathcal{D} \times \mathcal{D}_0$, which proves the measurability of (23).

CONVEXITY

We treat here a few facts about planar convex sets needed for Section 15. We assume the basic theory (support lines, convex hulls, and so on), and the treatment is somewhat informal.† Every convex set considered is assumed to be *compact.* Denote Euclidean distance by $d(x,y) = |x-y|$, and write $A^{+\epsilon} = (A^{\epsilon})^{-} = [x: d(x,A) \le \epsilon]$.

The *Hausdorff distance* between two compact sets is

$$h(A,B) = \Big(\sup_{x\in A}\inf_{y\in B}|x-y|\Big) \vee \Big(\sup_{y\in B}\inf_{x\in A}|x-y|\Big). \tag{26}$$

Equivalently, $h(A,B)$ is the infimum of those ϵ for which

$$A \subset B^{+\epsilon}, \qquad B \subset A^{+\epsilon}. \tag{27}$$

Let T be the space of (compact) convex subsets of the unit square $Q = [0,1]^2$. For the theory of Section 15, we must show that T is h-totally bounded, and in fact we need to estimate the size of the smallest ϵ-net.

M17. Preliminaries on Convexity. We first show that T *is h-complete*, which, once it has been shown to be h-totally bounded, will imply that T *is h-compact.* Suppose that $\{C_n\}$ is h-fundamental, and define C as the closed set $\bigcap_{i=n}^{\infty}(\bigcup_{j=i}^{\infty} C_j)^-$; note that, since the union decreases with i, the intersection is the same for all n. Suppose we have $h(C_n, C_i) < \epsilon$ for $i \ge n$. Then $\bigcup_{j=i}^{\infty} C_j \subset C_n^{\epsilon}$ for $i \ge n$, and hence $C \subset (\bigcup_{j=i}^{\infty} C_j)^- \subset C_n^{+\epsilon}$. On the other hand, if $x \in C_n$, then the disc $B(x,\epsilon)$ meets C_i for each $i \ge n$, and therefore $B(x,\epsilon)^- \cap (\bigcup_{j=i}^{\infty} C_j)^-$ is, for $i \ge n$, a decreasing sequence of nonempty, closed sets; it follows that the intersection $B(x,\epsilon)^- \cap \bigcap_{i=n}^{\infty}(\bigcup_{j=i}^{\infty} C_j)^- = B(x,\epsilon)^- \cap C$ is also nonempty. This means that $x \in C^{+\epsilon}$, and hence $C_n \subset C^{+\epsilon}$. Thus $h(C_n, C) \le \epsilon$, which shows that $h(C_n, C) \to 0$.

It remains to show that the closed set C is convex. Choose ϵ_n so that $h(C_n, C) < \epsilon_n \to 0$. Suppose that x and y lie in C and consider a convex combination $z = px + qy$. There are points x_n and y_n of C_n such that $|x - x_n| < \epsilon_n$ and $|y - y_n| < \epsilon_n$, and of course $z_n = px_n + qy_n \to z$. Also, z_n lies in C_n and hence lies within ϵ_n of a

---

† Eggleston [25], for example, has a detailed account of the general theory. He defines the Hausdorff distance as a sum rather than a maximum (which gives an equivalent metric), and he gives it no name.

point z_n' of C. This means that z_n' also converges to z, which proves that z lies in C. Therefore, the limit set C is convex, and T *is indeed complete.*

We also need this fact: Suppose that $C \subset C_{n+1} \subset C_n$ for all n. Then: $C_n \downarrow C$ *if and only if* $h(C, C_n) \downarrow 0$. Suppose first that $C_n \downarrow C$ but $h(C_n, C) > \epsilon$. There is then an x_n in C_n for which $d(x_n, C) \geq \epsilon$. Pass to a subsequence along which x_n converges to some x. Then $d(x, C) \geq \epsilon$, hence $x \notin C$, hence $x \notin C_{n_0}$ for some n_0, and hence $d(x, C_{n_0}) > 0$. But $x_n \in C_{n_0}$ for $n \geq n_0$, and so x_n cannot converge to x, a contradiction. Suppose on the other hand that $h(C, C_n) \to 0$. Of course, $C_n \downarrow A$ for some A, and $C \subset A \subset C_n \subset C^{+\epsilon}$ for large n; let $\epsilon \downarrow 0$: $A = C$.

We next prove that

$$\lambda C^{+\epsilon} - \lambda C = \begin{cases} l(\partial C)\epsilon + \pi\epsilon^2 & \text{if } C^\circ \neq \emptyset, \\ 2l(C)\epsilon + \pi\epsilon^2 & \text{if } C^\circ = \emptyset. \end{cases} \tag{28}$$

where λ denotes two-dimensional Legesgue measure and l denotes arc length. The case $C^\circ = \emptyset$ is easy, since then C is a closed line segment, $\partial C = C$, and $\lambda(C) = 0$. We show in the course of the proof that ∂C is rectifiable.

Suppose first that C is a convex polygon. On each side S_i of C, construct a rectangle R_i of height ϵ, as shown in Figure 1. Between R_i and R_{i+1} lies a fan-shaped region F_i, a sector of a disc of radius ϵ. If γ_i is the angle at the apex, then $\sum_i \gamma_i = 2\pi$, and so $\sum_i \lambda F_i = \pi\epsilon^2$. And of course $\sum_i \lambda R_i = l(\partial C)\epsilon$. Hence (28) for the polygonal case.

Next, take D to be a polygon inscribed in the general C, and apply (28) for $\epsilon = 1$: $l(\partial D) < \lambda D^{+1} - \lambda D \leq \lambda C^{+1}$. Thus $l(\partial D)$ is bounded, and ∂D is indeed rectifiable. And if $C \subset Q$, then $C^{+1} \subset [-1, 2]^2$, which gives

$$l(\partial C) \leq 9 \qquad \text{if } C \subset Q. \tag{29}$$

(It is intuitively clear that, in fact, $l(\partial C) \leq 4$.)

Before proving the general case of (28), we show that $\lambda(\partial C) = 0$. If $C^\circ = \emptyset$, this is easy. In the opposite case, translate C so that the origin is an interior point. The sets $\alpha(\partial C)$ are disjoint, and $\lambda(\alpha(\partial C)) = \alpha^2\lambda(\partial C)$, which is impossible if $\lambda(\partial C) > 0$.

To prove (28) for the general C with nonempty interior, note first that each point of C is the limit of points in C°: Given u in C, choose z_1 and z_2 in C° in such a way that u, z_1, z_2 are not colinear and hence are the vertices of a triangle V contained in C; but obviously, $V^\circ \subset C^\circ$ and hence $u \in \partial V^\circ$. Now choose inscribed polygons C_n in such a way that $C_n \uparrow A$, where $C^\circ \subset A \subset C$. Then $l(\partial C_n) \uparrow l(\partial C)$ (by the definition of arc length), and $\lambda C_n \uparrow \lambda A = \lambda C$ (since $\lambda(\partial C) = 0$). Also, $C_n^{+\epsilon} \subset C_{n+1}^{+\epsilon} \subset C^{+\epsilon}$. If $x \in C^\epsilon$, then $|x - y| < \epsilon$ for some $y \in C^\circ$. But then $y \in C_n$ for some n, so that $d(x, C_n) < \epsilon$ and $x \in C_n^{+\epsilon}$: $C^\epsilon \subset C_n^{+\epsilon}$. Thus $C_n^{+\epsilon} \uparrow B$, where $C^\epsilon \subset B \subset C^{+\epsilon}$, and so $\lambda C_n^{+\epsilon} \uparrow \lambda B = \lambda C^{+\epsilon}$ (since $\lambda(\partial C^{+\epsilon}) = 0$). Hence (28).

A further inequality:

$$\lambda(C_1 \Delta C_2) \leq 45h(C_1, C_2) \qquad \text{for } C_1, C_2 \in Q. \tag{30}$$

Suppose that $h(C_1, C_2) < h$. Then $C_2 \cap C_1^c \subset C_1^{+h} - C_1$ and $C_1 \cap C_2^c \subset C_2^{+h} - C_2$, and so, by (28), $\lambda(C_1 \Delta C_2) \leq 2(l(\partial C_1) + l(\partial C_2))h + 2\pi h^2$. And now (29) gives $\lambda(C_1 \Delta C_2) \leq 36h + 2\pi h^2$. Since $h(C_1, C_2) \leq \sqrt{2}$, letting h tend to $h(C_1, C_2)$ gives (30).

A boundary point of a convex set is *regular* if through it there runs only one support line for the set (it is not a "corner" of the boundary). Another fact we need is that, if C is convex, then every boundary point x of the larger convex set $C^{+\epsilon}$ is regular. Indeed, there is a unique point y of C such that $|x - y| = \epsilon$, and $B(y, \epsilon)^- \subset C^{+\epsilon}$; since any line through x that supports $C^{+\epsilon}$ must also support $B(y, \epsilon)^-$, there can be only one of them. Since $h(C, C^{+\epsilon}) = \epsilon$, it follows that every convex set can be approximated arbirarily closely in the Hausdorff metric by a larger convex set whose boundary points are all regular and whose interior is nonempty. (Although the boundary of the approximating set has no corners, it may contain straight-line segments.)

M18. The Size of ϵ-Nets. The basic estimate:

Theorem. *There is a constant A such that, for $0 < \epsilon < 1$, there exists in T an ϵ-net $T(\epsilon)$ consisting of $N(\epsilon)$ convex polygons, where*

$$\log N(\epsilon) \leq A\sqrt{\frac{1}{\epsilon}} \log \frac{1}{\epsilon}. \tag{31}$$

Further, for each C in T, there is a C' in $T(\epsilon)$ such that $C \subset C'$ and $h(C, C') < \epsilon$ (approximation from above).

It is possible to remove the logarithmic factor on the right in (31),[†] but the proof of the stronger inequality is both more complicated and

---

[†] Dudley [21], p. 62.

less intuitive. We prove only the inequality (31) as it stands, since this suffices for the application in Section 15.

PROOF. Let r be a positive integer, later to be taken of the order $\sqrt{1/\epsilon}$. Assume that $r \geq 8$. We want to approximate an element C of T by a convex polygon with a limited number of sides, and we proceed by constructing a sequence of approximations $C_1, C_2, \ldots$. These sets may not be contained in Q and hence may not be elements of T, but we can remedy this at the end by intersecting the last approximation with Q. First, we take C_1 to be a convex set that has only regular boundary points, has nonempty interior, and satisfies $C \subset C_1$ and $h(C, C_1) < 1/r^2$, as we know we can.

For x on ∂C_1, let $L(x)$ be the (unique) support line through x, and let $v(x)$ be the unit vector that starts at x, is normal to $L(x)$, and is directed away from C_1. If $p(x)$ is the endpoint of $v(x)$ after it has been translated to the origin, then $p(\cdot)$ maps ∂C_1 continuously onto the unit circle (it may not be one-to-one, since ∂C_1 may have flat places). For $0 \leq j < r$, let w_j be a point of ∂C_1 such that $p(w_j)$ has polar coordinates 1 and $2\pi j/r$. Since C_1 may not be contained in Q, (29) is not available as it stands. But $C_1 \subset [-1, 2]^2$, hence $C_1^{+1} \subset [-2, 3]^2$, and the argument leading to (29) gives $l(\partial C_1) \leq 25$. Therefore, there are points on ∂C_1, at most $25r$ of them, such that the distance along ∂C_1 from one to the next is at most $1/r$. Merge this set with set of the w_j. This gives points $x_0, x_1, \ldots, x_{k-1}$, ordered clockwise (say) around ∂C_1, such that, first, $|x_i - x_{i+1}| \leq 1/r$ ($i+1 = 0$ for $i = k-1$), second, the angle θ_i between the support lines $L(x_i)$ and $L(x_{i+1})$ satisfies $0 \leq \theta_i \leq 2\pi/r$, and third, $k \leq 26r$. Now $L(x_i)$ is the boundary of two closed half-planes; let $H(x_i)$ be the one containing C_1. Then $C_2 = \bigcap_{i=0}^{k-1} H(x_i)$ is a convex polygon which contains C_1.

Suppose at first that all the angles θ_i are positive, which implies that C_2 has exactly k sides. Consider successive points x_i and x_{i+1}, and let y_i be the point where the two lines $L(x_i)$ and $L(x_{i+1})$ intersect (Figure 2). If α_i and β_i are the angles at the vertices x_i and x_{i+1} of the triangle $x_i y_i x_{i+1}$, and if z_i is the foot of the perpendicular from y_i to the opposite side, then $\alpha_i + \beta_i = \theta_i$. From $r \geq 8$ follows $\theta_i \leq 2\pi/r \leq \pi/4$ and therefore,[†] $|y_i - z_i| = |x_i - z_i| \tan \alpha_i \leq |x_i - x_{i+1}| \tan \theta_i \leq (1/r)\tan(2\pi/r) \leq 4\pi/r^2$. This implies that each point of the segment from x_i to y_i is within distance $16/r^2$ of a point on ∂C_1. The same argument applies to the segment from x_i to the corresponding point

[†] If $0 \leq t \leq \pi/4$, then $\cos t \geq 1/\sqrt{2}$, and so $\tan t \leq t/\cos t \leq 2t$.

y_{i-1} on the other side. The segment from y_{i-1} to y_i is one of the k sides of C_2, and each of its points is within $16/r^2$ of ∂C_1. Since $C_1 \subset C_2$, it follows that $h(C_1, C_2) \leq 16/r^2$.

If a θ_i is 0, join the sides lying on $L(x_i)$ and $L(x_{i+1})$ (lines which coincide in this case) into one longer side. If we do this for each i such that $\theta_i = 0$, we see that C_2 now has $26r$ or fewer noncollinear sides ($k \leq 26r$), and the angle between each side and the next is at most $2\pi/r$.

There are infinitely many possibilities for the polygon C_2, because there is no restriction on the vertices. The next step is to replace the vertices of C_2 by elements of the lattice $\mathcal{L}_r$ of points of the form $(i/r^2, j/r^2)$ (i and j integers), which will lead to a convex polygon C_3 for which there are only finitely many possibilities.

Consider a vertex y_i joining adjacent sides S_i and S_{i+1} of C_2 (Figure 3). We want to move y_i to a nearby element of $\mathcal{L}_r$. Extend S_i and S_{i+1} past y_i to get rays R_i and R_i' from y_i; let A_i be the open region bounded by R_i and R_i'. Since $r > 4$, so that $2\pi/r < \pi/2$, the angle between R_i and R_i' is obtuse. It is clear that, for $\rho > 1/r^2$, any disc of radius ρ contains a point of $\mathcal{L}_r$ in its interior. Consider the disc of radius ρ that is tangent to the rays R_i and R_i'. Since the angle at y_i is obtuse, $|y_i - a| < \rho = |b - a|$, and so $|y_i - b| < 2\rho$. Take $\rho = 4/3r^2$: A_i contains a point y_i' that lies in $\mathcal{L}_r$ and satisfies $|y_i - y_i'| < 4/r^2$.

Let C_3 be the convex hull of the points y_i'. (Some of the y_i' may be interior to C_3.) Since y_i' lies in the open set A_i, a ray (Figure 4) starting from y_i' and passing through y_i will enter the interior of C_2 and will eventually pass between a pair y_j' and y_{j+1}' (or through one of them). This means that y_i lies in the triangle with vertices y_i', y_j', and y_{j+1}' and hence lies in the convex hull C_3. And since this is true for each vertex of the polygon C_2, it is a subset of C_3. Finally, since each y_j' is within $4/r^2$ of a point of C_2, we arrive at $h(C_2, C_3) < 4/r^2$.

The successive Haudorff distances from C to C_1 to C_2 to C_3 are less than $1/r^2$, $16/r^2$, and $4/r^2$, and therefore, $h(C, C_3) < 21/r^2$. And the sets are successively larger. How many of these convex polygons C_3 are there? Although C_3 may extend beyond Q, it is certainly contained in $[-1, 2]^2$. The number of points of $\mathcal{L}_r$ in this larger square is less than $9(r^2+1)^2$ and hence less than $36r^4$. Since each of the $26r$ or fewer vertices of C_3 is one of these points, the number of polygons C_3 is at most

$$M_r = \sum_{i=1}^{26r} \binom{36r^4}{i} \leq 26r \binom{36r^4}{26r} \leq 26r(36r^4)^{26r}.$$

(The first inequality holds if $26r$ is less than half of $36r^4$, which only requires $r > 1$.) Clearly, $\log M_r \le Br \log r$ for a constant B.

If $C_4 = Q \cap C_3$, then C_4 is a convex polygon containing C; then $h(C, C_4) < 21/r^2$, and M_r bounds the number of these special elements of T. Given an ϵ, take r an integer such that $r > \sqrt{21/\epsilon} \ge r - 1$. Then $21/r^2 < \epsilon$, and if ϵ is small enough, then $r \ge 8$ (a condition used in the construction). Since $r \le 2\sqrt{21/\epsilon}$, (31) follows. □

Suppose that C is in T, and choose a polygon C''_ϵ in $T(\epsilon)$ in such a way that $C \subset C''_\epsilon$ and $h(C, C''_\epsilon) < \epsilon$. Now let $C'_\epsilon = [x : d(x, (C''_\epsilon)^c) \ge \epsilon]$. Then C'_ϵ is convex and closed. And we can show that $C'_\epsilon \subset C$, that is, that $x \notin C$ implies $d(x, (C''_\epsilon)^c) < \epsilon$. Since this inequality certainly holds if $x \notin C''_\epsilon$, assume that $x \in C''_\epsilon - C$. If y is the point of C nearest to x (Figure 5), then C is supported by the line through y that is perpendicular to the segment from y to x. Extend this segment until it meets $\partial C''_\epsilon$ at the point z. Then y is the point of C nearest to z, and since $h(C, C''_\epsilon) < \epsilon$, we have $|y - z| < \epsilon$, and hence $d(x, (C''_\epsilon)^c) < \epsilon$. Therefore, $C'_\epsilon \subset C \subset C''_\epsilon$.

A simple argument shows that $\lambda(C''_\epsilon - C'_\epsilon) \le l(\partial C''_\epsilon)\epsilon \le 9\epsilon$: On each side of the convex polygon C''_ϵ construct a rectangle that has height ϵ and overlaps C''_ϵ (Figure 6). The extra factor of 9 can be eliminated by starting over with $\epsilon/9$ in place of ϵ and noting that replacing ϵ by $\epsilon/9$ in (31) gives a function of the same order.† (It is not true that $h(C'_\epsilon, C''_\epsilon)$ is always small: Take C''_ϵ to be an isosceles triangle with angle ϵ at the apex.)

To sum up: *If C is an element of T, there are an element C''_ϵ of $T(\epsilon)$ and a companion convex set C'_ϵ, such that*

$$(32) \qquad C'_\epsilon \subset C \subset C''_\epsilon, \qquad \lambda(C''_\epsilon - C'_\epsilon) < \epsilon, \qquad h(C, C''_\epsilon) < \epsilon.$$

Since C'_ϵ is completely determined by C''_ϵ, it follows that the number of possible pairs $(C'_\epsilon, C''_\epsilon)$ is $N(\epsilon)$.

Let T_0 be the countable class of finite intersections of sets contained in $\bigcup_n T(1/n)$. If $C_n = \bigcap_{k=1}^n C''_{1/k}$, then $C_n \in T_0$ and $h(C, C_n) \to 0$, and since $C \subset C_{n+1} \subset C_n$, it follows that $C_n \downarrow C$.

---

† In marking examination papers, Egorov (of Egorov's theorem) used to take off a point if a student ended up with 2ϵ instead of ϵ.

PROBABILITY

M19. Etemadi's Inequality. If $S_1, \ldots, S_n$ are sums of independent random variables, then

$$\mathsf{P}\Big[\max_{k\le n}|S_k|\ge 3\alpha\Big]\le 3\max_{k\le n}\mathsf{P}[|S_k|\ge\alpha].$$

To prove this, consider the sets B_k where $|S_k|\ge 3\alpha$ but $|S_j|<3\alpha$ for $j<k$. Since the B_k are disjoint,

$$\begin{aligned}
\mathsf{P}\Big[\max_{k\le n}|S_k|\ge 3\alpha\Big] &\le \mathsf{P}[|S_n|\ge\alpha]+\sum_{k\le n}\mathsf{P}(B_k\cap[|S_n|<\alpha])\\
&\le \mathsf{P}[|S_n|\ge\alpha]+\sum_{k\le n}\mathsf{P}(B_k\cap[|S_n-S_k|>2\alpha])\\
&= \mathsf{P}[|S_n|\ge\alpha]+\sum_{k\le n}\mathsf{P}B_k\cdot\mathsf{P}[|S_n-S_k|>2\alpha]\\
&\le \mathsf{P}[|S_n|\ge\alpha]+\max_{k\le n}\mathsf{P}[|S_n-S_k|\ge 2\alpha]\\
&\le \mathsf{P}[|S_n|\ge\alpha]+\max_{k\le n}(\mathsf{P}[|S_n|\ge\alpha]+\mathsf{P}[|S_k|\ge\alpha])\\
&\le 3\max_{k\le n}\mathsf{P}[|S_k|\ge\alpha].
\end{aligned}$$

M20. Bernstein's Inequality. Suppose that $\eta_1, \ldots, \eta_n$ are independent random variables, each with mean 0 and variance σ^2, and each bounded by 1. Let $S = \eta_1 + \cdots + \eta_n$; it has mean 0 and variance $s^2 = n\sigma^2$. Then

$$\mathsf{P}\Big[\Big|\frac{1}{\sqrt{n}}S\Big|\ge x\Big]\le 2\exp\Big[-\frac{x^2}{2(\sigma^2+x/\sqrt{n})}\Big] \qquad \text{for } x\ge 0. \tag{33}$$

This is a version of *Bernstein's inequality*.

To prove (33), suppose that $0<t<1$ and put $G(t)=1/(1-t)$. Then, since $\eta_i^r\le\eta_i^2$ for $r\ge 2$,

$$\mathsf{E}[e^{t\eta_i}]=1+\sum_{r=2}^{\infty}\frac{t^r\mathsf{E}[\eta_i^r]}{r!}\le 1+\frac{\sigma^2t^2}{2}G(t)\le\exp\Big[\frac{\sigma^2t^2}{2}G(t)\Big].$$

By independence,

$$\mathsf{E}\big[e^{tS}\big]\le\exp\Big[\frac{s^2t^2}{2}G(t)\Big],$$

and therefore, for positive y,

$$\mathsf{P}[S \geq y] = \mathsf{P}\big[e^{tS} \geq e^{ty}\big] \leq e^{-ty}\mathsf{E}[e^{tS}] \leq \exp\Big[\frac{t^2s^2}{2}G(t) - ty\Big].$$

We want to bound the probability on the left, and the method will be clear if at first we operate as though $G(t)$ were constant: $G(t) \equiv G$. This makes it easy to minimize the right side over t. The minimum is at

$$t = \frac{y}{s^2G}, \tag{34}$$

and substitution gives

$$\mathsf{P}[S \geq y] \leq \exp\Big[-\frac{y^2}{2s^2G}\Big]. \tag{35}$$

Although this argument is incorrect, a modification of it leads a good bound. Replace the G in (34) by $1/(1-t)$ and solve for t: $t = y/(s^2+y)$. Since this t is between 0 and 1, (35) does hold if we replace the G there by $1/(1-t) = (s^2+y)/s^2$:

$$\mathsf{P}[S \geq y] \leq \exp\Big[-\frac{y^2}{2(s^2+y)}\Big].$$

Replace y by $x\sqrt{n}$ and use $s^2 = n\sigma^2$; the resulting inequality, together with the symmetric inequality on the other side, gives (33).

M21. Rényi Mixing. Events $A_1, A_2, \ldots$ in $(\Omega, \mathcal{F}, \mathsf{P})$ are *mixing in the sense of Rényi* if

$$\mathsf{P}(A_n \cap E) \to \alpha\mathsf{P}(E) \tag{36}$$

for every E in $\mathcal{F}$. In this case, $\mathsf{P}(A_n) \to \alpha$. Suppose that $\mathcal{F}_0$ is a field in $\mathcal{F}$, and let $\mathcal{F}_1 = \sigma(\mathcal{F}_0)$: $\mathcal{F}_0 \subset \mathcal{F}_1 \subset \mathcal{F}$.

Theorem. *Suppose that* (36) *holds for every E in $\mathcal{F}_0$ and that $A_n \in \mathcal{F}_1$ for all n. Then*

$$\int_{A_n} g\,d\mathsf{P} \to \alpha \int g\,d\mathsf{P} \tag{37}$$

holds if g is $\mathcal{F}$-measurable and P-integrable; in particular, (36) *holds for all E in $\mathcal{F}$. And if P dominates P_0, a second probability measure on $\mathcal{F}$, then*

$$\mathsf{P}_0(A_n) \to \alpha. \tag{38}$$

PROOF. The class of $\mathcal{F}$-sets E satisfying (36) is a λ-system (use the M-test) which contains the field $\mathcal{F}_0$ and therefore [PM.42] contains $\mathcal{F}_1$. It follows that (37) holds if g is the indicator of an $\mathcal{F}_1$-set and hence if it is a simple, $\mathcal{F}_1$-measurable function. If g is $\mathcal{F}_1$-measurable and P-integrable, choose simple, $\mathcal{F}_1$-measurable functions g_k that satisfy $|g_k| \le |g|$ and $g_k \to g$. Now

$$\left| \int_{A_n} g\, d\mathsf{P} - \alpha \int g\, d\mathsf{P} \right| \le \left| \int_{A_n} g_k\, d\mathsf{P} - \alpha \int g_k\, d\mathsf{P} \right| + (1+|\alpha|)\mathsf{E}[|g-g_k|].$$

Let $n \to \infty$ and then let $k \to \infty$: (37) follows by the dominated convergence theorem.

Now suppose that g is $\mathcal{F}$-measurable and P-integrable and use conditional expected values. Since $A_n \in \mathcal{F}_1$ and $\mathsf{E}[g\|\mathcal{F}_1]$ satisfies (37), it follows that

$$\int_{A_n} g\, d\mathsf{P} = \int_{A_n} \mathsf{E}[g\|\mathcal{F}_1]\, d\mathsf{P} \to \alpha \int \mathsf{E}[g\|\mathcal{F}_1] d\mathsf{P} = \alpha \int g\, d\mathsf{P};$$

(37) again holds. To prove (38), take $g = d\mathsf{P}_0/d\mathsf{P}$. □

M22. The Shift Operator. Let $R^{+\infty}$ be the space of doubly infinite sequences $x = (\ldots, \xi_{-1}(x), \xi_0(x), \xi_1(x), \ldots)$ of real numbers; the ξ_k are the coordinate variables. Take $\mathcal{R}_0^{+\infty}$ to be the field of finite-dimensional sets (cylinders), and let $\mathcal{R}^{+\infty} = \sigma(\mathcal{R}_0^{+\infty})$. Let T be the (left) shift operator, defined by the equation $\xi_n(Tx) = \xi_{n+1}(x)$. If θ is a function on $R^{+\infty}$, define θT as the function with value $\theta(Tx)$ at x; for example, $\xi_n T = \xi_{n+1}$.

Define R^∞, $\mathcal{R}^\infty$, and $\mathcal{R}_0^\infty$ as in Example 1.2: $x = (\xi'_1(x), \xi'_2(x), \ldots)$. Let T' be the shift operator here: $\xi'_n(T'x) = \xi'_{n+1}(x)$ for $n \ge 1$.

Any stochastic process $\ldots \eta_{-1}, \eta_0, \eta_1, \ldots$ can be realized on the space $(R^{+\infty}, \mathcal{R}^{+\infty})$, in the sense that, by Kolmogorov's existence theorem, there is a probability measure P_η on $\mathcal{R}^{+\infty}$ under which the coordinate process $\{\xi_n\}$ has the same finite-dimensional distributions as $\{\eta_n\}$. Since all convergence-in-distributions results depend only on the finite-dimensional distributions, $\{\eta_n\}$ and $\{\xi_n\}$ are equivalent for our purposes. Now $\{\eta_n\}$ is stationary if and only if T preserves P_η on $R^{+\infty}$: $\mathsf{P}_\eta T^{-1} = \mathsf{P}_\eta$. And by definition, $\{\eta_n\}$ is ergodic if T is ergodic under P_η. Similarly, a one-sided process $(\eta_1, \eta_2, \ldots)$ can be realized on $(R^\infty, \mathcal{R}^\infty, \mathsf{P}'_\eta)$ for the appropriate P'_η; the original process is stationary if and only if T' preserves P'_η and is by definition ergodic if T' is ergodic under P_η.

But if $(\eta_1, \eta_2, \ldots)$ is stationary, it can also be represented on the space $(R^{+\infty}, \mathcal{R}^{+\infty}, \mathsf{P}_\eta)$ if, under P_η, $(\xi_u, \ldots, \xi_v)$ has the distribution of $(\eta_1, \ldots, \eta_{1+v-u})$ for all u and v. Thus a stationary one-sided process can be extended to a stationary two-sided process with the same finite-dimensional distributions. And T, T', and the two processes are all ergodic or none is. This is because ergodicity depends only on the finite-dimensional distributions: For example, if T is ergodic under P, then it follows by the ergodic theorem and the bounded convergence theorem that

$$(39) \qquad \frac{1}{n}\sum_{k=1}^{n} \mathsf{P}(A \cap T^{-k}B) \to \mathsf{P}(A)\mathsf{P}(B)$$

for all A and B in $\mathcal{R}^{+\infty}$. Conversely, this implies ergodicity: If $A = B$ is invariant, then $\mathsf{P}A$ is 0 or 1. And (39) holds in general if it holds for A and B in $\mathcal{R}_0^{+\infty}$: Given A and B in $\mathcal{R}^{+\infty}$ and an ϵ, choose sets A_ϵ and B_ϵ in $\mathcal{R}_0^{+\infty}$ in such a way that $\mathsf{P}(A \Delta A_\epsilon) < \epsilon$ and $\mathsf{P}(B \Delta B_\epsilon) < \epsilon$, and prove (39) by approximation.

If $\mathsf{E}[\xi_n \| \xi_1, \ldots, \xi_{n-1}] = 0$, so that $(\xi_1, \xi_2, \ldots)$ is a stationary martingale difference, then the two-sided extension satisfies the condidtion $\mathsf{E}[\xi_n \| \xi_{n-k}, \ldots, \xi_{n-1}] = 0$ for each $k \geq 1$, and it follows by a standard convergence theorem [PM.470] that $\mathsf{E}[\xi_n \| \ldots, \xi_{n-2}, \xi_{n-1}] = 0$.

If $\mathcal{F}$ is a σ-field in $R^{+\infty}$, let $T^{-1}\mathcal{F}$ be the σ-field of sets $T^{-1}A$ for $A \in \mathcal{F}$. Suppose that P is preserved by T. If θ (measurable $\mathcal{R}^{+\infty}$) is integrable and $A \in \mathcal{F}$, then, by change of variable,

$$\int_{T^{-1}A} \mathsf{E}[\theta T \| T^{-1}\mathcal{F}] d\mathsf{P} = \int_{T^{-1}A} \theta T \, d\mathsf{P} = \int_A \theta \, d\mathsf{P} = \int_A \mathsf{E}[\theta \| \mathcal{F}] \, d\mathsf{P}$$
$$= \int_{T^{-1}A} \mathsf{E}[\theta \| \mathcal{F}] T \, d\mathsf{P},$$

and therefore,

$$(40) \qquad \mathsf{E}[\theta \| \mathcal{F}]T = \mathsf{E}[\theta T \| T^{-1}\mathcal{F}].$$

If $\mathcal{F}_n = \sigma[\xi_k : k \leq n]$, then $T^{-1}\mathcal{F}_n = \mathcal{F}_{n+1}$, and if the coordinate variables ξ_i are integrable, then (40) gives

$$(41) \qquad \mathsf{E}[\xi_i \| \mathcal{F}_n]T = \mathsf{E}[\xi_{i+1} \| \mathcal{F}_{n+1}].$$

In particular,

$$(42) \qquad \mathsf{E}[\xi_0 \| \mathcal{F}_{-1}]T^n = \mathsf{E}[\xi_n \| \mathcal{F}_{n-1}].$$

For any θ, $\{\theta T^n : n = 0, \pm 1, \ldots\}$ is a stationary stochastic process, and [PM.495] if T is ergodic, so is this new process. Thus $\{\mathsf{E}[\xi_n \| \mathcal{F}_{n-1}]\}$ and $\{\mathsf{E}[\xi_n^2 \| \mathcal{F}_{n-1}]\}$ (for ξ_n^2 integrable) are ergodic processes.

M23. Uniform Mixing. Let $\| \cdot \|$ stand for the L^2 norm, and write $\xi \in \mathcal{A}$ if ξ is $\mathcal{A}$-measurable. There are three standard measures of dependence between σ-fields $\mathcal{A}$ and $\mathcal{B}$:

$$\alpha(\mathcal{A}, \mathcal{B}) = \sup[|\mathsf{P}(A \cap B) - \mathsf{P}A \cdot \mathsf{P}B| : A \in \mathcal{A},\ B \in \mathcal{B}], \tag{43}$$

$$\begin{aligned} \rho(\mathcal{A}, \mathcal{B}) = \sup[|\mathsf{E}[\xi\eta]| : \xi \in \mathcal{A},\ \mathsf{E}\xi = 0,\ \|\xi\| \le 1, \\ \eta \in \mathcal{B},\ \mathsf{E}\eta = 0,\ \|\eta\| \le 1], \end{aligned} \tag{44}$$

$$\varphi(\mathcal{A}, \mathcal{B}) = \sup[|\mathsf{P}(B|A) - \mathsf{P}B| : A \in \mathcal{A},\ \mathsf{P}A > 0,\ B \in \mathcal{B}]. \tag{45}$$

These measures are successively stronger; specifically,

$$\alpha(\mathcal{A}, \mathcal{B}) \le \rho(\mathcal{A}, \mathcal{B}) \le 2\sqrt{\varphi(\mathcal{A}, \mathcal{B})}. \tag{46}$$

To prove the left-hand inequality, simply take $\xi = I_A - \mathsf{P}A$ and $\eta = I_B - \mathsf{P}B$.

The right-hand inequality in (46) is harder to prove. It is enough to consider simple random variables $\xi = \sum_i u_i I_{A_i}$ and $\eta = \sum_j v_j I_{B_j}$, where $\{A_i\}$ and $\{B_j\}$ are finite decompositions of Ω into $\mathcal{A}$-sets and $\mathcal{B}$-sets, respectively. By Schwarz's inequality

$$\begin{aligned} |\mathsf{E}[\xi\eta]| &= \Big| \sum_i u_i \mathsf{P}^{1/2} A_i \Big[\mathsf{P}^{1/2} A_i \sum_j v_j (\mathsf{P}(B_j|A_i)\Big] \Big| \\ &\le \|\xi\| \Big\{ \sum_i \mathsf{P}A_i \,\Big|\, \sum_j v_j (\mathsf{P}(B_j|A_i) \Big|^2 \Big\}^{1/2}. \end{aligned}$$

Hence it is enough to prove that

$$\sum_i \mathsf{P}A_i \,\Big|\, \sum_j v_j \mathsf{P}(B_j|A_i) \Big|^2 \le 2^2 \varphi(\mathcal{A}, \mathcal{B}) \mathsf{E}[\eta^2]. \tag{47}$$

For each i, Schwarz's inequality gives (we assume $\mathsf{E}\xi = \mathsf{E}\eta = 0$)

$$\begin{aligned} \sum_j v_j \mathsf{P}(B_j|A_i) &= \sum_j v_j (\mathsf{P}(B_j|A_i) - \mathsf{P}B_j) \\ &\le \Big\{ \sum_j v_j^2 |\mathsf{P}(B_j|A_i) - \mathsf{P}B_j| \Big\}^{1/2} \Big\{ \sum_j |\mathsf{P}(B_j|A_i) - \mathsf{P}B_j| \Big\}^{1/2}. \end{aligned}$$

Since

$$\sum_i \mathsf{P}A_i \sum_j v_j^2 |\mathsf{P}(B_j|A_i) - \mathsf{P}B_j| \le 2\|\eta\|^2,$$

(47) will follow if we show that

$$\sum_j |\mathsf{P}(B_j|A_i) - \mathsf{P}B_j| \le 2\varphi(\mathcal{A}, \mathcal{B})$$

holds for each i. If C_i^+ $[C_i^-]$ is the union of those B_j for which the difference $\mathsf{P}(B_j|A_i) - \mathsf{P}B_j$ is positive [nonpositive], then C_i^+ and C_i^- lie in $\mathcal{B}$, and therefore,

$$\sum_j |\mathsf{P}(B_j|A_i) - \mathsf{P}B_j| = [\mathsf{P}(C_i^+|A_i) - \mathsf{P}C_i^+] + [\mathsf{P}C_i^- - \mathsf{P}(C_i^-|A_i)]$$
$$\le 2\varphi(\mathcal{A}, \mathcal{B}).$$

This proves the right-hand inequality in (46). (It is easy to strengthen the inequality between the first and third members of (46) to $\alpha(\mathcal{A}, \mathcal{B}) \le \varphi(\mathcal{A}, \mathcal{B})$.)

Finally, if η is $\mathcal{B}$-measurable and has a second moment, then

$$\|\mathsf{E}[\eta\|\mathcal{A}] - \mathsf{E}[\eta]\| \le \rho(\mathcal{A}, \mathcal{B})\|\eta - \mathsf{E}[\eta]\|. \tag{48}$$

We may assume that $\mathsf{E}[\eta] = 0$. By the definition (44), $|\mathsf{E}[\eta\|\mathcal{A}]\eta]| \le \rho(\mathcal{A}, \mathcal{B})\|\mathsf{E}[\eta\|\mathcal{A}]\| \cdot \|\eta\|$. But the left side here is $|\mathsf{E}[\mathsf{E}[\eta\|\mathcal{A}]\eta\|\mathcal{A}]]| = |\mathsf{E}[(\mathsf{E}[\eta\|\mathcal{A}])^2]| = \|\mathsf{E}[\eta\|\mathcal{A}]\|^2$, from which (48) now follows.

M24. The Gauss-Kuzmin-Lévy Theorem. The theorem is this:

Theorem. *Suppose that f_0 has two continuous derivatives on $[0, 1]$ and that $f_0(0) = 0$ and $f_0(1) = 1$. Define $f_1, f_2, \ldots$ by the recursion*

$$f_{n+1}(t) = \sum_{j=1}^{\infty} \left[f_n\left(\frac{1}{j}\right) - f_n\left(\frac{1}{j+t}\right) \right]. \tag{49}$$

Then, for $n \ge 0$ and $0 \le t \le 1$,

$$f_n'(t) = \frac{1}{\log 2} \frac{1}{1+t} + \delta_n(t), \tag{50}$$

and

$$f_n(t) = \frac{\log(1+t)}{\log 2} + \delta_n^\circ(t). \tag{51}$$

where $|\delta_n(t)| \vee |\delta_n^\circ(t)| \le K\theta^n$, $K > 0$, $0 < \theta < 1$.[†]

This can be used for a simple proof of the uniform mixing condition (19.38). Let

$$A = [x: a_i(x) = \alpha_i,\ 1 \le i \le k], \tag{52}$$

and define $f_n(t)$ as the conditional probability $\mathsf{P}([x: T^{k+n}x \le t]|A)$, where P is Gauss's measure (19.22). Since

$$[x: T^{k+n+1}x \le t] = \bigcup_{j=1}^{\infty}\left[x: \frac{1}{j+t} \le T^{k+n}x \le \frac{1}{j}\right],$$

it follows by countable additivity that the f_n satisfy (49). Let $\alpha(t)$ $[\beta(t)]$ be the smaller [larger] of $1\rfloor\overline{\alpha_1} + \cdots + 1\rfloor\overline{\alpha_k}$ and $1\rfloor\overline{\alpha_1} + \cdots + 1\rfloor\overline{\alpha_k + t}$. Then, since $x = 1\rfloor\overline{a_1(x)} + \cdots + 1\rfloor\overline{a_k(x) + T^k x}$, $A \cap [x: T^k x \le t]$ is the set of irrationals in $[\alpha(t), \beta(t)]$. Therefore,

$$f_0(t) = \frac{\log(1+\beta(t)) - \log(1+\alpha(t))}{\log 2 \cdot \mathsf{P}A},$$

and f_0 satisfies the conditions of the Gauss-Kuzmin-Lévy theorem.

Since $T^{k+n-1}x = 1\rfloor\overline{a_{k+n}(x)} + 1\rfloor\overline{a_{k+n+1}(x)} + \cdots$, it follows by the uniqueness of the partial quotients in an expansion that the general element of the σ-field $\mathcal{G}_{k+n} = \sigma(a_{k+n}, a_{k+n+1}, \ldots)$ has the form $B = [x: T^{k+n-1}x \in H]$ for $H \in \mathcal{R}^1$. Therefore, by (50),

$$\mathsf{P}(B|A) = \int_H f'_{n-1}(t)\,dt = \mathsf{P}H + \bar{\delta}_{n-1},$$

where $|\bar{\delta}_{n-1}| \le K\theta^{n-1}$. But $\mathsf{P}H = \mathsf{P}B$ by stationarity, and it follows that

$$|\mathsf{P}(A \cap B) - \mathsf{P}A \cdot \mathsf{P}B| \le K\theta^{n-1}\mathsf{P}A \tag{53}$$

holds if A has the form (52) and $B \in \mathcal{F}_{k+n}$. The class of sets A satisfying (53) is closed under the formation of countable disjoint unions, and each element of $\mathcal{H}_k = \sigma(a_1, \ldots, a_k)$ is such a union of sets (52). Therefore, (53) holds for $A \in \mathcal{H}_k$ and $B \in \mathcal{G}_{k+n}$, which proves (19.38).

[†] See Rockett & Szüsz [58], p. 152 for the proof. They do not give (50) (which obviously implies (51)) as part of the statement of the theorem, but it follows from their method of proof.

M25. Normal Tails. For $x > 0$,

$$\int_x^\infty e^{-u^2/2}du \le \int_x^\infty \frac{u}{x}e^{-u^2/2}du = \frac{1}{x}e^{-x^2/2}.$$

Therefore, for $x > 0$,

$$\mathsf{P}[N \ge x] \le \frac{1}{\sqrt{2\pi}}\frac{1}{x}e^{-x^2/2}. \tag{54}$$

Also, for $x \ge 0$,

$$\mathsf{P}[N \ge x] \le e^{-x^2/2}. \tag{55}$$

This obviously follows from (54) if $x \ge 1/\sqrt{2\pi}$. In the opposite case, we have $\mathsf{P}[N \ge x] \le \frac{1}{2} < e^{-1/4\pi} \le e^{-x^2/2}$.

For an inequality in the opposite direction, assume that $0 \le x < y$. Since the normal density exceeds $e^{-y^2/2}/\sqrt{2\pi}$ on (x, y), we have

$$\mathsf{P}[x < N < y] \ge \frac{y-x}{\sqrt{2\pi}}e^{-y^2/2}. \tag{56}$$

If $[B(s): 0 \le s \le t]$ is a Brownian motion and $\alpha \ge 0$, then, by symmetry and the reflection principle (see (8.20)),

$$\mathsf{P}[\sup_{s\le t}|B(s)| \ge \alpha] = 2\mathsf{P}[\sup_{s\le t} B(s) \ge \alpha] = 4\mathsf{P}[N \ge \alpha/\sqrt{t}].$$

Combine this with (54) to get

$$\mathsf{P}[\sup_{s\le t}|B(s)| \ge \alpha] \le \frac{4}{\sqrt{2\pi}}\frac{1}{\alpha/\sqrt{t}}e^{-\alpha^2/2t}, \quad \alpha > 0. \tag{57}$$

Combine it with (55) to get

$$\mathsf{P}[\sup_{s\le t}|B(s)| \ge \alpha] \le 4e^{-\alpha^2/2t}, \quad \alpha \ge 0. \tag{58}$$

SOME NOTES ON THE PROBLEMS

1.11. Given ϵ and a compact K, cover K by finitely many balls $B(x_i, r_i)$, $i \le k$, for which $x_i \in K$, $r_i < \epsilon/2$, and $B(x_i, r_i)^-$ is compact; take $G = \bigcup_i B(x_i, r_i)$ and $K_1 = G^-$. Choose δ so that $\delta < \epsilon/2$ and $\rho(K, G^c) > \delta$. Then $K \subset K^\delta \subset K_1 \subset K^\epsilon$. If $f(x) = (1 - \rho(x, K)/\delta)^+$, then $I_K \le f \le I_{K_1}$, and hence the uniformly continuous f has compact support. If $Pf = Qf$, then $PK \le Pf = Qf \le QK_1 \le QK^\epsilon$. It follows that $PK \le QK$ and (symmetry) $PK = QK$.

1.17. Suppose that L is locally compact and dense in the open ball B, and arrange that $B^- \cap L$ is compact; choose points x_n of B so that $\rho(x_m, x_n) \ge \epsilon > 0$ for $m \ne n$, then choose points y_n of $B^- \cap L$ so that $\rho(y_n, x_n) < \epsilon/3$, and conclude that $B^- \cap L$ is *not* compact.

1.18. Take $T = (0, 1]$. Let X be a Hamel basis for $[1, \infty)$ that contains 1 [PM.198], and take U to consist of the reciprocals of the elements of $X \backslash \{1\}$. Consider the elements of $C_b(T)$ defined by $x_u(t) = \sin(2\pi/ut)$ for $u \in U$. Choose α and β so that $0 < \alpha/2\pi < \beta/2\pi < 1/4$. Then, for each distinct u and v in U, there is (Hardy & Wright [36], Theorem 443) a t in T such that $(\{1/ut\}, \{1/vt\}) \in (0, \alpha/2\pi) \times (\beta/2\pi, \frac{1}{4})$, where the brackets denote fractional part. But then we have $|x_u(t) - x_v(t)| \ge \sin\beta - \sin\alpha > 0$. The $\|x_u - x_v\|$ are therefore bounded away from 0.

2.8. If $\{x_n\}$ contains a subsequence converging to x, then $P = \delta_x$ is easy to prove. Suppose on the other hand that $\{x_n\}$ contains no convergent subsequence. Then, for each k, $\{x_n\}$ contains no subsequence converging to x_k, hence there is an r_k such that $x_n \notin B(x_k, r_k)$ for infinitely many n, and hence $P(B(x_k, r_k)) \le \liminf_n \delta_{x_n}(B(x_k, r_k)) = 0$. But if F consists of the x_k and $G = F^c$, then F is closed (no sequence in F converges) and G is open, so that $PG \le \liminf_n \delta_{x_n} G = 0$. But now $S = G \cup \bigcup_k B(x_k, r_k)$, a countable union of sets of P-measure 0.

2.9. Let $\mathcal{F}_P$ be the field (Problem 2.5) of P-continuity sets. If F is closed, then $F^\epsilon \downarrow F$ and $F^\epsilon \in \mathcal{F}_P$ for all but countably many ϵ. The field $\mathcal{F}_P$ therefore generates $\mathcal{S}$, and [PM.169] for each A in $\mathcal{S}$ and each ϵ, $P(A\Delta B) < \epsilon$ for some $B \in \mathcal{F}_P$. But $P(B^- - B) = 0$: For each A in $\mathcal{S}$ and each ϵ, $P(A\Delta F) < \epsilon$ for some closed F. If $P|f| < \infty$, there is, for given ϵ, a simple $h_1 = \sum_{i=1}^k t_i I_{A_i}$ $(t_i \neq 0)$ such that $P|f - h_1| < \epsilon$. Choose closed sets F_i such that $P(A_i \Delta F_i) < \epsilon/|t_i|k$ and set $h_2 = \sum_{i=1}^k t_i I_{F_i}$; then $P|f - h_2| < 2\epsilon$. Now choose δ so that $PF_i \le PF_i^\delta < PF_i + \epsilon/|t_i|k$ for each i. Let $g_i(x) = (1 - \rho(x, F_i)/\delta)^+$ and $g = \sum t_i g_i$. Show that $P|f - g| < 3\epsilon$.

3.8. It would follow that $(S_{2n} - S_n)/\sqrt{n} = \sqrt{2}S_{2n}/\sqrt{2n} - S_n/\sqrt{n}$ converges in distribution to N and in probability to $(\sqrt{2} - 1)\eta$. But then η and $(\sqrt{2} - 1)\eta$ would have the same distribution as N, which is impossible.

4.2. Assume that $X^n \Rightarrow X$ (on Δ). To show that $(\hat{X}_1^n, \hat{X}_2^n) \Rightarrow (\hat{X}_1, \hat{X}_2)$, define $p_{ij}(x, y)$ as $xy/(1 - x)$ if $x \neq 1$; if $x = 1$, take it to be 1 or 0 as $j = i + 1$ or not. Let J consist of the pairs of distinct positive integers and let J_k consist of the (i, j) in J for which $i, j \le k$. Suppose that f is bounded and continuous on R^2. By Theorem 3.5, it will be enough to show that

$$
\begin{aligned}
(1) \qquad f(\hat{X}_1^n, \hat{X}_2^n) &= \sum\nolimits_J p_{ij}(X_i^n, X_j^n) f(X_i^n, X_j^n) \\
&\Rightarrow \sum\nolimits_J p_{ij}(X_i, X_j) f(X_i, X_j) = f(\hat{X}_1^n, \hat{X}_2).
\end{aligned}
$$

That this holds if J is replaced by J_k is clear. From $\sum_J p_{ij}(X_i^n, X_j^n) = \sum_J p_{ij}(X_i, X_j) = 1$, it follows that $\sum_{J_k^c} p_{ij}(X_i^n, X_j^n) \Rightarrow \sum_{J_k^c} p_{ij}(X_i, X_j)$. Therefore, for given ϵ and η, there exist k and n_0 such that $n \ge n_0$ implies $\mathsf{P}[\sum_{J_k^c} p_{ij}(X_i^n, X_j^n) > \epsilon] < \eta$. Now use the fact that f is bounded to derive (1).

4.3. Let G and Π have the GEM and Poisson-Dirichlet distributions, so that $\rho G =_d \Pi$. In the argument leading to Theorem 4.4, we showed that $\sigma Z^n \Rightarrow G$ and concluded by Theorem 4.1 that $\rho Z^n = \rho\sigma Z^n \Rightarrow \rho G =_d \Pi$. But now (see the preceding problem) $\sigma\rho Z^n \Rightarrow \sigma\Pi$ and $\sigma\rho Z^n =_d \sigma Z^n \Rightarrow G$, so that $\sigma\Pi =_d G$. And $\sigma G =_d \sigma\sigma\Pi =_d \sigma\Pi =_d G$.

5.12. **(a)** The class of Borel sets in Q has the power of the continuum [PM.36]. **(d)** If f is bounded and uniformly continuous, then

$$
|P_{x_n} f - P_x f| \le \int_0^1 |f(x_n, y) - f(x, y)|\, dy \le w_f(|x_n - x|) \to 0,
$$

where w_f is the modulus of continuity.

6.4. Consider the functions $\rho'(x, \cdot)$.

6.5. It is easy to show that $P_n \to P$ (weak*) implies $P_n \Rightarrow^\circ P_0$. Suppose now that $P_n \Rightarrow^\circ P_0$ and that f is bounded and continuous (and hence $\mathcal{S}$-measurable). If we can, for given η, construct a bounded, continuous, $\mathcal{S}_0$-measurable g such that $f \le g$ and $Pg \le Pf + \eta$, then $P_n \to P$ (weak*) will follow without difficulty. Assume that $0 < f < 1$, and choose points t_i such that $0 = t_0 < \cdots < t_k = 1$, $t_i - t_{i-1} < \eta$, and $P[f = t_i] = 0$. If $F_i = [t_{i-1} \le f \le t_i]$, then $f \le \sum_i t_i I_{F_i}$ and $\sum_i t_i PF_i \le Pf + \eta$. Choose ϵ so that $\sum_i t_i P(F_i^\epsilon) < \sum_i t_i PF_i + \eta$. Define g_i by (6.1) with F_i in place of F (M supports P), and take $g = \sum_i t_i g_i$. Then $f \le g$ and $P_g \le \sum_i t_i P(((F_i^\epsilon)^c \cap M)^c) = \sum_i t_i P(F_i^\epsilon) < \sum_i t_i PF_i + \eta \le Pf + 2\eta$.

7.2. See Dudley [23], p. 40.

8.1. Show that every locally compact subset of $C_0 = [x \in C : x(0) = 0]$ is nowhere dense and that each ball in C_0 has positive Wiener measure.

BIBLIOGRAPHICAL NOTES

Since this second edition is more a textbook than a research monograph, I think few historical notes are needed. In [6] (the first edition of this book) there are some notes of this kind; for more extensive accounts of the history, see Dudley [23], Ethier & Kurtz [26], Hall & Heyde [35], Jacod & Shiryaev [39], and Pollard [53].

Books devoted pimarily to weak convergence are Davidson [15] (for students of economics), Ethier & Kurtz [26], Jacod & Shiryaev [39], Parthasarathy [48], Pollard [53], Shorack & Wellner [60], and van der Waart & Wellner [65]. Books partly devoted to weak convergence are Dudley [23], Durrett [24], Hall & Heyde [35], Karatzas & Shreve [40], and Strook & Varadhan [63].

Section 4. Vervaat [66], p. 90, shows that (4.13) has (4.14) as its Laplace transform and connects it with limit theory for record values. Griffiths [33], p. 145, gives the joint moments for the density (4.15); this generalizes (4.17). For a different approach to Theorem 4.4, see Arratia, Barbour, & Tavaré [4]. For the case of cycle lengths ($\theta = 1$), the result is due to Shepp & Lloyd [59]. Theorem 4.5 was proved in [8]; the proof given here is that of Donnelly & Grimmet [17]. For the proof of Theorem 4.5 itself, the argument in [8] is the most straightforward, but the Donnelly-Grimmet approach is better because it puts the result in a very general and interesting context; see Arratia, Barbour, & Tavaré [3] and [4], and also Arratia [2]. For more on computation, see Griffiths [33] and [34]. The arguments at the end of Section 4 are laborious but require only calculus.

Section 5. Prohorov's fundamental paper is [56]. The proof of Theorem 5.1 given here is from [7]; for a similar construction on locally compact spaces, see Halmos [36], p. 231. The construction in Problem 5.12 is due to Léger; see Choquet [13]. In this connection, see also Preiss [55].

Section 6. "Weak°ly" can be pronounced "weak-circly."

Section 7. The second proof of Theorem 7.5 is due to Wichura [67].

Section 10. The maximal inequalities of this section are from [6]. The proofs here, much easier than those in [6], are due to Michael Wichura. See Bickel & Wichura [5] for extensions to multidimensional time.

Section 11. Theorem 11.1 is from a manuscript version of [6]; it was cut to keep the book short. For generalizations, see Gonzáles Villalobos [30]. For a different approach to functional limit theorems for lacunary series, see Philipp & Stout [52].

Section 12. Skorohod's basic paper is [61].

Section 14. The proofs here all follow the same pattern: We prove weak convergence of the finite-dimensional distributions and then verify tightness. Some patterns of proof circumvent the tightness issue. An application of Theorem 12.6, for example, requires no consideration of tightness, since that is effectively built into its hypothesis; see the proof of Theorem 17.3. Theorem 19.4 of [6] is another example. For a systematic treatment of this approach, see Ethier & Kurtz [26].

Section 15. Theorem 15.1 is from Dudley [20] and Theorem 15.2 from Bolthausen [11]. The proof in [11] uses the theory of analytic sets, but as the proof here shows, this is unnecessary. The theorem on the size of ϵ-nets [M18] is due to Dudley [21].

Section 16. The development is based on Lindvall [43] and Aldous [1].

Section 17. Theorem 17.2 is another result cut from [6] for reasons of space. For a different approach to these problems, see Philipp [49]. De Koninck and Galambos [16] have a result related to Theorem 17.3; they prove convergence of the finite-dimensional distributions but do not formulate a functional limit theorem.

Section 18. For statistical applications of martingale limit theorems, see Hall & Heyde [35].

Section 19. In [6], thirty pages (Sections 20 and 21) were necessary to cover what is achieved very simply and efficiently in this section by the use of Gordin's method of approximating stationary processes by martingales; see Gordin [31]. Although convergence to diffusions was briefly treated in [6] (see the comments on Section 14 above), it has become a subject of its own, and I omit any account of it here. See Ethier & Kurtz [26] for a systematic treatment; a good place to start is Durrett [24]. I have cut Section 22 of [6], on empirical distribution functions for dependent random variables.

Section 20. For a survey of various kinds of invariance principles, see Philipp [51]. Theorem 20.2 can be found in Breiman [12], p. 279, and in Freedman [29], p. 83.

Section 21. The second proof of Theorem 21.1 was shown to me by Walter Philipp; it is based on [50].

Section 22. The proof is from Strassen [62] and (on several points) Freedman [29].

BIBLIOGRAPHY

1. David Aldous: Stopping times and tightness, *Ann. Probab.* 6(1978) 335–340.
2. Richard Arratia: On the central role of scale invariant Poisson processes on $(0, \infty)$, *Discrete Math. Theoret. Comput. Sci.*, 41(1998) 21–41.
3. Richard Arratia, A.D. Barbour, and Simon Tavaré: Random combinatorial structures, *Notices Amer. Math. Soc.*, 44(1997)903–910.
4. Richard Arratia, A.D. Barbour, and Simon Tavaré: *Logarithmic Combinatorial Structures*, book in preparation.
5. P.J. Bickel and M.J. Wichura: Convergence criteria for multiparameter stochasic processes and some applications, *Ann. Math. Statist.*, 42(1971)1656–1670.
6. Patrick Billingsley: *Convergence of Probability Measures*, first edition, New York, Wiley, 1968.
7. Patrick Billingsley: *Weak Convergence of Measures: Applications in Probability*, Philadelphia, SIAM, 1971.
8. Patrick Billingsley: On the distribution of large prime divisors, *Periodica Mathematica Hungarica*, 2(1972)283–289.
9. Patrick Billingsley: *Probability and Measure*, third edition, New York, Wiley, 1995.
10. Harald Bohr: *Almost Periodic Functions*, New York, Chelsea, 1947.
11. Erwin Bolthausen: Weak convergence of an empirical process indexed by the closed convex subsets of I^2, *Z. Wahrscheinlichkeitstheorie* 43(1978)173–181.
12. Leo Breiman: *Probability*, Reading, Addison-Wesley, 1968.
13. G. Choquet: Sur les ensembles uniformément négligeables, *Sem. Choquet (Initiation à l'analyse)*, Paris, 9ᵉ année (1969/1970) no. 6.

14. J. Czipszer and L. Gehér: Extension of functions satisfying a Lipschitz condition, *Acta Math. Acad. Sci. Hungar.* 6(1955)213–220.

15. James Davidson: *Stochastic Limit Theory*, New York, Oxford University Press, 1994.

16. J.-M. De Konink and J. Galambos: The intermediate prime divisors of integers, *Proc. Amer. Math. Soc.* 101(1987)213–216.

17. Peter Donnelly and Geoffrey Grimmett: On the asymptotic distribution of large prime factors, *J. London Math. Soc.* (2), 47(1993) 395–404.

18. M. Donsker: An invariance principle for certain probability limit theorems, *Mem. Amer. Math. Soc.* 6(1951).

19. J.L. Doob: *Stochastic Processes*, New York, Wiley, 1953.

20. R.M.Dudley: Weak convergence of probabilities on nonseparable metric spaces and empirical measures on Euclidean spaces, *Illinois J. Math.* 10(1966)109–126.

21. R.M. Dudley: Metric entropy of some classes of sets with differentiable boundaries, *J. Approximation Th.* 10(1974)227-236; Correction, ibid. 26(1979)192–193.

22. R.M. Dudley: *A Course on Empirical Processes*, Springer Lectures Notes 1097(1984), New York, Springer.

23. Richard M. Dudley: *Real Analysis and Probability*, Pacific Grove, Wadsworth and Brooks/Cole, 1989.

24. Richard Durrett: *Stochastic Calculus*, Boca Raton, CRC Press, 1996.

25. H.G. Eggleston: *Convexity*, Cambridge, The University Press, 1958.

26. Stewart N. Ethier and Thomas G. Kurtz: *Markov Processes: Characterization and Convergence*, New York, Wiley, 1986.

27. W.J. Ewens: *Mathematical Population Genetics*, New York, Springer, 1979.

28. William Feller: *An Introduction to Probability Theory and Its Applications*, third edition, New York, Wiley, 1968.

29. David Freedman: *Brownian Motion and Diffusion*, San Francisco, Holden-Day, 1971.

30. Alvaro Gonzáles Villalobos: Some functional limit theorems for dependent random variables, *Ann. Probab.* 6(1974)1090–1107.

31. M.I. Gordin: The central limit theorem for stationary processes, *Soviet Math. Dokl.* 10(1969)1174-1176.

32. M.I. Gordin: On the behavior of the variances of sums of random variables forming a stationary process, *Theory Probab. Appl.* 16(1971)474–484.

33. R.C. Griffiths: On the distribution of allele frequencies in a diffusion model, *Theor. Pop. Biol.*, 15(1979)140–158.

34. R.C. Griffiths: On the distribution of points in a Poisson Dirichlet process, *J. Appl. Prob.*, 25(1988)336–345.

35. P. Hall and C.C. Heyde: *Martingale Limit Theory and Its Application*, New York, Academic Press, 1980.

36. Paul R. Halmos: *Measure Theory*, New York, Van Nostrand, 1950.

37. G.H. Hardy and E.M. Wright: *An Introduction to the Theory of Numbers*, fourth edition, London, Oxford University Press, 1960.

38. I.A. Ibragimov and Yu.V. Linnik: *Independent and Stationary Sequences of Random Variables*, Wolters-Noordhoff, Groningen, 1971.

39. Jean Jacod and Albert N. Shiryaev: *Limit Theorems for Stochastic Processes*, Berlin, Springer, 1987.

40. Ioannis Karatzas and Steven E. Shreve: *Brownian Motion and Stochastic Calculus*, New York, Springer, 1988.

41. John L. Kelley: *General Topology*, Princeton, Van Nostrand, 1955.

42. A. Ya. Khinchine: *Continued Fractions*, third edition (English translation), Chicago, University of Chicago Press, 1964.

43. Torgny Lindvall: Weak convergence in the function space $D[0,\infty)$, *J. Appl. Probab.* 10(1973)109–121.

44. L.A. Liusternik and V.J. Sobolev: *Elements of Functional Analysis*, New York, Frederick Ungar, 1961.

45. Péter Major: Approximation of partial sums of i.i.d.r.v.s when the summands have only two moments, *Z. Wahrscheinlichkeitstheorie*, 35(1976)221–229.

46. D. Monrad and W. Philipp: The problem of embedding vector-valued martingales in a Gaussian process, *Theory Probab. Appl.* 35(1990)374–377.

47. A.M. Mark: Some probability theorems, *Bull. Amer. Math. Soc.* 55(1949)885–900.

48. K.R. Parthasarathy: *Probability Measures on Metric Spaces*, New York, Academic Press, 1967.

49. Walter Philipp: Mixing sequences of random variables and probabilistic number theory, *Mem. Amer. Math. Soc.* 114(1971).

50. Walter Philipp: Weak and L^p-invariance principles for sums of B-valued random variables, *Ann. Probab.* 8(1980)68–82; Correction, ibid. 14(1986)1095–1101.

51. Walter Philipp: Invariance principles for independent and weakly dependent random variables, pp. 225–268 in *Dependence in Probability and Statistics*, Ernst Eberlein and Murad Taqqu, editors, Boston, Birkhaüser, 1986.

52. Walter Philipp and William Stout: Almost sure invariance principles for partial sums of weakly dependent random variables, *Mem. Amer. Math. Soc.*, 161(1975).

53. David Pollard: *Convergence of Stochastic Processes*, New York, Springer, 1984.

54. Jean-Pierre Portmanteau: Espoir pour l'ensemble vide? *Annales de l'Université de Felletin*, CXLI(1915)322–325.

55. David Preiss: Metric spaces in which Prohorov's theorem is not valid, *Z. Wahrscheinlichkeitstheorie und Verw. Gebiete* 27(1973)109–116.

56. Yu.V. Prohorov: Convergence of random processes and limit theorems in probability theory, *Theory Probab. Appl.*, 1(1956)157–214.

57. Philip Protter: *Stochastic Integration and Differential Equations*, New York, Springer, 1990.

58. Andrew M. Rockett and Peter Szüsz: *Continued Fractions*, New Jersey, World Scientific, 1992.

59. L.A. Shepp and S.P. Lloyd: Ordered cycle lengths in a random permutation, *Trans. Amer. Math. Soc.* 121(1966)340–357.

60. Galen R. Shorack and Jon A. Wellner: *Empirical Processes with Applications to Statistics*, New York, Wiley, 1986.

61. A.V. Skorohod: Limit theorems for stochastic processes, *Theory Probab. Appl.*, 1(1956)261–290.

62. V. Strassen: An invariance principle for the law of the iterated logarithm, *Z. Wahrscheinlichkeitstheorie und Verw. Gebiete*, 3(1964) 211–226.

63. Daniel W. Strook and S.R. Srinivasa Varadhan: *Multidimensional Diffusion Processes*, New York, Springer, 1979.

64. M. Talagrand: Les boules peuvent-elles engendrer la tribu borélienne d'un espace métrisable non separable? *Séminair Choquet*, Paris, 17e année: C5, 1978.

65. Aad W. van der Waart and Jon A. Wellner: *Weak Convergence and Empirical Processes*, New York, Springer, 1996.

66. W. Vervaat: *Success Epochs in Bernoulli trials,* Mathematical Centre Tracts, No. 41, Amsterdam, 1972.

67. M.J. Wichura: A note on the weak convergence of stochastic processes, *Ann. Math. Statist.* 42(1971)1769–1772.

INDEX

Greek symbols are alphabetized by their representation in roman letters (phi for ϕ, and so on). A reference [PM.n] is to page n of *Probability and Measure*, third edition.

WILEY SERIES IN PROBABILITY AND STATISTICS
ESTABLISHED BY WALTER A. SHEWHART AND SAMUEL S. WILKS

Probability and Statistics Section

*ANDERSON · The Statistical Analysis of Time Series
ARNOLD, BALAKRISHNAN, and NAGARAJA · A First Course in Order Statistics
ARNOLD, BALAKRISHNAN, and NAGARAJA · Records
BACCELLI, COHEN, OLSDER, and QUADRAT · Synchronization and Linearity: An Algebra for Discrete Event Systems
BASILEVSKY · Statistical Factor Analysis and Related Methods: Theory and Applications
BERNARDO and SMITH · Bayesian Statistical Concepts and Theory
BILLINGSLEY · Convergence of Probability Measures, *Second Edition*
BOROVKOV · Asymptotic Methods in Queuing Theory
BOROVKOV · Ergodicity and Stability of Stochastic Processes
BRANDT, FRANKEN, and LISEK · Stationary Stochastic Models
CAINES · Linear Stochastic Systems
CAIROLI and DALANG · Sequential Stochastic Optimization
CONSTANTINE · Combinatorial Theory and Statistical Design
COOK · Regression Graphics
COVER and THOMAS · Elements of Information Theory
CSÖRGŐ and HORVÁTH · Weighted Approximations in Probability Statistics
CSÖRGŐ and HORVÁTH · Limit Theorems in Change Point Analysis
DETTE and STUDDEN · The Theory of Canonical Moments with Applications in Statistics, Probability, and Analysis
DEY and MUKERJEE · Fractional Factorial Plans
*DOOB · Stochastic Processes
DRYDEN and MARDIA · Statistical Analysis of Shape
DUPUIS and ELLIS · A Weak Convergence Approach to the Theory of Large Deviations
ETHIER and KURTZ · Markov Processes: Characterization and Convergence
FELLER · An Introduction to Probability Theory and Its Applications, Volume 1, *Third Edition,* Revised; Volume II, *Second Edition*
FULLER · Introduction to Statistical Time Series, *Second Edition*
FULLER · Measurement Error Models
GHOSH, MUKHOPADHYAY, and SEN · Sequential Estimation
GIFI · Nonlinear Multivariate Analysis
GUTTORP · Statistical Inference for Branching Processes
HALL · Introduction to the Theory of Coverage Processes
HAMPEL · Robust Statistics: The Approach Based on Influence Functions
HANNAN and DEISTLER · The Statistical Theory of Linear Systems
HUBER · Robust Statistics
IMAN and CONOVER · A Modern Approach to Statistics
JUREK and MASON · Operator-Limit Distributions in Probability Theory
KASS and VOS · Geometrical Foundations of Asymptotic Inference

*Now available in a lower priced paperback edition in the Wiley Classics Library.

Probability and Statistics (Continued)

KAUFMAN and ROUSSEEUW · Finding Groups in Data: An Introduction to Cluster Analysis
KELLY · Probability, Statistics, and Optimization
LINDVALL · Lectures on the Coupling Method
McFADDEN · Management of Data in Clinical Trials
MANTON, WOODBURY, and TOLLEY · Statistical Applications Using Fuzzy Sets
MORGENTHALER and TUKEY · Configural Polysampling: A Route to Practical Robustness
MUIRHEAD · Aspects of Multivariate Statistical Theory
OLIVER and SMITH · Influence Diagrams, Belief Nets and Decision Analysis
*PARZEN · Modern Probability Theory and Its Applications
PRESS · Bayesian Statistics: Principles, Models, and Applications
PUKELSHEIM · Optimal Experimental Design
RAO · Asymptotic Theory of Statistical Inference
RAO · Linear Statistical Inference and Its Applications, *Second Edition*
RAO and SHANBHAG · Choquet-Deny Type Functional Equations with Applications to Stochastic Models
ROBERTSON, WRIGHT, and DYKSTRA · Order Restricted Statistical Inference
ROGERS and WILLIAMS · Diffusions, Markov Processes, and Martingales, Volume I: Foundations, *Second Edition;* Volume II: Îto Calculus
RUBINSTEIN and SHAPIRO · Discrete Event Systems: Sensitivity Analysis and Stochastic Optimization by the Score Function Method
RUZSA and SZEKELY · Algebraic Probability Theory
SCHEFFE · The Analysis of Variance
SEBER · Linear Regression Analysis
SEBER · Multivariate Observations
SEBER and WILD · Nonlinear Regression
SERFLING · Approximation Theorems of Mathematical Statistics
SHORACK and WELLNER · Empirical Processes with Applications to Statistics
SMALL and McLEISH · Hilbert Space Methods in Probability and Statistical Inference
STAPLETON · Linear Statistical Models
STAUDTE and SHEATHER · Robust Estimation and Testing
STOYANOV · Counterexamples in Probability
TANAKA · Time Series Analysis: Nonstationary and Noninvertible Distribution Theory
THOMPSON and SEBER · Adaptive Sampling
WELSH · Aspects of Statistical Inference
WHITTAKER · Graphical Models in Applied Multivariate Statistics
YANG · The Construction Theory of Denumerable Markov Processes

Applied Probability and Statistics Section

ABRAHAM and LEDOLTER · Statistical Methods for Forecasting
AGRESTI · Analysis of Ordinal Categorical Data
AGRESTI · Categorical Data Analysis
ANDERSON, AUQUIER, HAUCK, OAKES, VANDAELE, and WEISBERG · Statistical Methods for Comparative Studies
ARMITAGE and DAVID (editors) · Advances in Biometry
*ARTHANARI and DODGE · Mathematical Programming in Statistics
ASMUSSEN · Applied Probability and Queues
*BAILEY · The Elements of Stochastic Processes with Applications to the Natural Sciences
BARNETT and LEWIS · Outliers in Statistical Data, *Third Edition*

*Now available in a lower priced paperback edition in the Wiley Classics Library.

Applied Probability and Statistics (Continued)

BARTHOLOMEW, FORBES, and McLEAN · Statistical Techniques for Manpower Planning, *Second Edition*
BATES and WATTS · Nonlinear Regression Analysis and Its Applications
BECHHOFER, SANTNER, and GOLDSMAN · Design and Analysis of Experiments for Statistical Selection, Screening, and Multiple Comparisons
BELSLEY · Conditioning Diagnostics: Collinearity and Weak Data in Regression
BELSLEY, KUH, and WELSCH · Regression Diagnostics: Identifying Influential Data and Sources of Collinearity
BHAT · Elements of Applied Stochastic Processes, *Second Edition*
BHATTACHARYA and WAYMIRE · Stochastic Processes with Applications
BIRKES and DODGE · Alternative Methods of Regression
BLOOMFIELD · Fourier Analysis of Time Series: An Introduction
BOLLEN · Structural Equations with Latent Variables
BOULEAU · Numerical Methods for Stochastic Processes
BOX · Bayesian Inference in Statistical Analysis
BOX and DRAPER · Empirical Model-Building and Response Surfaces
BOX and DRAPER · Evolutionary Operation: A Statistical Method for Process Improvement
BUCKLEW · Large Deviation Techniques in Decision, Simulation, and Estimation
BUNKE and BUNKE · Nonlinear Regression, Functional Relations and Robust Methods: Statistical Methods of Model Building
CHATTERJEE and HADI · Sensitivity Analysis in Linear Regression
CHILÈS and DELFINER · Geostatistics: Modeling Spatial Uncertainty
CHOW and LIU · Design and Analysis of Clinical Trials: Concepts and Methodologies
CLARKE and DISNEY · Probability and Random Processes: A First Course with Applications, *Second Edition*
*COCHRAN and COX · Experimental Designs, *Second Edition*
CONOVER · Practical Nonparametric Statistics, *Second Edition*
CORNELL · Experiments with Mixtures, Designs, Models, and the Analysis of Mixture Data, *Second Edition*
*COX · Planning of Experiments
CRESSIE · Statistics for Spatial Data, *Revised Edition*
DANIEL · Applications of Statistics to Industrial Experimentation
DANIEL · Biostatistics: A Foundation for Analysis in the Health Sciences, *Sixth Edition*
DAVID · Order Statistics, *Second Edition*
*DEGROOT, FIENBERG, and KADANE · Statistics and the Law
DODGE · Alternative Methods of Regression
DOWDY and WEARDEN · Statistics for Research, *Second Edition*
DRYDEN and MARDIA · Statistical Shape Analysis
DUNN and CLARK · Applied Statistics: Analysis of Variance and Regression, *Second Edition*
ELANDT-JOHNSON and JOHNSON · Survival Models and Data Analysis
EVANS, PEACOCK, and HASTINGS · Statistical Distributions, *Second Edition*
FLEISS · The Design and Analysis of Clinical Experiments
FLEISS · Statistical Methods for Rates and Proportions, *Second Edition*
FLEMING and HARRINGTON · Counting Processes and Survival Analysis
GALLANT · Nonlinear Statistical Models
GLASSERMAN and YAO · Monotone Structure in Discrete-Event Systems
GNANADESIKAN · Methods for Statistical Data Analysis of Multivariate Observations, *Second Edition*
GOLDSTEIN and LEWIS · Assessment: Problems, Development, and Statistical Issues
GREENWOOD and NIKULIN · A Guide to Chi-Squared Testing
*HAHN · Statistical Models in Engineering

*Now available in a lower priced paperback edition in the Wiley Classics Library.

Applied Probability and Statistics (Continued)

HAHN and MEEKER · Statistical Intervals: A Guide for Practitioners
HAND · Construction and Assessment of Classification Rules
HAND · Discrimination and Classification
HEIBERGER · Computation for the Analysis of Designed Experiments
HINKELMAN and KEMPTHORNE: · Design and Analysis of Experiments, Volume 1: Introduction to Experimental Design
HOAGLIN, MOSTELLER, and TUKEY · Exploratory Approach to Analysis of Variance
HOAGLIN, MOSTELLER, and TUKEY · Exploring Data Tables, Trends and Shapes
HOAGLIN, MOSTELLER, and TUKEY · Understanding Robust and Exploratory Data Analysis
HOCHBERG and TAMHANE · Multiple Comparison Procedures
HOCKING · Methods and Applications of Linear Models: Regression and the Analysis of Variables
HOGG and KLUGMAN · Loss Distributions
HOSMER and LEMESHOW · Applied Logistic Regression
HØYLAND and RAUSAND · System Reliability Theory: Models and Statistical Methods
HUBERTY · Applied Discriminant Analysis
JACKSON · A User's Guide to Principle Components
JOHN · Statistical Methods in Engineering and Quality Assurance
JOHNSON · Multivariate Statistical Simulation
JOHNSON and KOTZ · Distributions in Statistics
Continuous Multivariate Distributions
JOHNSON, KOTZ, and BALAKRISHNAN · Continuous Univariate Distributions, Volume 1, *Second Edition*
JOHNSON, KOTZ, and BALAKRISHNAN · Continuous Univariate Distributions, Volume 2, *Second Edition*
JOHNSON, KOTZ, and BALAKRISHNAN · Discrete Multivariate Distributions
JOHNSON, KOTZ, and KEMP · Univariate Discrete Distributions, *Second Edition*
JUREČKOVÁ and SEN · Robust Statistical Procedures: Aymptotics and Interrelations
KADANE · Bayesian Methods and Ethics in a Clinical Trial Design
KADANE AND SCHUM · A Probabilistic Analysis of the Sacco and Vanzetti Evidence
KALBFLEISCH and PRENTICE · The Statistical Analysis of Failure Time Data
KELLY · Reversability and Stochastic Networks
KHURI, MATHEW, and SINHA · Statistical Tests for Mixed Linear Models
KLUGMAN, PANJER, and WILLMOT · Loss Models: From Data to Decisions
KLUGMAN, PANJER, and WILLMOT · Solutions Manual to Accompany Loss Models: From Data to Decisions
KOVALENKO, KUZNETZOV, and PEGG · Mathematical Theory of Reliability of Time-Dependent Systems with Practical Applications
LAD · Operational Subjective Statistical Methods: A Mathematical, Philosophical, and Historical Introduction
LANGE, RYAN, BILLARD, BRILLINGER, CONQUEST, and GREENHOUSE · Case Studies in Biometry
LAWLESS · Statistical Models and Methods for Lifetime Data
LEE · Statistical Methods for Survival Data Analysis, *Second Edition*
LEPAGE and BILLARD · Exploring the Limits of Bootstrap
LINHART and ZUCCHINI · Model Selection
LITTLE and RUBIN · Statistical Analysis with Missing Data
LLOYD · The Statistical Analysis of Categorical Data
MAGNUS and NEUDECKER · Matrix Differential Calculus with Applications in Statistics and Econometrics
MALLER and ZHOU · Survival Analysis with Long Term Survivors
MANN, SCHAFER, and SINGPURWALLA · Methods for Statistical Analysis of Reliability and Life Data

*Now available in a lower priced paperback edition in the Wiley Classics Library.

Applied Probability and Statistics (Continued)

McLACHLAN and KRISHNAN · The EM Algorithm and Extensions
McLACHLAN · Discriminant Analysis and Statistical Pattern Recognition
McNEIL · Epidemiological Research Methods
MEEKER and ESCOBAR · Statistical Methods for Reliability Data
MILLER · Survival Analysis
MONTGOMERY and PECK · Introduction to Linear Regression Analysis, *Second Edition*
MYERS and MONTGOMERY · Response Surface Methodology: Process and Product in Optimization Using Designed Experiments
NELSON · Accelerated Testing, Statistical Models, Test Plans, and Data Analyses
NELSON · Applied Life Data Analysis
OCHI · Applied Probability and Stochastic Processes in Engineering and Physical Sciences
OKABE, BOOTS, and SUGIHARA · Spatial Tesselations: Concepts and Applications of Voronoi Diagrams
PANKRATZ · Forecasting with Dynamic Regression Models
PANKRATZ · Forecasting with Univariate Box-Jenkins Models: Concepts and Cases
PIANTADOSI · Clinical Trials: A Methodologic Perspective
PORT · Theoretical Probability for Applications
PUTERMAN · Markov Decision Processes: Discrete Stochastic Dynamic Programming
RACHEV · Probability Metrics and the Stability of Stochastic Models
RÉNYI · A Diary on Information Theory
RIPLEY · Spatial Statistics
RIPLEY · Stochastic Simulation
ROUSSEEUW and LEROY · Robust Regression and Outlier Detection
RUBIN · Multiple Imputation for Nonresponse in Surveys
RUBINSTEIN · Simulation and the Monte Carlo Method
RUBINSTEIN and MELAMED · Modern Simulation and Modeling
RYAN · Statistical Methods for Quality Improvement
SCHUSS · Theory and Applications of Stochastic Differential Equations
SCOTT · Multivariate Density Estimation: Theory, Practice, and Visualization
*SEARLE · Linear Models
SEARLE · Linear Models for Unbalanced Data
SEARLE, CASELLA, and McCULLOCH · Variance Components
SENNOTT · Stochastic Dynamic Programming and the Control of Queueing Systems
STOYAN, KENDALL, and MECKE · Stochastic Geometry and Its Applications, *Second Edition*
STOYAN and STOYAN · Fractals, Random Shapes and Point Fields: Methods of Geometrical Statistics
THOMPSON · Empirical Model Building
THOMPSON · Sampling
TIJMS · Stochastic Modeling and Analysis: A Computational Approach
TIJMS · Stochastic Models: An Algorithmic Approach
TITTERINGTON, SMITH, and MAKOV · Statistical Analysis of Finite Mixture Distributions
UPTON and FINGLETON · Spatial Data Analysis by Example, Volume 1: Point Pattern and Quantitative Data
UPTON and FINGLETON · Spatial Data Analysis by Example, Volume II: Categorical and Directional Data
VAN RIJCKEVORSEL and DE LEEUW · Component and Correspondence Analysis
VIDAKOVIC · Statistical Modeling by Wavelets
WEISBERG · Applied Linear Regression, *Second Edition*
WESTFALL and YOUNG · Resampling-Based Multiple Testing: Examples and Methods for p-Value Adjustment
WHITTLE · Systems in Stochastic Equilibrium

*Now available in a lower priced paperback edition in the Wiley Classics Library.

Applied Probability and Statistics (Continued)

WOODING · Planning Pharmaceutical Clinical Trials: Basic Statistical Principles
WOOLSON · Statistical Methods for the Analysis of Biomedical Data
*ZELLNER · An Introduction to Bayesian Inference in Econometrics

Texts and References Section

AGRESTI · An Introduction to Categorical Data Analysis
ANDERSON · An Introduction to Multivariate Statistical Analysis, *Second Edition*
ANDERSON and LOYNES · The Teaching of Practical Statistics
ARMITAGE and COLTON · Encyclopedia of Biostatistics: Volumes 1 to 6 with Index
BARTOSZYNSKI and NIEWIADOMSKA-BUGAJ · Probability and Statistical Inference
BERRY, CHALONER, and GEWEKE · Bayesian Analysis in Statistics and Econometrics: Essays in Honor of Arnold Zellner
BHATTACHARYA and JOHNSON · Statistical Concepts and Methods
BILLINGSLEY · Probability and Measure, *Second Edition*
BOX · R. A. Fisher, the Life of a Scientist
BOX, HUNTER, and HUNTER · Statistics for Experimenters: An Introduction to Design, Data Analysis, and Model Building
BOX and LUCEÑO · Statistical Control by Monitoring and Feedback Adjustment
BROWN and HOLLANDER · Statistics: A Biomedical Introduction
CHATTERJEE and PRICE · Regression Analysis by Example, *Second Edition*
COOK and WEISBERG · An Introduction to Regression Graphics
COX · A Handbook of Introductory Statistical Methods
DILLON and GOLDSTEIN · Multivariate Analysis: Methods and Applications
DODGE and ROMIG · Sampling Inspection Tables, *Second Edition*
DRAPER and SMITH · Applied Regression Analysis, *Third Edition*
DUDEWICZ and MISHRA · Modern Mathematical Statistics
DUNN · Basic Statistics: A Primer for the Biomedical Sciences, *Second Edition*
FISHER and VAN BELLE · Biostatistics: A Methodology for the Health Sciences
FREEMAN and SMITH · Aspects of Uncertainty: A Tribute to D. V. Lindley
GROSS and HARRIS · Fundamentals of Queueing Theory, *Third Edition*
HALD · A History of Probability and Statistics and their Applications Before 1750
HALD · A History of Mathematical Statistics from 1750 to 1930
HELLER · MACSYMA for Statisticians
HOEL · Introduction to Mathematical Statistics, *Fifth Edition*
HOLLANDER and WOLFE · Nonparametric Statistical Methods, *Second Edition*
HOSMER and LEMESHOW · Applied Survival Analysis: Regression Modeling of Time to Event Data
JOHNSON and BALAKRISHNAN · Advances in the Theory and Practice of Statistics: A Volume in Honor of Samuel Kotz
JOHNSON and KOTZ (editors) · Leading Personalities in Statistical Sciences: From the Seventeenth Century to the Present
JUDGE, GRIFFITHS, HILL, LÜTKEPOHL, and LEE · The Theory and Practice of Econometrics, *Second Edition*
KHURI · Advanced Calculus with Applications in Statistics
KOTZ and JOHNSON (editors) · Encyclopedia of Statistical Sciences: Volumes 1 to 9 wtih Index
KOTZ and JOHNSON (editors) · Encyclopedia of Statistical Sciences: Supplement Volume
KOTZ, REED, and BANKS (editors) · Encyclopedia of Statistical Sciences: Update Volume 1
KOTZ, REED, and BANKS (editors) · Encyclopedia of Statistical Sciences: Update Volume 2

*Now available in a lower priced paperback edition in the Wiley Classics Library.

Texts and References (Continued)

LAMPERTI · Probability: A Survey of the Mathematical Theory, *Second Edition*
LARSON · Introduction to Probability Theory and Statistical Inference, *Third Edition*
LE · Applied Categorical Data Analysis
LE · Applied Survival Analysis
MALLOWS · Design, Data, and Analysis by Some Friends of Cuthbert Daniel
MARDIA · The Art of Statistical Science: A Tribute to G. S. Watson
MASON, GUNST, and HESS · Statistical Design and Analysis of Experiments with Applications to Engineering and Science
MURRAY · X-STAT 2.0 Statistical Experimentation, Design Data Analysis, and Nonlinear Optimization
PURI, VILAPLANA, and WERTZ · New Perspectives in Theoretical and Applied Statistics
RENCHER · Methods of Multivariate Analysis
RENCHER · Multivariate Statistical Inference with Applications
ROSS · Introduction to Probability and Statistics for Engineers and Scientists
ROHATGI · An Introduction to Probability Theory and Mathematical Statistics
RYAN · Modern Regression Methods
SCHOTT · Matrix Analysis for Statistics
SEARLE · Matrix Algebra Useful for Statistics
STYAN · The Collected Papers of T. W. Anderson: 1943–1985
TIERNEY · LISP-STAT: An Object-Oriented Environment for Statistical Computing and Dynamic Graphics
WONNACOTT and WONNACOTT · Econometrics, *Second Edition*

WILEY SERIES IN PROBABILITY AND STATISTICS

ESTABLISHED BY WALTER A. SHEWHART AND SAMUEL S. WILKS

Editors
Robert M. Groves, Graham Kalton, J. N. K. Rao, Norbert Schwarz, Christopher Skinner

Survey Methodology Section

BIEMER, GROVES, LYBERG, MATHIOWETZ, and SUDMAN · Measurement Errors in Surveys
COCHRAN · Sampling Techniques, *Third Edition*
COUPER, BAKER, BETHLEHEM, CLARK, MARTIN, NICHOLLS, and O'REILLY (editors) · Computer Assisted Survey Information Collection
COX, BINDER, CHINNAPPA, CHRISTIANSON, COLLEDGE, and KOTT (editors) · Business Survey Methods
*DEMING · Sample Design in Business Research
DILLMAN · Mail and Telephone Surveys: The Total Design Method
GROVES and COUPER · Nonresponse in Household Interview Surveys
GROVES · Survey Errors and Survey Costs
GROVES, BIEMER, LYBERG, MASSEY, NICHOLLS, and WAKSBERG · Telephone Survey Methodology
*HANSEN, HURWITZ, and MADOW · Sample Survey Methods and Theory, Volume 1: Methods and Applications

*Now available in a lower priced paperback edition in the Wiley Classics Library.